ZYGMUNT ZAWIRSKI: HIS LIFE AND WORK

# BOSTON STUDIES IN THE PHILOSOPHY OF SCIENCE




VOLUME 157

Zygmunt Zawirski (1882–1948)

IRENA SZUMILEWICZ-LACHMAN

# ZYGMUNT ZAWIRSKI: HIS LIFE AND WORK

## With Selected Writings on Time, Logic and The Methodology of Science

Translations by
Feliks Lachman

Edited by
ROBERT S. COHEN
*Boston University*

With the Assistance of
Bettina Bergo

SPRINGER SCIENCE+BUSINESS MEDIA, B.V.

**Library of Congress Cataloging-in-Publication Data**

Zawirski, Zygmunt, 1882–1948.

Zygmunt Zawirski : his life and work with selected writings on time, logic & the methodology of science / edited by Irena Szumilewicz-Lachman ; translations by Feliks Lachman ; with editorial assistance of Bettina Bergo and Robert S. Cohen.

p. cm. -- (Boston studies in the philosophy of science ; 157)

Includes index.

ISBN 978-94-010-4329-8 ISBN 978-94-011-0768-6 (eBook)

DOI 10.1007/978-94-011-0768-6

1. Zawirski, Zygmunt, 1882–1948. 2. Science--Methodology. 3. Science--Philosophy. 4. Science--Poland--History. 5. Scientists--Poland--Biography. I. Zsumilewicz-Lachman, Irena. II. Title. III. Series.

Q143.Z35Z38 1994

502.8--dc20 93–38214

ISBN 978-94-010-4329-8

*Printed on acid-free paper*

In memory of my beloved husband Feliks

*Irena Szumilewicz-Lachman*

# CONTENTS

# PREFACE

Among the extraordinary Polish philosophers of the past one hundred years, Zygmunt Zawirski deserves to be given particular attention for his fusion of analytic and historical scholarship. Strikingly versatile, and contributing original work in all his fields of competence, Zawirski thought through issues in the philosophical aspects of relativity theory, on the claims of intuitionalistic foundations of mathematics, on the nature and usefulness of many-value Logics, and on the calculus of probability, on the axiomatic method in science and in the philosophy of science, on the genesis and development of scientific and philosophical concepts, and in his crowning achievement, the conceptual history of notions of time.

His work has been too little known in English despite the respect which has been so clearly shown by his Polish colleagues and students. In this generous selection from his papers and from his great critical study *L'Evolution de la Notion du Temps*, Dr. Irena Szumilewicz-Lachman has provided a clear account of Zawirski's achievement; and she has written a fine comprehensive introductory essay which provides both the personal and historical context of his work and a systematic survey of his principal publications.

I wish to note several matters of particular interest. First, Zawirski conceived the philosophical task broadly and he did not reject consideration of conceptions which have inadequate empirical support; but while not anti-metaphysical, he was indeed apprehensive when it came to undisciplined speculation. His task was to explore the application of the axiomatic method, of hypothetico-deductive logic, to metaphysics. Just as scientific theories, so metaphysical theories might undergo that 'rational reconstruction' so dear to the Vienna Circle of philosophical scientists whom he admired. He wrote in 1947,

Polish philosophers . . . were treated by representatives of logical positivism as coming close to their standpoint. That was right to some extent, but not very much, because Polish scientific philosophy did not share the most important point of old and new positivism. A radically anti-metaphysical attitude is the essence of positivism. Yet Polish scientific philosophy did not preclude the possibility that at least some issues of traditional metaphysics . . . should be treated in a scientific manner.

Second, he was clear in his mind about the difficulties posed for the understanding of time by the twofold origin of that understanding, in private, 'subjective' and in public, 'objective' experience. We see in this volume his discussion of the conceptual puzzles of continuity, infinity, and irreversibility. Thus, for example, Zawirski, physicist as well as philosopher, considered that empirical science would decide the issue of the infinity of time and space whereas he saw that issue to be, for the Vienna Circle, a matter of logical syntax. His exchange with Tarski in 1936 emphasized this disagreement, not only about space–time infinity but more generally about method in philosophical work.

I am particularly pleased to have Zawirski and his work introduced in English by Dr. Irena Szumilewicz-Lachman, and to recognize also the rigorous assistance that she received from the late Dr. Feliks Lachman, a man of humane erudition. This book has also benefited from the philosophically sensitive contributions of Bettina Bergo to the translation from the French.

Irena Szumilewicz-Lachman devoted much of her own scientific work to the nature of time. Her doctoral thesis was on 'Thermodynamical death of the universe', and included an important reference to Zawirski's essay on the 'eternal return of worlds'. In her work for Habilitation entitled 'The Direction of Time' she dealt with Zawirski's major work. We met in Poland in 1962, ever since to be colleagues and friends. At that time I discovered Zawirski's book on time and bought a copy for myself in a rare book store. I received special permission to take this copy out of Poland; the appropriate ministry had determined that it was a national treasure. So it was, and so it still is.

*Robert S. Cohen*
*July 1994*

ACKNOWLEDGEMENTS

First of all I wish to thank John Watkins, without whose help, hospitality, friendship and guidance I would not have been able to write this book.

I wholeheartedly thank Robert S. Cohen for his advice, friendly criticism and arduous work while editing this book.

I wish to stress that while a Research Associate of the LSE Department of Philosophy, Logic and Scientific Method since 1980, I took part in the faculty seminars, and profited immensely from the stimulating discussions with my colleagues, Elie Zahar, Colin Howson, Peter Urbach and John Worrall.

I am most grateful to the British Academy (London) and the Fritz Thyssen Stiftung (Köln, West Germany) for their help, which enabled me not only to work but also to travel to Poland a few times – where I collected material for my research – and for covering my expenses for typing and some translations. I am also most grateful to Barbara Wisznieniski for her friendship and help in correcting the proofs.

And, last but not least, I thank my husband Feliks Lachman for his devotion and love. Being an accomplished linguist, he translated most of the articles from Polish and French and helped me with his encyclopaedic knowledge of history, literature and science.

I. S.-L.

# PART I

# A HISTORICO-CRITICAL INTRODUCTION

CHAPTER 1

# SOME DATA

At the turn of the century a new, highly interesting and original philosophical school arose in Poland, known under the name of the Lwów–Warsaw School (LWS).

The founder of this School was Kazimierz Twardowski, an outstanding philosopher. The school came into being at the end of the 19th century when Twardowski was offered a chair at the Lwów University (1895); it lasted to the outbreak of World War II. The influence of the School upon the development of philosophy and science in Poland during the inter-war period, and even later, cannot be overestimated. Zygmunt Zawirski was one of the outstanding members of that School. In order to understand his role and work, one has to get acquainted with the historical background and characteristic features of Kazimierz Twardowski, and with the outlines of the philosophy of his school. The latter task is the more interesting since the LWS played an important role in the development of philosophy not only in Poland, but also in the entire world, especially in the fields of logic, mathematical logic and semiotics. It seems to be quite widely accepted today that LWS was one of many philosophical centres initiated by the Vienna Circle. Nothing could be more erroneous! This statement is not only at variance with the historic truth, but it is also particularly exasperating for the representatives of the LWS. Professor Izydora Dąmbska repeated this opinion many a time when talking to me. I remember her well when, a few weeks before her death in 1983, I went to Cracow to collect material for my work on Zawirski. She stressed that it was imperative to clear up this misunderstanding, so unfair to Poland. This had been, so to say, her last will and testament which I am anxious to execute.

At the end of the 19th century Poland was far from being an important cultural centre. The question arises, what factors contributed to the fact that a philosophical school characterized by a great *élan* and many brilliant ideas, a school to which many outstanding and famous scholars belonged, was born in a country of, so to say, parochial character. I shall attempt to stress at least some factors which played a role here.

*Partitions of Poland*

In the 18th century Poland found itself in the state of potential danger, nay, of virtual foreign occupation. This can be illustrated by the fact of constant marches-past of Russian troops across Poland (during the Seven-Years War). When the last Polish King was elected in Warsaw in the year 1764, the capital of Poland had a strong Russian garrison.

This was a period of economic decay and political debacle; private interest prevailed over patriotic feelings. The robber-neighbours understood that Poland had become an easy prey. Three partitions of the Polish territories, among Prussia, Austria and Russia, took place between 1772 and 1795; the second and third partitions (1793 and 1795) occurred during the early years of the French Revolution which was virtually saved by the mere facts of the dismemberment of Poland.

Thus, in 1795 Poland ceased to exist as an independent state. This horrible lesson of history became the beginning of national rebirth. The Polish nation closed, as it were, her ranks and began the long, bloody and heroic struggle for the right of national existence. This struggle took place on battlefields in various countries. Polish nationals could be found in almost any country where struggle against tyranny took place.[1]

The final victory in the Polish struggle was the recovery of national independence in 1918.

The fate of the Polish population varied from one partition power to another. *Russia* and *Prussia* used all available means to denationalize the Poles. The language of the partition state was obligatory in the offices, schools, universities, even in churches (in Prussia). The difference between Russia and Prussia consisted mainly in that Russia crushed any resistance by using prisons, torturing and deportation to Siberia, while Prussia rather used the methods of economic and political pressure. However, force engendered resistance. The effort of the partitioning powers was found to be a Sisyphean labour. The history of the resistance offered by the society is illustrated in the Polish literature (The novel *The Sisyphean Labours*, by the eminent Polish author Stefan Żeromski, describes the life of the Polish youth in Russian schools).

In addition to denationalizing measures, economic pressure was applied by Prussia. A semi-official anti-Polish organization '*Ostmarkenverein*' (H.K.T.) was founded in 1894. Its aim was to force Polish peasant-farmers to sell their farms to Germans only. A law was passed in Prussia in 1904, which not only forbade the sale of any landed estate to a non-German person, but also forbade introducing any improvements into a Polish-owned estate. When a Polish peasant-farmer, Michał Drzymala, did not get a permit to build a new cottage on his land, he constructed a 'gypsy wagon' (caravan) and lived there with his family. 'Drzymala's wagon' became the symbol of the Polish fight against economic and cultural discrimination and oppression.

This state of affairs also found its reflection in Boleslaw Prus's short story *Placòwka* (*The Outpost*), dealing with the heroic part played by Polish peasants in Prussia, who were defending the 'outposts' of national culture. Maria Konopnicka, a popular poetess, wrote an anti-Prussian anthem which exhorted Polish people, especially children, to struggle against Germanization.

*Austria*: "Pre-war [pre-World War I] Austria," writes Bronislaw Malinowski in his preface to *Cassubian Civilization* (Faber & Faber, London 1935, by F. Lorentz, A. Fischer and T. Lehr-Splawinski), "presented . . . a sound solution to all minority problems." During the period of ca. 50 years ending with the collapse of Austro-Hungarian empire in 1918, the Polish Province of Austria (Galicia) enjoyed a Home Rule (Polish Governor-General, Polish Diet, Self-Government and Education). Of course not all national minorities in Austria enjoyed the same privileges. This is easily explained by the proverbial policy *divide et impera*. Still, the fact remains that Poles in Galicia enjoyed civil liberties to a much greater extent than did their compatriots under Prussian or Russian rule.

There were two Polish universities in Galicia: the first and most ancient, the Jagellonian University in Cracow founded in 1364 (one year earlier than the Vienna University), and the University of Lwów (1681). The Lwów Technological University was founded in 1877.

It should be particularly noted that the Austrian cultural climate was much closer to the Polish tradition than the Prussian Protestant jackboot or the Russian Orthodox-Greek-Christian autocracy. Likewise one should not forget that Austria was the only Roman Catholic state among the partition powers.

The existence of three academic schools and the relative freedom were the factors which favoured the development of a vigorous centre of Polish culture in Galicia, that radiated across the whole of Poland.

No wonder that many Polish scholars and scientists made for Galicia (especially for Lwów). Some of them had previously been forced to look for a kind of asylum beyond the frontiers of Russia and Prussia in

European cultural centres such as Paris, Milan or Vienna. Polish scientists, who frequently gained recognition in foreign countries, were very well acquainted with the main directions of the cultural development all over the world. They began now to concentrate at Lwów. Their immense contribution to the development of the Lwów center was based not only on individual achievements, but also on importing cultural traditions of countries where they lived before. Thus Lwów became heir to many European cultural trends.

These scholars usually had a very good command of the principal European languages. This enabled them not only to have an easier access to European culture, but also to publish the results of their research in foreign languages, thus surmounting the language barrier which had separated the Poles from the rest of Europe for many years.

The professors were soon followed by young Poles who were attracted by the names of famous scholars and by the atmosphere of relative freedom which prevailed in Lwów.

One ought to mention that humanities and such branches of science as mathematics and logic enjoyed special popularity among the youth. This is easily explained. Unlike natural sciences, humanities provide a much more profound and wide world outlook. The Poles had been treated by 'history' with an utmost 'injustice', hence the trend towards the understanding of the structure of the society, the mechanism of justice, and the role played by moral systems.

At the same time patriotic tendencies, which are very strong in the Polish nation, encouraged students to become acquainted with the treasures of their national culture, its past and present state. Besides, more pedestrian reasons were involved here. Scientific research in technological fields requires, as a rule, greater financial resources than the non-technological research. No wonder that, if one wanted not to fall below the European standards, it was easier to cultivate such disciplines as mathematics, logic, and humanities, which do not require too much financial expenditure.

As a result of this state of affairs, the elite of Polish youth chose humanities, and especially philosophy, as the subject of studies. The first disciples of Twardowski included people whose names soon became famous: Wladyslaw Witwicki, Jan Łukasiewicz, Tadeusz Kotarbiński, Tadeusz Czeżowski and Zygmunt Zawirski. One reckons among the alumni of the School also Stanislaw Leśniewski who established contact with Twardowski prior to his doctorate, and Wladyslaw Tatarkiewicz. In addition to philosophers *sensu stricto*, the seminar of Twardowski produced numerous scholars: psychologists, philologists, educationists, mathematicians, historians of medicine and of art.

CHAPTER 2

# KAZIMIERZ TWARDOWSKI (1866–1938), THE CREATOR OF THE LWÓW–WARSAW SCHOOL (LWS)

Prior to dealing with the LWS some attention should be paid to its creator.

Kazimierz Twardowski was born in Vienna on October 20th 1866, as the third child of Pius and Malwina née Kuhn. In 1885, he passed the secondary school-leaving examinations in the Vienna Theresianum *cum laude*. Next he studied in the Philosophical Faculty of Vienna University where he attended Franz Brentano's lectures with special enthusiasm. In 1891 he was granted the degree of Doctor of Philosophy on the basis of the thesis *Idee und Perzeption. Eine erkenntnistheoretische Untersuchung aus Descartes.* At the same time he passed examinations in philosophy, classical philology, mathematics and physics. In 1894 he was granted *veniam legendi* in philosophy, having submitted his habilitation thesis *Zur Lehre vom Inhalt und Gegenstand der Vorstellungen.* In 1895 he was invited to a professorial chair in Lwów where he lectured for 35 years.

He retired in 1930, but this did not reduce his activity. As professor *honoris causa* at the Lwów University, he took part in university life, and worked at the university as a member of many committees. He also took a lively part in the philosophical life of the country. He devoted much time to reading theses and articles which were forwarded to him from various quarters for appraisal and analysis, especially by his former students.

Twardowski was not only the creator of a new philosophical school but also the organizer of intellectual life in Poland. He created the institutional framework for philosophy without which the development of philosophy would not have been possible on such a scale. He was a highly-talented organizer. *Inter alia* he was responsible for the founding of many cultural institutions such as: *Seminarium Filozofiozne*(The Philosophical Seminar), *Polskie Towarzystwo Filozofiozny* (The Polish Philosophical Society), and *Przegląd Filozofiozny* (*The Philosophical Journal*). Many of these institutions have outlived Twardowski by many years. Right to the end of his life, Twardowski was respected, revered and adored as philosopher, man and teacher. He died on February 11, 1938, leaving behind numerous friends and disciples in deep sorrow.

Much as the importance of Twardowski in the development of Polish philosophy is undeniably related to the founding of the LWS, which covered the name of Poland with glory, one ought not to forget that Twardowski is marked by his own important and original contribution to the development of contemporary philosophy, especially the philosophy of mind.

Twardowski's philosophical conception, especially during the first stage of his activity, was mainly influenced by the currents and trends prevalent in Austria. Austrian philosophy, in particular in the 19th century, developed in a different way from, and even in opposition to, the German idealism of Immanuel Kant (1724–1804), J.G. Fichte (1762–1814), F.W.J. Schelling (1775–1854) and G.W.F. Hegel (1770–1841). B. Bolzano (1781–1848) and Franz Brentano (1838–1917), played the most important part in creating the realistic trend in Austrian philosophy.

It was undoubtedly Franz Brentano who played a decisive role in the development of Austrian philosophical thought. His influence on the students, and beyond their circle, was greater than that of any of his predecessors. His disciples, both in and outside Austria, initiated new philosophical schools whose significance for contemporary philosophy can hardly be overestimated.

We do not intend to present here Brentano's most interesting philosophy in full. He was a realist and an empiricist. It is from this standpoint that he fought against German idealistic metaphysics and against the tendency to build new philosophical systems. His criticism was directed especially against Kant. He wrote that the entire Kantian philosophy was an error which had resulted in still greater errors, and led ultimately to a complete chaos in philosophy.

Kazimierz Twardowski took over Brentano's epistemological realism and his aversion to speculative metaphysics and Kant's philosophy. He developed some of Brentano's ideas in an important manner. Let us cite an example:

Just as Brentano, Twardowski chose as his starting-point mental phenomena, which had finally made it possible to find a basis for realism, and to oppose subjective idealism. Brentano's realism was based on a clear distinction between the mental act and the object toward which the act is directed. The objects exist independent of any observer. Mental acts have a specific ability to direct themselves towards something which is different from them. This specific relation between a mental act and its object is called 'intentional'. However, there was a weak point in

Brentano's approach: he did not discriminate clearly between the content of a mental act and the object.

Twardowski eliminated this weak point and, what is more important, developed his teacher's ideas and made them specific. He was the first to discriminate distinctly between the three elements of the process of perception: the *content* of the mental act, the *act* itself, and the *object*.

The discrimination between the act, its content and the object of perception was found to be of special importance. . . . This discrimination determines the relation with the content and object. This took place with most of the empiricist philosophers and with Kant, for whom the content was considered to be the representative of the object, which, so to say, shielded the object and appeared instead of it. In contrast, Twardowski grasped the matter in such a way that the content renders the object accessible, as if making it present in a certain perspective, so that various contents provide information about various sides of the object. Such an understanding of the relation between the content of the act and the object presents the background of epistemological realism and permits us to establish a clear boundary between the objects which are directly cognizable by the contents and the hypothetical objects. (p. IX). Signed: Polish Philosophical Society, [Preface to: *Selected Writings of Kazimierz Twardowski* (Kazimierz Twardowski, *Wybrane Pisma Filozoficzne*), PWN, Warsaw, 1965.]

It is considered that Twardowski was the first to discriminate clearly the *act,* the *content* and the *object*. The question of priority in Brentano's school has been controversial up to now. Discrimination between the object and the content of the presentation was introduced a few years earlier in Höfler's and Meinong's *Logic*. Marty is of the opinion that this distinction had been known and applied at an earlier date. The followers of Brentano, however, used a new approach to consciousness, viz. not as a set of contents, but as a set of acts. In this context the work done by Twardowski receives a novel meaning [cf. Jan Woleński, *Szkola Lwowsko–Warszawska* (*The Lwów–Warsaw School*) Wydawnictwo Wyższej Szkoly Pedagogicznej Rzeszowie, Rzeszow, 1983].

The discrimination between act and content is understood by Twardowski as the distinction between an action and the product of this action. The thought is the product of the action of cognition. It can be fixed in the form of writing, as a work of art or technology. These permanent products are components of culture and present the object of humanistic research. The same applies to objects which are dealt with in mathematics; they are fixed psycho-physical products whose meaning consists in mathematical mental contents; they are neither inscriptions having definite forms nor non-sensual ideal objects. Twardowski stresses discrimination between both sides of the fixed

fixed psycho-physical object, viz. its *form* (e.g. inscription) and its *meaning*. He stresses that the lack of such a discrimination leads to epistemological relativism and subjectivism.

Thus, starting from epistemological analysis of psychic phenomena, Twardowski finally arrives at epistemological realism and objectivism:

. . . i.e. at the position that cognition, through its content, reaches as far as an object which differs from it, and that the truth contained in the propositions as well as in ethical principles is absolute and objective, i.e. it remains true regardless of who and under what circumstances expresses it (*ibid*).

Twardowski took over from Brentano not only his main idea in the field of the philosophy of mind, which enabled him to support his realistic approach, but also his aversion to metaphysics in its traditional sense. Metaphysics in this sense was connected with the cultivation of speculative philosophy, and its main offence was lack of clarity in the formulation of thought. This vagueness was often explained by the alleged difficulty and intricacy of the problem treated. Now Twardowski considered that this vagueness of style and lack of precision in the formulation of problems conceal the obscurity in the thoughts of the philosopher in question and are, to a large extent, responsible for the lack of progress in the field of philosophy in the course of centuries.

The initial condition for the healing of the old philosophy and for the cultivation of good philosophy was said to consist, *inter alia*, in a clear formulation and presentation of the problems. Every idea ought to be presented so as to be understood by anybody endowed with common sense and a secondary education. A thought which is vaguely formulated is not worth the trouble of understanding: ". . . the lack of clarity in the style of some philosophers is not the inevitable consequence of factors inherent in the object of their reasoning, but has its source in the obscurity and lack of clarity in their method of thinking . . . if the above-mentioned notes are correct, they release us from the duty of puzzling about what a philosophical author who uses an unclear style really thinks . . . we can safely admit that an author who cannot express his thoughts clearly is also unable to think clearly, and hence his thoughts are not worth the effort connected with their assessment. 'O Jasnym i Niejasnym Stylu Filozoficznym' (The Clear and Unclear Philosophical Style) Kazimierz Twardowski (*ibid.*, pp. 346–348).

The concern for a clear and logical formulation of one's thoughts was not an aim in itself for Twardowski. He did not admit the slogan *art for art's sake.* In every argument – writes Twardowski – one should always think of the problems under consideration. The clarity of the style of reasoning is the *means* but not the *aim*.

This discrimination between means and the aim to be achieved was often stressed by Twardowski. He comes out against some defects, not of the old metaphysics, but of the new philosophy based on the use of mathematical and logical symbols. In his excellent and balanced paper 'Symbolomania i Pragmatofobia' ('Symbolomania and Pragmatophobia'), K. Twardowski, [*Ibid.*, pp. 354–364], while stressing the tremendous advantage gained from the use of symbols, points strongly to the danger connected with the misuse of symbols. The paper starts with the enumeration of advantages obtained by the use of symbols in scientific research.

When used as conventional signs in certain sciences instead of linguistic expressions, the symbols rendered uncommon service to these sciences; moreover, some sciences simply would not be able to make a single step forward had they not used the system of symbols right from the beginning; therefore the development of such sciences proceeds *pari passu* with the development of their system of symbols . . . when making use of symbols and manipulating them, one has to take into account always and very conscientiously that they play the role of an agent . . . Therefore the system of symbols to use this expression, for the sake of brevity, to denote both the symbols and the operations carried out with them should be adapted as thoroughly as possible to the symbolized notions and objects; they should be, time and again, subjected to strict control, in order not to risk any difficulties in the final interpretation of the results obtained by means of the system of symbols expressed in its language . . . for various reasons symbolism can lead to false results, e.g. because it uses non-univocal symbols or because its operations are not governed by sufficiently strict rules or because its assumptions are inadequate, etc. (*ibid.*)

Twardowski also writes that there are people who do not realize that symbols are merely means,

. . . they forget that symbols symbolize something . . . they do not attempt to interpret the results clothed in the symbolic attire; both the symbols and the operations carried out with them, which originally had been means to attain an aim, become for them the aim in itself, the object of ardent love and the source of a great intellectual pleasure.

Such love, exuberant in a certain direction, when it takes possession of the entire method of somebody's thinking and behaviour is called a mania. One speaks of graphomania, erotomania, etc. why should it be impossible to speak of symbolomania?

On the other hand, what is related to symbolomania and always accompanies it, as a kind of negative complement, is *pragmatophobia*, if one is permitted to give this name to an aversion to objects so characteristic of symbolomaniacs i.e. aversion to what is symbolized by symbols, which are the signs of objects in the broadest meaning of the word. (*Ibid.*)

The danger of symbolomania and pragmatophobia, writes Twardowski, is especially great in the field of logistics. This state of affairs is encouraged by two moments.

(a) The new logic, which opposes traditional logic and looks at it with a certain disrespect, has some 'imperialistic tendencies'. The tendencies are related to over-estimation of new methods and to the belief that all problems will be solvable with their aid.

(b) The mathematical character of logic and of its symbolism is the other factor favoring symbolomania. "The symbols are not mathematical, since neither do the individual logical symbols have a mathematical meaning nor are the logical operations mathematical operations. Both the one and the other can only happen when a special interpretation is applied".

When one limits oneself to operation with symbols but fails to see any other aim beside it says Twardowski one is changed into a machine and ceases to be a *homo sapiens* (K. Twardowski, *ibid.*).

Little can be added to Twardowski's remarks. Even now, these remarks have not become out of date.

Twardowski himself tried not only to exhort people to philosophizing in accordance with the principles of clarity and logic, he also served as the best example of such behavior. His articles were always prepared with great thoughtfulness; they were always formulated with clarity and limpidity, and presented an example of good philosophy.

Twardowski followed the same rules in his lectures. He was not only a teacher, imparting knowledge to his students, but an educator, who influenced the character and morale of his students, as well. No student was ever allowed to be late for his lectures which started at 6.30 a.m. Tadeusz Kotarbinski, one of his most eminent disciples (and, later, colleagues), writes about Twardowski's lectures:

> . . . they were always prepared with utmost care, constructed with clarity, and intelligible to every attentive listener. The mental effort of Twardowski was always and indivisibly coupled with the concern for a clear and distinct expression; the thinker's task was combined with that of a teacher. [Preface to *Selected Writings of Kazimierz Twardowski* (*Wybrane Pisma Filozoficzne*), PWN, Warsaw, 1965.]

Even Twardowski's appearance evoked admiration and confidence. "He was a good-looking man of imposing stature. He inspired respect, when he paced lonely, slowly and quietly, along the streets of the town, or the long corridors of the University . . . And it was this man . . . who almost became a motherly protector, a gently and cordially close elder friend of those who sought his friendship as close disciples, who of their own accord agreed to become his followers" [*ibid,* pp. V–VI]

Twardowski was a *Dei gratia* teacher. He subordinated not only his creative work but also his private life to his pedagogical work.

He was beloved and adored by his disciples. He was their personal ideal. With many disciples, he maintained relations all his life, rejoicing at their successes, advising them at difficult junctures, and especially following their scientific development assiduously. And yet Twardowski was never a forbearing man. He required many things both from himself and from other people.

Highly interesting are Twardowski's views regarding the role of philosophy and its relation to science.

Twardowski had an especially great esteem for the natural sciences and their development. He believed that philosophy was developing much less vigorously, and sometimes, when studying its history, one had the impression that it did not develop at all. However, the new philosophy, based on the achievements of science and using clear language and logic, has a great future. Twardowski thought that the place of philosophy is to be found in a central position, being bounded by science, art and religion. The philosophical aspect of the universe and life supplies the sciences with specific ideas and thoughts indispensable for the shaping of the scientific outlook upon life. Scientific discoveries and theories affect the formation of a philosophical outlook. With the passage of time the process of the mutual approach of these two outlooks does take place. This is a process in which its two members, philosophy and science, undergo changes, a process which will never reach an end.

In agreement with this broad conception of philosophy, Twardowski understood the ideal of a philosopher as a sage with a fundamental knowledge of sciences, and with a good understanding of the outlook on life. A philosopher ought to be a man who not only proclaims certain principles, but who firmly and consistently applies them in practice. Twardowski realized this ideal in his own life.

He remained, in the memory and hearts of people who knew him, an absolutely honest man and thinker, a Stoic in the Roman manner: *vir constans, integer, scelerisque purus.*

CHAPTER 3

# THE LWÓW – WARSAW SCHOOL (1895 – 1939)

When Poland regained independence in 1918, conditions for the development of philosophy improved dramatically.

The number of universities increased to six. Four philosophical periodicals were regularly published: in addition to the *Przegląd Filozoficzny* (*Philosophical Review*) and *Ruch Filozoficzny* (*Philosophical Movement*), the quarterly *Kwartalnik Filozoficzny* (1922) and *Studia Philosophica* (1935) were established Branches of the philosophical societies came into existence in all university towns. Periodic conferences were held, which not only deepened contact and co-operation between philosophers, but also enabled the important achievements of science and philosophy to be discussed.

Having obtained a suitable framework, Polish philosophy began to develop exuberantly and in many directions as never before. The characteristic feature of this period was the transition from general philosophy to its special branches in such fields as the philosophical problems of physics, mathematics, medicine, law, etc. Similarly, research into language resulted in the development of semiotics (semantics and syntax) which, in turn, led to such highly important philosophical problems as, e.g., the definition of truth, being solved.

Polish philosophy was made famous mostly by the development of logic, especially of mathematical and many-valued logic (Jan Łukasiewicz). The set of ontological problems, such as the problem of determinism, acquired a new explanation as soon as many-valued logic appeared. It opened a new chapter in the history of logic. For example, the invention of three-valued logic, based on the analysis of modal sentences and the new interpretation of Aristotle by Łukasiewicz just at the time of difficulties in the interpretation of quantum mechanics, was of utmost importance. It inspired a long and fruitful discussion in which, *inter alia,* Zygmunt Zawirski took part. Zawirski tried to adapt many-valued logic to the interpretation of statistical laws. To achieve this, he tried to change many-valued logic so as to make it consistent with two-valued logic and to contain two-valued logic as a limiting case. He wanted the postulate of reduction to be fulfilled. Generally speaking, Polish philosophy could be found on the main track of scientific and philosophical discussions.

There was hardly any important development in philosophy that could not be discussed in Poland. Polish scientists were in touch with other world centers of philosophy.

Just as before, during the partition era, so also following the regaining of independence, the predominant role in Polish intellectual life was played by the members of the LWS. Twardowski's disciples and adherents could be found in the fields of philosophy, logic, mathematics and psychology all over the Polish universities. The individual university centers not only became stronger, and independent of Lwów, but they also became specialized in various fields of philosophy. For example, the Lwów centre was mostly engaged, as before, in the analysis of ideas, as a basic task to be carried out profoundly and exactly. Twardowski himself was working at the boundary between logic and psychology. The Warsaw center's work, with Kotarbiński, Leśniewski and Łukasiewicz, was mainly centred on research in the field of logic, especially mathematical logic, and attained very important results in this field. Besides, research in the field of semiotics was also energetically developed there.

In Cracow, the main stress was laid upon the methodology of science and upon history, viz. both the history of philosophy and the history of science (cf. again, the work of Zygmunt Zawirski).

The pedagogical and philosophical ideas of Twardowski were continued in all Polish universities except the Lublin University. Just before World War II there were more than 50 active philosophers and scholars in Poland who were publishing their works. The philosophical life of the country was highly animated.

An answer remains to be given to the question whether the LWS was a philosophical school *sensu stricto.* As we noted, during the inter-war period the individual academic centers in Poland were developing research in many different directions. The multipolar and multidirectional character of Polish philosophy, its open character and (as frequently stressed by Twardowski) the principle of uncommitted philosophy, had the result that philosophers who belonged to the LWS differed sometimes in their views concerning some basic philosophical problems. Thus, there belonged to the LWS: materialists (Kotarbinski) and idealists (Łukasiewicz), nominalists (Lesniewski) and realists (Ajdukiewicz), neo-scholastics (Sałamucha and Swieżawski) and scientists (Zawirski), the quasi-phenomenologists (Blaustein) and quasi-positivists (Kokoszynska–Lutmanowa). (Cf. Woleński, *Ibid.*).

*In this situation it is very difficult to provide a characteristic of the School on the basis of a set of answers to the basic questions of the philosophy. Is it possible therefore to talk of a philosophical school? Yet I think that in fact, LWS was a philosophical school because, notwithstanding differences in views, the basic method of philosophizing was similar. Two fundamental postulates, stressed so emphatically by Twardowski himself, were generally accepted: the dislike of speculative metaphysics, and the care for precision and clarity in the formulation of problems. The method of philosophizing and especially the postulate of clarity meant in practice the broad use of logic, the chariness of words and terms, and precision in their formulation. The philosophical method was based on a profound knowledge of the philosophical workshop and, if the object of research was a specific field of science, a good technical acquaintance with that field. This was the anti-irrationalist method, characteristic of analytical philosophy. The principle of scientific criticism, good work and the pursuit of perfection was binding.*

*Twardowski's pupils, like their master, were characterized by a realistic and, on the whole, anti-idealistic approach.*

*Their critical attitude with respect to old philosophy did not mean at all a tendency towards its complete elimination. The need for metaphysics was admitted, and it was understood that old philosophy has its place in science and philosophy, and that some of its problems are highly important for mankind.*

*As mentioned before, Twardowski voiced the principle of an open philosophy and the right to have personal convictions. He writes: "... everybody has the right to adopt such an outlook upon the world and life as suits him, provided this outlook is free from internal contradictions, consistent with science and intelligible", 'Speech on the occasion of the twenty-fifth anniversary of The Polish Philosophical Society',* (Księga Pamiątkowa Polskiego Towarzystwa Filozoficznego we Lwowie), Lwów pp. 7–5, 1931.

To be in accord with Twardowski's opinions, a scientist ought to search for truth and have the courage to preach it. His behaviour ought to be in full accord with the opinions he voiced. A philosopher ought to be not only a sage, but also a teacher and educator. To his disciples, the disciples' disciples and to his friends, Kazimierz Twardowski was, and remains, the personal example of the philosopher, teacher and man. I had the privilege of meeting his disciples, Tadeusz Czeżowski, Izydora Dąmbska and Janina Kotarbińska, and my contact with them taught me very much.

CHAPTER 4

# THE LWS AND THE VIENNA CIRCLE (VC)

The birth of the Vienna Circle (VC) is connected with the seminar conducted in Vienna from 1923, by Moritz Schlick. In 1929 members of the seminar proclaimed their *credo* in the manifesto: "The scientific conception of the world 'The Vienna Circle'". When VC came into being, a lively and fruitful co-operation developed between the LWS and the VC which was younger by a quarter of a century. The cooperation was based on the similarity of aims and methods.

Both schools belonged to the trend of analytical philosophy in a broader sense: they referred to the tradition of Austrian philosophy, especially of Franz Brentano (1838–1917) in opposition to German idealism, especially to Kant's idealism. Their predecessors were the philosophers of the second positivism and of analytical scientists (Poincaré, Duhem, Einstein), and of the logicians (Whitehead and Russell).

Both schools rejected the speculative philosophy, were devoted to logic and natural sciences, and advocated methodological empiricism and philosophical realism. The admiration of natural sciences was complementary to the dislike of speculative philosophy.

The lack of progress in the field of philosophy, the haziness of the language used by the old philosophers, these were opposed to the brilliant achievements of physics and to the demand to define its terms in a clear-cut, unambiguous and empirically perceptible manner.

The members of the LWS believed that a philosopher ought to complete his education by sound studies in the field of science.

In the Polish School, mathematics and/or physics were studied by Kazimierz Ajdukiewicz, Jan Łukasiewicz, Stanislaw Leśniewski, Jerzy Slupecki, Alfred Tarski, Zygmunt Zawirski, and many other philosophers belonging to the younger generation.

VC was the collective work of learned specialists who had come to philosophy mostly without previous preparation in this field. These initial shortcomings were of course compensated with the passage of time.

Moritz Schlick, the initiator of the Circle, was originally a physicist. Rudolf Carnap, Hans Hahn, Karl Menger and Kurt Gödel were mathematicians.

The situation in science was changing dramatically at that time. The crisis in mathematics, related to the discovery of antinomies in the theory of sets, gave inspiration to look for a deep explanation by way of research into the foundations of mathematics. The physicists, e.g. Philipp Frank, were attracted to philosophy by the revolution in the realm of physics, connected with the theories of relativity and quantum mechanics.

A close collaboration and a lively exchange of ideas, based on real understanding and personal contacts, started between the Polish and Austrian philosophers. Zygmunt Zawirski, an eminent representative of LWS, described the special relations between the two schools as follows:

> It is known that the Vienna Circle, which represents logical positivism, felt a certain spiritual unity with contemporary Polish logicians. Suffice it to mention Carnap's visit to Warsaw to deliver lectures, and the invitation of Tarski to Vienna about two years earlier. In recent times the Vienna Circle proposed an idea of organizing an International Congress devoted to the Philosophy of Science in Paris in 1935, and with that end in view they invited Polish logicians to a preliminary conference in Prague, which took place two days before the Congress [of Philosophy] (August 31 and September 1). Quite a number of lectures were delivered at that conference for the purpose of the mutual exchange of ideas and of establishing a closer acquaintance. On the Polish side lectures were delivered by Professors Ajdukiewicz and Łukasiewicz and Dr. Tarski. Ajdukiewicz presented his philosophical position in his lecture 'On the Scientific Perspectives of the World', and in another lecture '*On Positivism in Poland*' he turned his attention, however, not so much to the older scientific workers who, in Poland, remained under the influence of Avenarius, Mach, etc., but rather to the circle of contemporary logicians, and in concordance with the present interests of the Vienna Circle.
>
> Professor Łukasiewicz delivered a highly interesting lecture 'On the History of the Calculus of Sentences Starting with the Ancient Times', while Dr. Tarski had a highly specialized lecture on the 'Methodological Research into the Definability of Ideas'. In addition to Poles and representatives of the Circle there were the Frenchman Rougier, the American Morris, the Dane Jörgensen and others. Our representation turned out to be excellent and we were able to observe that the richness of our achievements called forth general admiration and amazement among people who heretofore had not known us more closely (cf. in this book 'An Account of the Eighth International Philosophical Congress', pp. 99–100).

At the 8th International Philosophical Congress which took place between September 2nd and 7th in Prague, the VC and LWS philosophers represented the same trend.

"Perhaps a more important phenomenon, likewise significant for the Congress, was the appearance of a novel philosophical trend represented vigorously in part by the Vienna Circle and in part by most of the Polish philosophers who attended the Congress. The traditions of Mach's positivism merged in the Vienna Circle with contemporary trends of mathematical logic toward the revision of philosophical problems in such a manner both that the Polish logicians came closer to that group of philosophers and that, in the section of the Congress on 'logical analysis of cognition' they play the dominating role along with the representatives of the Vienna Circle." (Ibid.)

Co-operation between the two Schools became closer and closer with the passage of time. Plans were made for the publication of an *International Encylopaedia of Unified Science*. Besides, several congresses had been organized, dealing with the unity of the sciences. Hitler's victories, and World War II in general, interrupted this highly promising cooperation.

The similarity between the purposes and methods of both schools of thought is probably best reflected in the pronouncement of Jan Łukasiewicz (who had formulated his programme in 1927 at the 2nd Congress of Polish Philosophy):

The scientific philosophy of the future must build its structure from new beginnings, from new foundations. This means, first of all, the reviewing of philosophical problems and the choosing of only those that can be comprehensively formulated; the rest should be rejected. Mathematical logic can be useful even at this pre-introductory stage by helping us to establish the meanings of expressions which belong to philosophy. Next, we should try to answer those questions which can be comprehensively formulated. The most suitable method for this purpose seems to be that of mathematical logic, deductive and axiomatic . . .

. . . Results obtained in this way should be constantly checked against the data of intuition and experience, and also against findings in other sciences, particularly the natural sciences . . . We should uninterruptedly maintain contact with reality in order not to create mythical beings like Platonic ideas or the things-in-themselves, but to try to comprehend the essence and the structure of the real world in which we live, in which we act, and which we want to transform into a better and more perfect one.

In this work we should behave as if nothing had been done in philosophy. Any return to Aristotle, or Leibniz, or Kant will be of no use; it will rather be a drawback because we become influenced by these great names and acquire bad mental habits.[2]

If we compare this early statement of Łukasiewicz with the Manifesto of the VC, the similarity becomes striking. The main watchword here is condemnation of speculative philosophy and a move toward the creation of a new scientific philosophy. This scientific philosophy is to be based on natural sciences and on the connection with the real world in which we live. The methods of research ought to be based on mathematical logic as on an omnipotent panacea.

Notwithstanding considerable similarities between both schools, there also were serious differences. One finds that sufficient attention is not given to this in the literature. The representatives of LWS themselves were especially sensitive to these distinctions, and they often gave voice to them.

Kazimierz Ajdukiewicz put it quite explicitly at the Prague Congress. He said: ". . . none of the Polish philosophers recognizes the VC thesis as his own."

The same point has been brought to my attention more than once in conversations with Izydora Dąmbska, one of the last survivors of the LWS (the last such conversation took place in May, 1983, a few weeks prior to her death).

The differences between the Polish School and the Vienna Circle were caused, *inter alia*, by the fact that the Polish School developed in a multipolar manner. As mentioned before, there existed in Poland several centers connected with various university towns. Each of these centers possessed its own specific lines of research and of interest. This state of affairs favoured the openness of Polish philosophy. In the Vienna Circle there was only one center, the Moritz Schlick seminar. The result of this state of affairs was that, at least during the first years of its existence, the VC had been much more 'dogmatic' than the LWS.

The LWS was much more mature than the VC, if only because it was a quarter of a century older, and it had besides achieved a lot in the fields of logic, methodology and history at the time when the VC came into being. The Polish philosophers had a profound philosophical training in the history of philosophy and in contemporary trends in science and philosophy. The VC philosophers were mostly scientists without such preparation. Consequently, though the Poles had very often the same aims, they did not come to such extremes as their Viennese counterparts, and so did not have to withdraw from those extreme positions later.

The difference between both schools was reflected in a rather critical view adopted by the LWS, especially in relation to some early radical ideas of the VC. The criticism was based on profound studies into, and on their brilliant development of, logic, semantics and syntax.

In particular, many 'immature' ideas of the Viennese thinkers were analysed by the Polish school, who often pointed to the incompleteness of these first 'world-shaking' and radical ideas. The Poles sometimes referred to their own earlier solutions, which were often reached only much later, and generally independently, by the Vienna thinkers. This took place, not infrequently, owing to the fact that some Polish works had been inaccessible because of the language in which they had been published. The Polish community was like an older, more experienced brother, more mature and less radical. The Vienna philosophers on the other hand, used to propose new, interesting and bold ideas which were discussed by the Polish School in a lively interchange; and the Poles were extremely helpful in clarifying some obscure and difficult problems, often because of their sharp and radical form.

Let me give some examples.

(a) No Polish philosopher approved of 'physicalism', a view according to which the most perfect system of concepts is that developed by physics. According to some of the Vienna thinkers, physicalism was to yield a unity of sciences by reducing the languages of all sciences to the space-time language of physics. The Poles pointed to the difficulties involved in the realization of such a reduction, if only by referring to the example psychology.

(b) Common to both Schools, the Polish as well as the Vienna one, was a condemnation of traditional speculative metaphysics. This condemnation was understood and interpreted in Vienna differently than it was in Poland.

Much as the Polish philosophers like Kazimierz Twardowski were opposed to fruitless speculation, they did nonetheless esteem philosophy. According to Twardowski, philosophy is indispensable for the intellectual development of mankind. There was, there is, and there will be, a mutual link and influence between science and philosophy – of great importance for both sides. The philosophical aspect of life, nature and the universe supplies science with many fruitful ideas indispensable for its development (e.g., the idea of the atom).

On the other hand not only the difficulties and problems science comes across, but also its foundations are the subject of philosophical investigation. In this connection, the new philosophy, based on the achievements of science and logic and using a clear language, has a great future, according to Twardowski.

The VC philosophers were much more radical. For them, the old philosophy was not only valueless, but outright harmful.

According to the VC speculative philosophy was objected to because some of its sentences were either without sense or else referred to pseudo-problems.

A language consists of a vocabulary and a syntax, i.e. a set of words which have meanings and rules of sentence formation. These rules indicate how sentences may be formed out of the various sorts of words. Accordingly, there are two kinds of pseudo-statements: either they contain a word which is erroneously believed to have meaning, or the constituent words are meaningful, yet are put together in a counter-syntactical way, so that they do not yield a meaningful statement. We shall show in terms of examples that pseudo-statements of both kinds occur in metaphysics. Later we shall have to inquire into the reasons that support our contention that metaphysics in its entirety consists of such pseudo-statements. (Rudolf Carnap, 'The Elimination of Metaphysics through Logic Analysis of Language', in A.J. Ayer, ed., *Logical Positivism*, Illinois, 1959, p. 61)

To make such an annihilating appraisal of metaphysics, one has, first, to establish what conditions must be fulfilled by a proposition if one admits that it does have a sense. The answer to this question was that for a proposition to be scientific, two conditions must be fulfilled: all of its terms must have a definite meaning and it ought to be constructed in accordance with the rules of syntax.

The former condition is met when, for a given language, criteria have been established which enable one to decide, by means of a simplest judgement, whether a certain name can be given to a certain object.

Leibniz was the first to criticize the Newtonian concept of absolute motion, in accordance with this criterion. The term 'absolute motion', he wrote, is devoid of a sense since the problem as to whether the body is in 'absolute motion' cannot be decided on the basis of any experiment. We observe always only the relative motion of one body with respect to another.

Einstein, using a similar presupposition, criticized the term 'simultaneous' in the frame of classical physics. The point is that classical physics does not enable us to decide whether two distant events are simultaneous. The VC philosophers followed Einstein. According to their view only such statements are meaningful which are supported by experience: "A (pseudo) statement which cannot in principle be supported by an experience, and which therefore does not have any factual content would not express any conceivable state of affairs and therefore would not be a statement, but only a conglomeration of meaningless marks or noises." Rudolf Carnap, *The Logical Structure of the World and Pseudo Problems in Philosophy,* Routledge & Kegan Paul, London, 1967, p. 328.

The Vienna philosophers illustrated their rigorous negative appraisal of speculative philosophy by examples taken from leading philosophers, and especially from Hegel.

In addition to illustrations taken from philosophical works, the Vienna thinkers supported their criticism of metaphysics with arguments of a general nature, that the propositions of metaphysics are senseless because there exists no method of deciding whether they are true or false.

A decision cannot be reached by using the *a priori* method since the propositions are not analytical. They deal with reality. In accordance with the foundations of logical empiricism, there exist no synthetic *a priori* propositions in the Kantian sense. Hence the *a priori* method does not enable one to decide on the truth or falsity of metaphysical propositions. The VC excludes as well the existence of empirical metaphysics. The empirical method cannot be applied to philosophical propositions according to VC. This devastating appraisal of metaphysics embraces also the normative disciplines: ethics and aesthetics.

The philosophical sciences remaining after the expulsion of metaphysics, ethics and aesthetics include only the logical and the objective propositions. (The logical propositions comprise the utterances relating to language constructions, while the objective propositions refer to constructions other than the linguistic ones.)

When engaged in this task of 'cleansing' philosophy, Carnap proposed to reduce all philosophical questions to questions of language (see *The Logical Syntax of Language*, London, 1951). He presented his point of view at the First International Congress of Scientific Philosophy (Paris, September 15–23, 1935). Here is a description of Carnap's speech by one of the members of the Congress [this volume, p. 103]

Monday morning there spoke, among others, the chief representatives of the Vienna Circle, Carnap and Neurath. Carnap presented three phases which can be distinguished in the evolution of the V.C. programme. The first phase is characterized by the anti-metaphysical attitude inherited from Mach, i.e. the conviction that metaphysics, *qua* science, is impossible. In this phase, the theory of knowledge appeared to them to be the fundamental philosophical science. In the following phase, the representatives of the Circle become conscious of the fact that, not only can metaphysics not enter into the composition of the philosophy of science, but also the traditional theory of cognition must be excluded from this role, frequently based on metaphysical foundations as it has been, and too such a theory of *a priori* cognition as that created by Kant in his transcendental philosophy. The third phase is distinguished precisely by the transition from such a theory of cognition to the logic of the sciences (*Wissenschaftslogik*) as the proper aim of philosophy, differing from the psychological or sociological treatment of philosophical problems. In a philosophy thus understood, its role is reduced to the investigation of the formal structure of the languages of scientific theories.

Lukasiewicz was absent at this Congress, but his answer appeared almost immediately in a paper entitled 'Logistyka i Filozofia' (Logistics and Philosophy). Jan Lukasiewicz expresses the greatest approval as regards the precision and clarity of Carnap's formulations. Yet he rejects Carnap's idea emphatically. He writes:

Among questions which can be determined on the basis of language, I would include only questions of the type whether all bodies are extended, assuming that, by a 'body', I understand something that can be extended. Only such statements are analytical and only such statements are, according to my opinion, capable of being discussed on the basis of a language. I cannot comprehend, however, how the question, whether the world is finite or infinite can be discussed on that basis. By the word 'world' I mean neither something that is finite nor something that is infinite, so I am concerned here not with an analytical statement, but with a synthetic one. Furthermore, I know that being finite is something different from being infinite, and that only the one or the other can be the case, and what is really the case does not depend at all upon us nor does it depend upon our linguistic rules. The same refers to the questions of causality and determinism. Either causal necessity rules supremely in the world, or not, and everything is either determined *a priori* or not, and for this again one cannot depend on any of our rules of syntax of language. These questions I regard as questions of facts; these are 'subject matter' questions and are objective, and have to do with more than formal linguistics. I have strong and unyielding objections to the way in which Carnap attempts to reduce objective questions to linguistic ones . . . Carnap's considerations of this matter I regard as hazardous philosophical speculation which will pass away just as similar speculations have done. [*Przegląd Filozoficzny*, (*Philosophical Review*) 39, 1936 pp. 126].

It seems that his opinion does not require any complements. The prediction was fulfilled.

The same problem was tackled by Kazimierz Ajdukiewicz in a different way. He wrote that among the objective propositions one finds also such as do not satisfy the VC condition that they have sense. It turns out, however, that some propositions of this kind are only apparently senseless. They acquire a sense as soon as they become correctly formulated, for instance, when one passes from the objective to the formal way of speaking. To explain this matter, Kazimierz Ajdukiewicz uses as an example the proposition: 'quadrilaterality is the essential feature of a square'. This proposition seems to be senseless because it contains the term 'essential feature', which is senseless according to the criteria of the Vienna thinkers. Now what is to be expressed by the above proposition does not at all deal with quadrilaterality or with the square, but it refers to the names 'quadrilateral' and 'square'. The point is, namely, that one can predicate the word 'quadrilateral' about the word 'square' when one is given the analytical proposition: 'the square is quadrilateral'.

Such a transition from the objective manner of speaking to the formal one is called a paraphrase. It presents a highly important step in the development of metaphysics. Ajdukiewicz writes on this subject:

It is by means of this paraphrase that problems which have been the immemorial ballast of philosophy. . . find their way to incontestable solution . . . However, one should make it clear that the author of the idea of this paraphrase and its application to numerous problems was a Pole, professor St. Leniewski, and that, many years before the Vienna thinkers, this idea had been used in the Polish philosophy for solving many difficulties. ['O. Tzw. Neopozytywizmie', ('About The So-Called Neopositivism'), *Myśl Wspólczesna*, (*The Contemporary Ideas*), Nr. 6–7, pp. 155–176, 1946].

Unfortunately, the ideas of Leśniewski had been unknown abroad. No wonder the Vienna thinkers had to solve this problem quite independently.

The role played by Polish philosophers in the very important discussion of the problem of truth and semantics can serve as a further example.

For such a notion as truth it is characteristic that it deals with language expressions on the one side and with reality outside the language on the other. All attempts to construct a definition of truth according to the classical definition resulted in insuperable difficulties and antinomies. According to the VC philosophers, as mentioned above, the only aim in scientific philosophy is to deal with the syntax of scientific language. But it was very difficult, or even impossible, to reduce the notion of truth to pure syntax.

The outcome of this situation was that the term 'truth' (cf. Carnap) was to be expelled from scientific philosophy.

In 1933 a paper by Alfred Tarski 'The Concept of Truth in the Languages of Deductive Sciences' appeared in *Sprawozdania Towarzystwa Naukowego Warszawskiego*. This paper was later published in German in the journal *Studia Philosophica* (1935) by the Polish Philosophical Society at Lwów (*Polskie Towarzystwo Filozoficzne we Lwowie*). In the same year the author had read a paper on the same subject at the First International Congress of Scientific Philosophy (Paris, 1935). Tarski's lecture brought about a fundamental turn in the discussion. He demonstrated that the classical notion of truth as: *veritas est adaequatio rei et intellectus* can be defined correctly only in a meta-language (ML), i.e. in a language which is more comprehensive than the language (L) to which the defined idea of truth refers. Tarski proved that in formal languages of a finite order, the meaning of truth could be defined quite properly with the help of some morphological terms. Difficulties and anomalies emerge only within universalistic languages, e.g. within the language of everyday, because the latter contains its own semantics and is able to talk about itself. In order to avoid difficulties, a language (L) should be separated from the meta-language (ML). Only in the ML is it possible to formulate a definition of a true sentence which belongs to L.

The definition proposed by Tarski was in full harmony with our intuition and with the classical definition of truth. The definition reads as follows: '*x* is a true sentence if, and only if, p', where *x* denotes the name of a sentence in L. The name *x* belongs to the vocabulary of L, and *p* presents the translation of the sentence *x* into L. This enables us to avoid antinomies which had prompted Carnap to expel the term 'truth' from scientific philosophy. The definition of a true sentence may properly be formulated, according to Tarski, only for languages which either contain only expressions of a finite degree (in these languages all variables belong to the same semantical category), languages where the number of semantical categories is finite, or those languages where the number of semantical categories is infinite but the number of classes of categories is finite.

For formalized languages with an infinite number of classes it is impossible to formulate a definition of a true sentence; the only thing that could be done is to establish rules of a proper use of the notion of truth. In everyday language it is not only impossible to construct a definition of a true sentence, but we are even unable to establish rules for the proper use of the notion of truth.

Tarski's paper, an outline of which has now been briefly given above, was commonly accepted as being the most important event of the Congress.

In a Congress report, one can find an interesting description of the discussion which took place after Tarski had presented his paper:

. . . Carnap's relation to the content of the paper was more factual [than Neurath's, I.L.]; as a scientist, he expressed his readiness to accept all corrections and complements to the lacunae in his system, provided they were duly motivated. Though usually silent, Russell spoke up too, pointing to the fact that the work of Tarski in a way carries into effect his previsions which he expressed in the foreword to the work by Wittgenstein *Tractatus Logico-Philosophicus* of 1922. The point is that, while Wittgenstein was sceptical with regard to the scientific nature of sentences of the scientific language, Russell, in the foreword to the work of his disciple, had already noticed then that the difficulties could be removed by creating a language of a higher order for the preceding language . . . At the same time Russell paid compliments to the Vienna Circle, stating that he liked very much especially the physicalism of the Circle . . . [Cf. Impressions from the 1st International Congress of Scientific Philosophy, this volume, pp. 109–110].

During the same Congress, another Polish philosopher presented a paper supporting Tarski's proposition: Mrs. Lutmanowa-Kokoszyńska pointed to some shortcomings in Carnap's book. She said that syntax could be practiced in two different ways, but Carnap limited his investigation to one

way only. Mrs. Lutmanowa-Kokoszyńska asserted that there exists:
(a) pure syntax using only logical and mathematical terms, and
(b) descriptive syntax using some descriptive terms as well.
The descriptive syntax (b) could be divided into Syntax Proper, using only syntactical terms, and Extended Syntax, using all descriptive terms of the language under investigation. Carnap investigates Syntax Proper only.

Since the notion of truth did not fit into that kind of syntax, Carnap declared it to be unscientific, and, as such, decided to expel the notion of truth from science altogether. In her closing remarks, Mrs. Lutmanowa-Kokoszyńska referred to Tarski's investigation which, though giving us the definition of a true sentence only in relation to certain languages, still does appeal to the classical definition of truth and is satisfactory from the logical point of view.

Both VC and LWS are commonly considered to be historical formations whose existence came to an end with the outbreak of World War II. Many representatives of these Schools lived for a long time after the War; their opinions, of course, underwent evolution, yet the method of philosophizing and the perfection in the application of logical analysis remained in principle unchanged.

May I present here, in a rather incomplete manner, an example from the post-war period. I have in mind the profound and excellent criticism of Karl Popper's falsificationism (*The Logic of Scientific Discovery*, 1959), contained in Janina Kotarbińska's paper 'Kontrowersja dedukcjonizm/indukcjonizm' (The controversy of deductivism versus inductivism'), *Studia Filozoficzne 1* (22) 1961.

Popper's conception, writes Kotarbińska, is directed against the inductionism of inductive sciences. Induction is considered here to be an unscientific and deceptive method; according to Popper it ought to be replaced by the deductive method of the criticism of hypotheses by falsification. The heart of the problem consists in finding which of these two methods makes it possible to justify general statements on the basis of certain distinct sentences, i.e. the 'basic' singular statements.

According to Popper, it is for two reasons that induction does not lend itself as a method of justification.

First, because the inductive way of reasoning goes in a direction opposite to that of justification. It is the case that the observations forestall the hypotheses. However in the reasoning of justification, one begins with a hypothesis.

Secondly, the conclusion by induction cannot be proved on the basis of the premises; it can even be the case that it is found to be false when all of the premises are true.

Now according to Popper his method is supposed to be immune to all of these objections.

Janina Kotarbińska demonstrates that Popper here attacks induction in the traditional sense, viz. as the passage from individual sentences to universal ones. However, if we understand that induction is a process of thinking, based on an hypothetical adoption of a nomological statement and on singular statements which are its special cases, then this interpretation cannot be subjected to Popper's objections. 'Moreover when induction is so understood, the method of criticism of hypotheses is only a variant of inductive reasoning. Thus the controversy becomes merely verbal.'

According to Kotarbińska, another serious sin of Popper's is his departure from empiricism, since basic statements are accepted in a conventional way. What is more, says Professor Kotarbińska, Popper's falsifiability of a system is open to some serious criticism. Firstly, for many reasons, no 'basic' statement is certain. Secondly, we never test one single hypothesis, but always a whole set of hypotheses, so we never know which hypothesis was falsified, as Poincaré had rightly pointed out. (Unfortunately, I must limit myself to these examples.)

It would also be worthwhile to clarify the source of the opinion prevailing after World War II, according to which the achievements of the LWS have been belittled. This seems to be due to several factors:

1. A certain part in the underestimation of the LWS has come from the Polish philosopher Ingarden, who wrote about it in his work *Glówne kierunki polskiej filozofii (The Main Trends in Polish Philosophy)*. According to him the LWS was only a variant of neo-positivism. However, in Poland, a decisive part was played by the first critical review by the Marxists (A. Schaff, *Poglądy filozoficzne K. Ajdukiewicza (K. Ajdukiewicz's Philosophical Views),* Warsaw, 1952; and B. Baczko, *O Poglądach filozoficznych T. Kotarbińskiego (T. Kotarbinski's Philosophical Views)*. These reviews, which were also repeated in the USSR, became the signal for reprisals against the representatives of the LWS. Many older scholars were pensioned prematurely, while the younger ones were simply dismissed from the university. A certain mitigation took place after 1956, and several scholars were reinstated (Toruń, Cracow, Poznań). Notwithstanding the

attacks directed against the LWS, many of its eminent representatives retained their academic chairs, thus giving dimension and significance to Polish philosophy after World War II. It is worth mentioning that they even influenced their Marxist adversaries who, unlike Marxists in other countries, had to master semiotics and logic in order to be able to start a discussion on the merits of the subject.

2. A considerable role on the international scale was played by the incomplete acquaintance with the achievements of LWS, due to the language barrier.
3. As mentioned before, the LWS belonged to the broader current of analytical philosophy. The principle of honest and profound analysis was binding here and, consequently, a disinclination to broad syntheses prevailed. Bold generalizations usually attract attention to a much greater extent than closely focussed work.

It ought to be stressed that the opinion which depreciated the achievements of LWS was held neither by the members of VC, nor by specialists in various fields of logic. Many eminent members of the VC had a great regard for Polish philosophy, especially in the field of logic. Here are some examples:

Carnap visited Warsaw in 1930 to give lectures on metaphysics, physicalism and the logical analysis of reasoning. Many years later, in his 'Intellectual Autobiography' (Schilpp, *The Philosophy of Rudolf Carnap,* La Salle, p. 31), Carnap writes:

> I found that the Polish philosophers had done a great deal of thoroughgoing and fruitful work in the field of logic and its applications to foundation problems, in particular to foundations of mathematics, and in the theory of knowledge and the general theory of language, the results of which were almost unknown to philosophers in other countries. I left Warsaw grateful for many stimulating suggestions and the fruitful exchange of ideas which I had enjoyed.

Carnap's opinion was not isolated; it was shared by other representatives of the VC, whom I met in Pittsburgh in 1980, such as Ernest Nagel, Carl Hempel and Herbert Feigl.

It is also worthwhile to mention Karl Popper here much as he distanced himself from VC. He not only dedicated his book, *Objective Knowledge*, (Oxford University Press, 1972) to Alfred Tarski, but also devoted an entire chapter to him, in which we read:

> Although Tarski was only a little older than I, and although we were, in those days, on terms of considerable intimacy, I look upon him as my teacher in philosophy. I have never learned so much from anybody else. (*Ibid*, p. 322).

As regards the fact that the Polish Logical School, which came into being in the course of one generation, was an extraordinary phenomenon, it is worthwhile to cite the opinion expressed in the work of A. Fraenkel, Y. Bar-Hillel, A. Levy, O. van Dalen, *Foundations of Set Theory*, (Amsterdam, 1973, p. 200):

There exists probably no other country which, in proportion to the number of the population, would achieve so much for mathematical logic as Poland did.

Some interesting examples can be found in the new edition of the *Encyclopedia Britannica* (15th Edition, Macropedia, 1974) (EB). In the section of EB devoted to logic, the achievements of Polish logicians, like Stanislaw Leśniewski, Jan Lukasiewicz, Alfred Tarski, Jerzy Slupecki and many others are quoted at length. (See Appendix)

In conclusion I should like to stress that my picture of the LWS might appear to be too flattering. This is partly due to my aim of demonstrating that the opinion which depreciates the role and the achievements of the LWS is inconsistent with historical truth. However, in order not to run the risk of being charged with bias, I should like also to stress that I duly appreciate the role and importance of the philosophical formation represented by VC; however, since this is commonly accepted, it does not need to be emphasized here. The ingenuity, *élan*, creativeness and energy of the Vienna scholars played a tremendous role in the development of our present-day philosophy. Suffice it to mention the change of the face of philosophy, brought about by the splendid work of Rudolf Carnap, Hans Reichenbach, Herbert Feigl, Carl Hempel, Philipp Frank, Karl Popper and others whose influence on today's philosophy of science, theory of knowledge, logic, linguistics, history of ideas, foundations of various branches of science, etc., cannot be overestimated. It is worth mentioning that Karl Popper often stressed that he never did belong to the VC, yet he was quite close to many of their ideas. His appearance at the first International Congress was described as follows:

The representatives of that Circle formed a rather homogeneous and coherent union; a certain opposition (a 'legal' one according to Neurath) was represented within the Circle by the young Viennese Popper who, in agreement with the content of his recent book *Logik der Forschung*, denied any logical character to the so-called inductive conclusion-inference (cf. *supra*).

CHAPTER 5

# ZYGMUNT ZAWIRSKI: A SHORT BIO-BIBLIOGRAPHY (28.IX.1882 – 2.IV.1948)

Zygmunt Michal Zawirski was born at Berezowica Mała, in the Polish (then Austrian) province of Tarnopol.

His father, Jan, an ardent Polish patriot, took part in the 1863 uprising against the Tsarist Empire. After the suppression of this uprising Jan found shelter in Galicia. For reasons of security he changed his surname from Buchowiecki to Zawirski, and his first name from Jan to Józef. Zawirski had five brothers and three sisters; the family was very closely knit, and the atmosphere at home was highly patriotic.

Zygmunt Zawirski completed his studies and was granted the doctor's degree at the Lwów University in 1910. He was fortunate enough to have, as an academic sponsor, the most eminent Polish philosopher, Kazimierz Twardowski. Zawirski's love and respect for Twardowski lasted all his life. After the death of his master, he wrote an obituary in which he expressed his emotional and intellectual debt to Twardowski ('Kazimierz Twardowski (1866–1938)', Kraków, 1938, Biblioteczka *Czasu*).

In the meantime, in order to make his living, Zawirski became an assistant schoolmaster in 1906. In 1907 he passed exams (in the propaedeutics of philosophy, mathematics and physics) thus becoming fully eligible to teach in secondary schools. During almost a quarter of a century, he worked *con amore* as a schoolmaster. Much as his work was time-consuming, he did not neglect his studies, and published quite a lot during that period (except during the first world war). His tenure in this position was an excellent preparation for his future career, especially from the pedagogical point of view. When Zawirski started his academic career, one of his great merits (always appreciated by both students and colleagues) was the excellent construction of his lectures. They were expressed in a clear-cut manner, and were delivered in an interesting and attractive form; no wonder that a lot of students, not only from the Department of Philosophy, attended his lectures. Based on this experience, Kazimierz Twardowski encouraged his alumni especially as one who wanted to embrace the academic career to begin with the schoolteacher's practice. During the period 1909–1910 Zawirski went abroad in order to continue his studies in Berlin and Paris.

A brief outline of his university career, based on the university questionnaire answered by Zawirski himself, is shown below:

| Where | Post | When |
|---|---|---|
| Inspector General of Education, Lwów | assistant teacher | 10.9.1906–31.8.1907 |
| | schoolmaster | 1.9.1907–31.10.1918 |
| University, Poznán | assistant professor | 1.10.1928–31.7.1929 |
| | professor extraordinary | 1.8.1929–31.10.1934 |
| | professor ordinary | 1.11.1934–31.12.1936 |
| Jagiellonian University (Cracow) | professor ordinary, permanent | 1.1.1937–6.11.1939 |
| | (break for WW2) | |
| | continued | 20.1.1945 |

The above list ought to be completed as it does not include some details of his career. Zawirski's promoter in his examination to obtain the right of academic teaching was one of the most eminent Polish scholars, professor W. Heinrich (1924). Soon after this examination Zawirski started to lecture in philosophy in the general department of the Lwów Technological University. In 1928 the University of Poznán invited him to the chair of the Theory and Methodology of Sciences in the Department of Mathematics and Natural Sciences, where he worked until 1937 when he was asked to accept the Chair of Philosophy in Cracow. This was undoubtedly evidence of the high evaluation of Zawirski's scientific activity. The Jagellonian University was not only the oldest but also the most esteemed Polish scientific institution. Before World War II and immediately after the expulsion of the Germans, Zawirski held the office of Dean in the Department of Mathematics and Natural Sciences. On the memorable day of November 6th, 1939, when all of the Polish professors were deported to German concentration camps, Zawirski remained safe because, fortunately, he was not in Cracow.

Zawirski was an extremely industrious man, engrossed in scientific work, devoted utterly to it. All the same, he always managed to find time for social work. From 1934, he had been the co-editor, and after the war editor, of the highly respected periodical *Philosophical Quarterly* (*Kwartalnik Filozoficzny*). He was the president of the Polish Philosophical Society in Cracow from 1938. Likewise, he was a member of the Philosophical Committee of the Polish Academy of Sciences.

Between 1939 and 1945 Zawirski took part in clandestine educational activities on an academic level.

Zawirski established personal contacts with many eminent scientists not only in Poland, but also abroad. Some factors were important in this respect. He had a perfect command of English, French and German, and travelled a lot. He visited Britain, France, Germany, Italy, Belgium, Holland, Denmark, USSR, Austria, Hungary, Czechoslovakia and Yugoslavia, and some countries, like France, many times.

Thanks to these factors Zawirski always had first-hand information in his field of interest: philosophy, logic and physics. He was not only very well-informed and well-read in these fields, but also took an active and creative part in the discussions on all important problems of his time. He participated not only in Polish congresses but in the international ones as well, such as the congresses of scientific philosophy right from their very beginning, and the general philosophical congresses, always delivering papers on interesting and lively problems. He presented papers at the VIIth Congress (Oxford), the VIIIth in Prague, the IXth in Paris; for the Xth Congress (Amsterdam, August, 1948) he had already sent his paper. The proofs arrived at Cracow, from Amsterdam, one day after his death.

It is also worth mentioning that Zawirski played a very important role as a communicant and interpreter of novel ideas for Polish readers, on the one hand, and to foreign scientists about Polish achievements, on the other. His scholarly output is astounding from the point of view of erudition and breadth. It has a tremendous historical and substantive value.

Prior to passing to the detailed characteristics of his purely scientific activity, let us first establish that his main subjects were the philosophy of physics, and formal logic. Logic interested him both from the points of view of methodology and of the application to natural sciences, as well as that of logic itself. He was also concerned with ontological-epistemological problems, especially dealing with problems of time. In addition to these, he was attracted by problems related to the construction of the philosophical *Weltanschauung*. He prefaced almost all of his works with a historical approach, and only then did he pass to an analysis in the contemporary context.

Being a pupil of Twardowski, he began with descriptive psychology (which, at that time, formed an integral part of philosophy). Soon he passed to investigations conducted in Cracow, under the guidance of professor W. Heinrich, and his disciples Boleslaw Gawecki and Joachim Metallman. The gist of these investigations consisted in the logical analysis of the epistemological foundations of physics and in the philosophical consequences of discoveries in modern physics. Particularly characteristic of Zawirski is the fact that his considerations of the consequences of the theory of relativity and the quantum theory are accompanied by the tendency to maintain the former 'epistemological foundations of knowledge of natural history' ie, the principle of causality, the principle of continuity and the determinist approach. At the same time Zawirski tried to apply to physics the achievements of mathematical logic (e.g. axiomatization of theory, application of multi-valued logic to the quantum theory, etc.). His imposing scholarly production was probably richer by two papers whose manuscripts have been lost, viz. a university-level textbook of the methodology of natural sciences, and a work in the field of the history of philosophy or religion (probably in connection with his studies on Origen). In 1947, Zawirski showed to professor Gawecki an almost complete manuscript of his methodology. During his trip to the Polish mountain resort of Zakopane, his suitcase with manuscripts was stolen. He never had time to reconstruct the lost works.

During the first period of his scientific career, Zawirski inherited from Twardowski his scientific interest. His doctoral thesis was on the border of psychology and philosophy: 'Ilość praw kojarzenia przedstawień' (On The Laws of association of images, and their number), Rzeszów, 1910. In the introduction he presents the historical background of the relevant problems, and then the present state of the subject. In view of the ambiguity of ideas, Zawirski proposes to reduce all the laws of association to one single law that of contiguity. The last term is defined rather perfunctorily; it is to mean association in time, in space, by way of contrast, etc. This is the only work of Zawirski in the field of psychology; it does not contain many original ideas and it would be hardly worthwhile to mention it, were it not for the tremendous range of information in the fields of mathematics, logic and history of philosophy, which is found in the work of this 23-year-old author. It is this mastery of the field of science, as well as the conscientiousness and perspicacity that constitutes the essential factor in Zawirski's future successes.

What also deserves notice was his decisive call for an empirical approach which, by and large, he supported right to the end. During this early period, the tendency to adopt as scientific only ideas that could be examined experimentally, might have been pushed too far. In the name of a somewhat simplified empiricism, in accordance with the 19th-century approach (according to which the essential characteristic of the psyche ought to be consciousness), the idea of an unconscious psyche seemed to involve a contradiction. Zawirski writes: '. . . whatever takes place in subconsciousness will not be the subject of our inquiries.' (p. 59). One might accept as a mitigating factor the circumstance that Zawirski's work was written in 1905, when depth psychology was still in its initial stage. However, Freud's *Die Traumdeutung* (Interpretation of Dreams) was published in 1900, and immediately aroused a great interest. This seems to bear out the generally accepted opinion that Freud failed to be 'a prophet in his own country'. (At that time Galicia was still a part of Austria). Much as he gained the greatest fame in the whole world, yet in Austria he remained, right to the end, only a professor extraordinary (not a full professor, without a chair).

We shall pass now to a brief survey of the most important of Zawirski's works. One may divide Zawirski's achievements into three main groups:

(a) Methodological problems of science (causality, determinism, axiomatic method, *experimentum crucis*)
(b) Logic (multi-valued and intuitionistic logic) application of multi-valued logic to quantum physics, logic and methodology.
(c) The set of ontological questions (problems of time, space and cosmology).

## 5.1 THE METHODOLOGICAL PROBLEMS OF SCIENCE

The first group comprises consecutively: *Causality and Functional Relation* (1912, prize work); 'Zwiazek Zasady Przyczynowości z Zasadą Względności' ('Relation Between the Principle of Causality and the Principle of Relativity'), *Kwartalnik Filozoficzny,*[2] 1924, 'Teoria Kwantów a Zasada Przyczynowosci' ('The Quantum Theory and the Principle of Causality'), *Przegląd Filozoficzny 33*, 1930; *Considerations on the Method of Natural Sciences* Lwów, 1931, Książnica Atlas with Zawirski's paper 'W Sprawie Indeterminizmu Fizyki Kwantowej ('On the Indeterminism in Quantum Physics), [Commemorative Book on the occasion of the 25th Anniversary of the *Polskie Towarzystwo Filozoficzne* (*Polish Philosophical Society*)], Lwów, 1931.

In the above works, which cover a period of twenty years, Zawirski defends the principle of causality against successive attacks directed against this principle, first as one of the fundamental presuppositions of the mechanist viewpoint, next in connection with the statistical character of the laws which triumphantly entered into physics, then in relation to the theory of relativity, and ending with difficulties in connection with quantum mechanics.

One of the philosophical presuppositions of mechanism was determinism based on the principles of the uniformity of nature and of causality. Towards the end of the 19th century, the principle of causality, its value and the limits of its applicability, became the subject of lively discussions in the light of new developments in science. Mach and Avenarius launched a strong attack on causality as an obscure, metaphysical, anthropomorphic ingredient of science. According to them, causal relation should be replaced by functional relations. In addition to not being metaphysical, they are empirically observable and can be treated mathematically in a clear-cut way; hence a functional relation is superior to a causal one. According to Zawirski, Mach's idea is open to some objections. The main arguments adduced by Zawirski are as follows.
(a) *The metaphysical character of causality.* Mach states that the principle of causality smacks of metaphysics. The intermediate proof in this respect is allegedly the fact that only philosophers show an interest in causality, but not the physicists whose voice is, after all, decisive as far as the development of physics is concerned.

Zawirski answers both objections in an accurate and penetrating way:

'. . . with the causality principle . . . Maxwell, Einstein, as well as Laue, and Minkowski, who are physicists of no mean stature, are concerned . . .'

The objection that the causality principle is a metaphysical term is not of slightest consequence. Science cannot do without metaphysical assumptions; otherwise it would cease to exist as science. All generalizations are of metaphysical character. Well, what of it? They are indispensable for the progress of science. Even considerations which are useless or speculative according to some scientists, might play, or have played, an enormous role in the advance of science. Let us take, for instance, says Zawirski, the fruitless search for solutions of certain problems. 'Fruitless,' says he, was the search for a *perpetum mobile*, or for the squaring of the circle. These apparently fruitless efforts had enriched science enormously. The search for an inertial frame helped in the formulation of the general theory of relativity (GTR); the pursuit of *perpetuum mobile* provided foundations for the principle of the conservation of energy;

while studies relating to the squaring of the circle made it possible to understand the role and importance of irrational (transcendental) numbers such as $\pi$, the role played by metaphysics in science is indispensable.

(b) *The heuristic power of causality.* A strong argument in favour of causality is its heuristic power. To support this claim, Zawirski referred to the context of discovery of GTR. Einstein claimed that his confidence in the causality principle was most helpful in formulating an important axiom of GTR, viz. that the laws of nature are the same irrespective of the frame of reference. Einstein's interpretation of the causality principle was in the spirit of a radical empiricism:

'. . . the causality principle is of value as an assessment of the real state of affairs if, and only if, cause and effect appear to be, in the last instance, observable facts' (*Grundlagen der allgemeinen Relativitätstheorie*).

Einstein claimed that non-observance of this advice could result in serious difficulties. As an illustration he gave the example of a defective explanation which was reached on the basis of Classical Mechanics (CM) and of the Special Theory of Relativity (STR).

Let us imagine two identical liquid bodies $S_1$ and $S_2$, rotating about an axis passing through their centers. At the same time we assume that these bodies are so far from one another and from other bodies that any external action can be disregarded. We observe that $S_1$ assumes a spherical shape, while $S_2$ the form of an ellipsoid of rotation. To explain this state of affairs, CM refers to the fact that laws of mechanics are in force in the space $R_1$ with respect to which $S_1$ is at rest, while different laws are in force in the space $R_2$ with respect to which $S_2$ is at rest.

Einstein believes that the above explanation is epistemologically defective. The point is that there exists no empirically perceptible difference between the spaces $R_1$ and $R_2$. It is therefore difficult to admit that one of them is privileged compared with the other one. Einstein proposes to accept the principle that the same laws are valid in all spaces and to give up the rules of CM.

(c) *Some ontological arguments.* The cause-effect relation is asymmetrical, for the cause must precede the effect in time. This follows from the assumptions of relativistic physics. Had the cause-effect relation been symmetrical these events ought to have been simultaneous, which is at variance with the assumptions of GTR.

According to GTR all actions are propagated with a finite velocity (the upper limit of which is the velocity of light in vacuum). Therefore a cause cannot be simultaneous with the effect which is brought about by its action. Even if the time interval between them is infinitely small, it must exist. A cause should precede its effect. The causal relation is asymmetrical, because of the structure of space-time in reality. A functional relation could be either symmetrical (in some cases) or asymmetrical, but a causal relation is always asymmetrical. Therefore causality cannot be crammed into the Procrustean bed of a functional relationship.

Mach would be able to propose a re-formulation of the function idea so as to make it possible clearly to distinguish reversible processes from the irreversible ones. But, according to Zawirski, such a way out could not be considered satisfactory: ' . . . However, one can find that the entire controversy will only boil down to words.'

The irreversible processes which run only in one direction constitute an essential feature of reality. This directional sense of events finds its justification in the description of reality as supplied by relativistic physics.

All events, says Zawirski, can be divided, according to Minkowski's conception, into two classes: those which can be linked causally, i.e. in which the temporal order cannot be reversed, and those in which such a reversal is possible. Minkowski constructed a light cone the surface of which separates the area where the temporal order cannot be reversed (within the cone) from that in which it can be reversed. Hilbert submitted the condition that must be satisfied by the components of the gravitational potential if the temporal order of events is to be preserved in accordance with the GTR for the time-space continuum. In accordance with classical physics and with the *actio-in-distans* principle (momentary propagation of disturbances) the entire universe would influence every event. The relativistic physics, writes Zawirski, has introduced the principle of space-time continuum, thus facilitating the method of discovering the causes of phenomena as they occur. It suffices, according to Zawirski, to enclose this phenomena within a limiting surface, and only observe changes within a certain time across this surface. All events must pass through all intermediate places, and through a certain limiting surface. For every event, it is the immediately preceding state that is decisive, so also for space it is the processes that occur in close vicinity that are decisive.

It can be seen that the development of science did provide additional arguments to the adherents of the causality principle. Yet, as happens frequently, unforeseen difficulties appeared at the same time. What Zawirski had in mind was the inhomogeneity of the space-time continuum, as opposed to the homogeneity of the absolute time and absolute space of classical physics.

The inhomogeneity of space makes it impossible to apply the principle of the uniformity of nature. The formulation of the causality principle stipulates that the same effects follow when the same causes are repeated.

In ancient times one followed Aristotle's belief that space was heterogeneous. The contrast between top and bottom was of an absolute character. For light bodies the 'natural place' to which they tended was the top, while the analogous role was played by the bottom for heavy bodies. Since Newton's times, both time and space had been considered, for nearly three centuries, to be absolute, independent of the processes occurring, and homogeneous. In relativistic physics the time-space has a definite structure, which depends on mass distribution. It is only the empty time-space that resembles the Newtonian one (it is considered to be pseudo-Euclidean). Consequently, the question arises whether it is possible to verify empirically the causality principle, whether 'the same causes' are likely to occur in different places and at different times. In this matter Zawirski took a position similar to that expressed by Russell (cf. his paper 'On the Notion of Cause', p. 17). He draws attention to the fact that the repetition of the same facts ($F_1, F_2 \ldots$) in their entire specificity and individuality does not seem to be possible since every fact (F) is a resultant of actions of a very large number of different factors. One should therefore introduce, as opposed to absolute theoretical sameness, an approximate, i.e. practical, identity which takes place when differences (between $F_1$ and $F_2$) are not greater than the errors of measurement.

The time-space inhomogeneity which results from the theory of relativity can be replaced by the approximate homogeneity in limited, not too large, ranges where, for practical purposes, one can assume the occurrence of homogeneity in space-time.

The analysis of the role and place of the causality principle in science (in the context of the principle of relativity) is of particular interest in connection with Zawirski's empirical approach.

While duly appreciating both the necessity, in physics, of grasping the problems in a clear-cut way, and the role played by the mathematical approach to relationships, Zawirski opposes Mach's claim to ban the causal relationship from science. The causal relationship, he writes, characterizes definite links which occur in reality itself, and that is why it is indispensable to keep this relationship in science.

Zawirski not only defends the place of the causality principle, but he also opposes the excessively radical interpretation of the requirement advanced by Einstein.

Zawirski points to the difficulties encountered when one applies the principle of 'empirical observability', recommended by Einstein. Thus it is not clear what should be understood by 'empirical observability': is this the sensual observability or observability by means of instruments?

A further problem is that of interpolation and extrapolation of empirical data, which go beyond experiments. However, no generalization can be carried out without these measures. And, without generalization, science would be only a sum of disconnected facts, thus being unable to fulfill the fundamental functions of explanation and prediction.

There is a difference between purely metaphysical hypotheses and extrapolation or interpolation based on facts. When assuming the propagation of changes in a continuous way we carry out an interpolation and often an extrapolation of empirical data. Then we submit our mental construction to an experimental control. In the case of metaphysical hypotheses such a test is out of question.

The postulate of 'empirical observability' had been advanced many times before Einstein, e.g. by Hume and Kant. Even earlier by Leibniz when he formulated his principle of observability: 'I answer, motion does not indeed depend upon being observed; but it does depend upon being possible to be observed.' ('Je réponds, que le mouvement est indépendant de l'observation, mais qu'il n'est point indépendant de l'observabilité') 5$^{th}$ letter to Clarke, *The Leibniz-Clarke Correspondence*, H.G. Alexander (Manchester Press, 1956, p. 74).

Before long the principle of causality again came under scrutiny, mainly in connection with quantum mechanics and with the Heisenberg indeterminacy principle. Zawirski took part in the heated discussion of this subject. During the Oxford Congress, he delivered a lecture in French 'The quantum theory and the principle of causality'. He also published a paper on the same subject in the Polish periodical *Przegląd Filozoficzny,* vol. *33*, 1930. Here again he defends the place and role of the causal principle in science. He compares the evolution of views on time and space with the change of opinions on the causality principle:

> . . . the adoption of absolute space and time made it more difficult to understand the formulae of relativistic physics. Consequently it became necessary to revise and transform fundamentally views on this subject. In the case of the causal relationship, the situation is completely different. The development of physics confirmed the existence of causal relations as specific ones in reality itself.

The advent of quantum physics did not undermine decisively the existence of causal relationships, though science had to face serious problems. According to Zawirski, the preservation of the causality principle does not introduce new difficulties, but, on the other hand, its rejection puts us up against insoluble riddles. According to Zawirski it is possible for the so-called duality phenomena to be approached using the (Łukasiewicz) three-valued, perhaps multi-valued, logic. Zawirski takes up studies of the connection between multi-valued logic and the probability calculus. He thinks that the 'probabilistic' logic would constitute a possible solution of this difficult situation. In this approach, the causal relationship would be connected with the calculus of probability and would be of a probabilistic character (we shall revert to this question in connection with Zawirski's logical works).

The final conclusion drawn from the above considerations consists in leaving the question, so to say, open. The solution, writes Zawirski, is likely to emerge with the further development of science.

### 5.2. LOGIC AND ITS APPLICATIONS TO PHYSICS

For a very long time logic had been considered to be a domain which had been completed once and for all time by Aristotle.

This was gradually changed not earlier than at the end of the 19th and the beginning of the 20th centuries. I am alluding here to the pioneering work of De Morgan in the field of syllogisms, and, in the 20th century, to the development of three- and many-valued logic, as well as of intuitionist logic.

Just as the invention of non-Euclidean geometries undermined the philosophical interpretation of Euclidean geometry as synthetic judgement *a priori* and the only possible description of our space, so also the creation of multi-valued logics called in question the prevailing opinion, that logic constitutes the description of the laws of human thinking.

At present the classical logic of Aristotle represents a small fraction of modern logic whose basic part is the so-called sentential calculus. The development of logic enabled the work of Aristotle to be understood better and more profoundly.

In the earlier centuries the wrong interpretation of Aristotle had its main root in an insufficient knowledge of logic, and especially of the structure and function of deductive systems. The discovery of the key for the understanding of traditional logic became an inspiration for the further impressive development of logic.

Jan Łukasiwicz devoted many years to his studies of Aristotle's syllogisms. The results of these studies were presented (in June, 1939) at the session of the Department of History and Philosophy of the Polish Academy of Sciences and Letters, in Cracow. Lukasiewicz demonstrates *inter alia* that Aristotle's syllogisms, contrary to earlier interpretations which presented them in the form of inference patterns, were formulated by Aristotle as logical theses or laws in the form of implications, where the antecedent is produced by the conjunction of premises, and the consequent by their conclusion. Aristotle's syllogism is a deductive system constructed by the axiomatic method. Lukasiewicz completes this system and accurately defines its theorems by using the modern formal method.

The development of investigations of Aristotle's logic and the studies of his original texts have also made it possible to discover new problems in the domain of modal sentences. This, in turn, has played a motivating role in the creation of multi-valued logics. The point is that Aristotle was extremely close to attributing a third logical value to modal sentences relating to future events. We have here in mind in the first place a chapter in which we read (19a 23–19a 39):

> Now that which is must needs be when it is, and that which is not must needs not be when it is not. Yet it cannot be said without qualification that all existence and non-existence is the outcome of necessity. For there is a difference between saying that which is, when it is, must needs be, and simply saying that all that is must needs be, and similarly in the case of that which is not. In the case, also, of two contradictory propositions this holds good. Everything must either be or not be, whether in the present or in the future, but it is not always possible to distinguish and state determinately which of these alternatives must necessarily come about.
>
> Let me illustrate. A sea-fight must either take place to-morrow or not, but it is not necessary that it should take place to-morrow, neither is it necessary that it should not take place, yet it is necessary that it either should or should not take place tomorrow. Since propositions correspond with facts, it is evident that when in future events there is a real alternative, and a potentiality in contrary directions, the corresponding affirmation and denial have the same character.
>
> This is the case with regard to that which is not always existent or not always non-existent. One of the two propositions in such instances must be true and the other false, but we cannot say determinately that this or that is false, but must leave the alternative undecided. One may indeed be more likely to be true than the other, but it cannot be either actually true or actually false. It is therefore plain that it is not necessary that of an affirmation and a denial one should be true and the other false. (*On Interpretation* 19a 23–19b 2: tr. E.M. Edghill)

Aristotle also dealt with modal sentences, on two occasions, in *Analytica* 2,8–22 and in *Peri Hermeneias* 12–13. In these texts two different meanings of the term 'is possible' and 'it happens' appear. In *Peri Hermeneias*, $p$ is possible when $-p$ is not necessary or if $p$ is not impossible. On the other hand, in accordance with some passages of the *First*

*Analytica*, it does not suffice to admit that *p* is possible for *p* not to be impossible; *p* must be, in addition, not necessary. Thus, what is not possible is either necessary, or impossible. (Aristotle also uses multivocally the terms 'necessary' and 'accidental'.)

Opinions differ as to Aristotle's views. Some philosophers accuse him of a certain inconsistency. In his *Metaphysics*, Aristotle claims that the principle of the excluded middle is of general value, whereas in the text adduced above he gives up this idea with respect to sentences relating to future possible events.

The problem of modal sentences and the possibility of attributing to some such sentences a different value was the subject of a lively discussion at the beginning of the 20th century. Zawirski followed this discussion quite assiduously, and took part in it.

In 1914 he published a discourse 'On the Modality of Judgements', (PTF, Lwów). This discourse is of a historical significance for today's readers. It is a kind of a critical analysis of the discussion. Zawirski presents in it, in a systematic way, the development of the problems involved, starting from ancient times. Next he introduces the reader, as objectively as possible, to the course of the modern discussion, taking into consideration its most important representatives: Sigwart, Windelband, Wundt, Twardowski and Husserl. It is only on this basis that Zawirski presents his own point of view, declaring decidedly for the two-valued logic. (We shall see that later he ceased to be so sure.)

The proposition, as an expression of a certain conviction, or as possessing an objective true value, does not allow any further differentiation as regards the factor which forms its essence. The proposition can be false, or unfounded, it can refer to a state of uncertainty and only state uncertainty, but neither itself nor the truth contained therein can be weaker or stronger, more or less intensive, more or less certain. A doubtful or uncertain proposition is a *contradictio in adiecto* (p. 107).

The publication of the discourse on the 'Modality of Judgements' was followed in the creative work of Zawirski by a longer, almost seven-year break, caused by the war.

In the meantime Jan Łukasiewicz published a momentous exposition. The studies of the Aristotelean logic and the analysis of modal sentences inspired Lukasiewicz to formulate a new three-valued logic. In addition to true (1) and false (0) values, he introduces a third value (½).

Łukasiewicz attributes the logical value (½) to modal propositions relating to the future, of the type 'it is possible that *p* will take place'. Such propositions are neither true (1) nor false (0). This logic is sometimes called non-Aristotelean. If anything, it ought to be called non-Chrysippan. The Stoic Chrysippus stuck to the principle of two values. Aristotle (as can be seen above) did not apply this principle in the case of assertions relating to possible future events. Lukasiewicz presented his idea for the first time at one of the sessions of the Lwów Philosophical Society (1920, proceedings in *Ruch Filozoficzny,* 5, p. 170). He treated the same problem in more detail in 1930 in an article: 'Philosophische Bemerkungen zu mehrwertigen Systemen des Aussagenkalküls' ('Proceedings of Warsaw Scientific Society') Towarzystwo Naukowe Warszawskie, 1930. The bold new idea of Lukasiewicz gave rise to a lively discussion. Zawirski took part in it by publishing a series of papers in the field of logic.[3]

The main subject he is interested in is the possibility of using multivalued logics for solving the numerous difficulties which appeared with the development of physics (quantum mechanics) and mathematics.

Zawirski sees the merits of Łukasiewicz's idea in that it makes it possible to eliminate certain logical antinomies and difficulties arising in mathematics and in quantum mechanics:

(a) The paradoxicality of antinomies in which one arrives at the equivalence of a pair of contradictory sentences disappears if one attributes to the antinomian sentences this third logical value which is neither true nor false.

However, Zawirski believes that a better method of removing these antinomies is the theory of types as suggested by Russell (were it only because it does not go beyond two-valued logic).

(b) . . . some propositions of the theory of sets are called in question because the method of proving them is not obvious (e.g. the so-called diagonal procedure) . . . Since the three-valued logic removes these suspect methods of demonstration not directly, there occurs likewise here . . . the possibility of using this logic for revising the correctness of demonstrating some propositions in the theory of sets (p. 3).

(c) The new logic can be applied to quantum mechanics. 'The indeterminism of modern physics appears to supply a field of applicability to the new logic'. 'The three-valued logic provides the only possible way of understanding the complementary theory' in micro-physics.

While appreciating duly the values of Łukasiewicz's suggestion, Zawirski realizes the numerous difficulties associated with the relinquishment of the two-valued logic.

The difficulties one finds in the application of the three-valued logic compared with the traditional way of handling marks are mainly associated with compound sentences. Thus, for instance, the negation of a sentence with a logical value of '½' likewise obtains the value of '½'. Thus, from among the tautologies of traditional logic, it is only certain tautologies that retain the value of truth '1'. Laws of logic, which are as important as the law of contradiction and the law of excluded middle, assume the value of '½'. They are merely possible. But this is not the only problem.

Zawirski considers that the main shortcoming of Lukasiewicz's logic consists in that it does not satisfy the postulate of correspondence. Roughly speaking, the postulate of correspondence represents the requirement that later theories ($T_n$) ought to remain in correspondence with the earlier theories ($T_{n-1}$). The correspondence ought to consist in that the equations of the theory $T_n$ pass into equations of the theory $T_{n-1}$ in the limiting cases, i.e. when certain parameters occurring in both theories tend either to zero or to infinity, or to a certain critical value.

The authority which was enjoyed by the principle of correspondence among scientists was due to the conviction that this principle provided a kind of guarantee that nothing will be lost from what is valuable in the achievement of the development of science. At the same time this principle makes it possible to conduct a critical analysis and to understand the shortcomings in the preceding stage of the development of science, from a new, more mature point of view.

Zawirski writes that the renouncement of the principle of correspondence would make it more difficult or even impossible to understand the mechanism of the development of science:

Once upon a time we adopted the postulate that, whenever science is forced to renounce a certain epistemological principle which was assumed to be intuitively certain, one ought to admit in its place such a new principle that the former principle would be valid at least as a limiting case of the new one: otherwise it would have been incomprehensible in what way principles which are intuitively certain in science could result in arriving at a stage that would force us to go beyond these intuitive principles ('The Indeterminism of Quantum Physics', p. 25).

| p | q | $p \supset q$ (Łukasiewicz) | $p \supset q$ (Brouwer-Heyting) |
|---|---|---|---|
| 1) 1 | 1 | 1 | 1 |
| 2) 0 | 0 | 1 | 1 |
| 3) 0 | 1 | 1 | 1 |
| 4) 1 | 0 | 0 | 0 |
| 5) 0 | ½ | 1 | 1 |
| 6) ½ | ½ | 1 | 1 |
| 7) 1 | ½ | ½ | ½ |
| 8) ½ | 0 | ½ | 0 |
| 9) ½ | 1 | 1 | 1 |

It is from the point of view of satisfying the correspondence postulate that Zawirski compares Łukasiewicz's logic with that of Brouwer. Now the table on page 47 is the comparative table for the implications '$p \supset q$' in both logics.

We can see from the table that the intuitionist logic comes nearer to the traditional one. When the antecedent has a higher value than the consequent, the implication is false in the traditional logic. In Lukasiewicz's logic it is possible, '½'. In Brouwer's logic, however, it is false (no. 8 in the table).

A still weightier argument pointing to the advantage of the intuitionist logic is the fact that, in the two-valued logic, there are no propositions that ought to be rejected in Brouwer's logic as being false; only some of them are possible. The situation appears to be different in Lukasiewicz's logic. E.g., the law of negation of the equivalence of two contradictory sentences $-(p \equiv -p)$ is false in Lukasiewicz's logic, since for $p = ½, -p = ½$. In Brouwer-Heyting's logic this law continues to be true owing to a different interpretation of the negation of the negation.[4]

The main problem of paramount importance for Zawirski was the application of multi-valued logic to the task of interpreting the probabilistic laws of quantum mechanics. Lukasiewicz's logic had only three values ('1', '0' and '½'). This was not enough for the interpretation of the calculus of probability.

Zawirski tries to effect the necessary alterations and works out his own variant of a multi-valued logic which is to fulfill the following conditions; it should:

(1) be in the relation of correspondence with the traditional logic,
(2) link the truth value of a proposition to its probability
(3) with reference to the compound propositions (such as the logical sum and product) link their value, in a univocal manner, to the truth value of the arguments.

To this set of problems he devotes two following works: 'Significance of Multi-Valued Logic' and 'The Relation of Multi-Valued Logic to the Calculus of Probability'. The task which Zawirski had set for himself was far from being an easy one: '. . . the junction of logic and probability calculus is impossible *a priori* since in the logic of propositions, the values of the sum and the product are one-to-one functions of the values of their propositional arguments, while the probabilities of the sum and the product are not one-to-one functions of the probabilities of arguments. Different probability values can correspond to the same argument values.

Hence the former values cannot be considered as new logical values (apart from '0' and '1').

The order of introduction of arguments is here extremely important.

To achieve a situation where the logical product and sum are one-to-one functions of their arguments Zawirski had to introduce new logical factors:

'. . . we have increased the number of logical operators of the sum and of the product in such a way that only one logical value corresponds to each of these factors. Let us take, for example, the logic of five values symbolized by the series of numbers 0, 1/4, 2/4, 3/4, 1. It is true that, in the logic of Lukasiewicz and in that of Post, the logical sum acquires only one value; for the values of the arguments v(p)[5] = 2/4, and v(q) = 2/4, the sum obtains the value 2/4, while the calculus requires for the same values the value 2/4 as well as 3/4 or even 1. However, we have assumed for each of these cases another definition of the sum, thus we obtained three different formulae, and the values of each of them became a one-to-one function of the value of their arguments.

This innovation was not a whimsical work; we were authorized to adopt it for very important reasons. Two possible definitions of the logical sum (and of the product) pvq = (p ⊃ q) ⊃ q, or pvq = ~ pvq, which are equivalent in ordinary logic, are not equivalent in the many-valued logics. For v(p) = 2/4 and v(q) = 2/4, the former sum obtains the value 2/4, while the second yields 1. It was therefore necessary to introduce these two formulae into the new logic.

The same result can be achieved by considering each proposition of the n-valued logic as one class of (n–1)-valued propositions'.

In our example of the logic with five values, the different formulae of the sum are justified by the difference in the order in which the true propositions follow false propositions in these classes. If v(p) = 2/4 and v(q) = 2/4, the corresponding classes, containing two true propositions for four true or false ones, can have the forms (the true propositions being denoted by 1 and the false propositions by 0): p = (0, 0, 1, 1), q = (0, 0, 1, 1), or else p = (1, 1, 0, 0), q = (0, 0, 1, 1), or again p = (0, 0, 1, 1), q = (1, 0, 1, 0). Let us now interpret the logical sum p   q   as a number of two-valued logical sums formed by joining the first member of the first series with the first member of the second series, the second member of the first series with the second member of the second series, etc., and we shall obtain in the former case p   q = (0 0, 00, 1 1, 1 1) = (0, 0, 1, 1) the value of the sum will be 2/4; in the second case p   q = ( 1 0, 1 0, 0 1, 0 1) = (1, 1, 1, 1) it will be 1, the case of the maximum value for the sum; finally, in the last case p   q = (0 1, 0 0, 1 1, 1 0) = (1, 0, 1, 1), the value of the sum will be 3/4. (*Ibid.*)

The results achieved were presented by Zawirski at the Prague Conference in 1934, and then at the session specially devoted to this subject during the First International Congress of Scientific Philosophy in Paris (September, 1935). During the same session the main speaker on this subject was Hans Reichenbach, who completely independently of Zawirski worked on the logic of probability.

The discussion of both papers (Reichenbach's and Zawirski's) took place jointly.

A fundamental objection was raised against both conceptions: why does one create a novel multi-valued logic when '. . . the theory of probability can be comprised within the framework of ordinary (two-valued) logic?'

Zawirski writes that it would appear that only an apparent illusion of many-valued logic is created if one must substitute, for each many-valued proposition, a class of two-valued propositions. One situation is different, namely one ought to distinguish the proposition '*in suppositione formali*' from the proposition '*in suppositione materiali*':

> The probability is related to propositions, hence it is not the propositions in *suppositione formali* that here play the role of arguments, but rather the names of these propositions, propositions in *suppositione materiali*. We set forth the propositions about the probability of the mentioned propositions, so to say in inverted commas. If the names of the propositions ought to be divided into more than two classes, nothing follows for the propositions themselves and nothing entitles us to look for more than two values of the latter ones. (*Ibid.*)

Both during the Congress itself, and still more distinctly in his 'report' of the Congress ('Impressions from the First International Congress of Scientific Philosophy'), Zawirski stresses the differences between his own idea *re* multi-valued logic and that of Hans Reichenbach, differences to which nobody paid sufficient attention. These differences are as follows:

(1a) Reichenbach interpreted some expressions obtained directly from the probability calculus as a kind of generalized logic.

(1b) Zawirski obtained a certain parallelism between the formulae of the calculus of probability and those of Łukasiewicz's and Post's many-valued logic, after having carried out some changes in that logic. Thus a certain formal agreement was achieved between many-valued logic and the calculus of probability.

(2a) According to Reichenbach many-valued logic was valueless. He believed that one cannot, in a suitable manner, attribute more than two values to a proposition.

(2b) For Zawirski, the many-valued logic is of great significance: He achieved a formal agreement between many-valued logic ('topological' or 'modal' according to Reichenbach's terminology) and the calculus of probability. According to Zawirski, one has to treat the many-valued logic and the probability calculus as two different systems, one of which provides the empirical basis for the other:

The formulae of the calculus of probability, based undoubtedly on the two-valued logic, create an empirical foundation for new propositions relating to the future, which themselves already belong to the many-valued modal, or topological, logic.

With time, Reichenbach changed his opinion, and so did Zawirski. Reichenbach gradually ceased to consider the topological, or qualitative interpretation to be utterly valueless. Zawirski, on the other hand, arrived at a better understanding of the role played by the definition of probability, which was sometimes the cause of misunderstandings.

It appeared to us then that our adversaries, who reject any links between the calculus of probability and the many-valued logic, frequently simplify the problem when accepting the classical definition of the measure of probability; it is only then that one can accept the plain answer that probability concerns the names of the sentences about which we pronounce the sentences of two-valued logic. However, if we accept the frequency definition, then the matter will not be as simple as that, because the probability sentence will be found to be also probable, and so *ad infinitum*, and therefore we shall never reach the sentence of two-valued logic.[6]

In 1938 Zawirski published a highly interesting paper (in Polish) on 'The Importance of Logical and Semantical Investigations for the Theories of Contemporary Physics' (*Przegląd Filozoficzny 41*, 1938, pp. 15–30). A brief summary of the paper is given here:

The relationship between logic and science is a mutual one. This can be seen from both sides. The new developments in the field of deductive systems are of great importance for the enlightenment and understanding of the theories in science. On the other hand, new achievements in physics, new theories, such as quantum mechanics or the theory of relativity, provide a strong stimulus for the development of logico-semantical investigations.

Zawirski supported his claim by some examples.

The problem of finding a demarcation line between science and non-science was a very hot one in the Thirties. One of the most discussed problems was the question: when could a proposition be called 'sensible' or 'scientific'. The Vienna Circle philosophers proposed calling a proposition a 'scientific' one only if its truth value could be decided directly by experience, or by reference to the so-called protocol sentences (observational or otherwise basic sentences of a certain distinctive type; the character of protocol sentences was a subject of lively discussions). Zawirski wrote:

It is said that a proposition makes sense, or is a scientific one if it can be verified or falsified (by experience) [*op. cit.*, p. 26].

When analysing this proposition, Zawirski put forward the question whether sensibility can be identified with decidability (of the true value) of a proposition.

In the first place the description of a sensible proposition, suggested above, cannot be applied to deductive systems. In such systems the language depends on our arbitrary choice. It is commonly accepted that in such systems a proposition makes sense if it can be expressed in the language of the system. Yet such a proposition, though making sense according to the above description, could appear to be undecidable. The recent investigation of deductive systems has shown that, apart from the so-called complete systems, where every proposition expressed in the proper language is either a thesis of that system (true) or else contradicts it (false), there exist also incomplete systems lacking this property. In incomplete systems there are propositions which, though expressed in the language of the system, are neither true nor false. Such propositions are undecidable. For example, to the surprise of the scientific community, arithmetic was found to be an incomplete deductive system. Fermat's theorem added to arithmetic does not lead to a contradiction, but all attempts to obtain a proof of this theorem in a general form have failed. Moreover, Kurt Gödel invented a general method of constructing 'undecidable' propositions in arithmetic and all systems which are superstructures of arithmetic.

This result is of great importance – not only for deductive systems, *in abstracto*, but for physical theories as well. Physical theories are superstructures of arithmetic: 'Therefore every physical theory carries in itself, in the quality of applied arithmetic, problems which are, in a sense, undecidable.'

When analysing the status of physical theories one has to remember that they are based not only on arithmetic, but also on logic.

New developments in logic and, especially, in the many valued and intuitionistic logic gave rise to some important problems. For example, in the so-called intuitionistic logic of Brouwer and Heyting the law of excluded middle is not valid. In this logic, apart from the true and false propositions, there also exist undecidable propositions.

Consequently, Zawirski concludes very rightly: '. . . sensibility must not be identified with decidability.' (*op. cit.* p. 26).

The next problem (the importance of which could not be overestimated), initiated by Popper's famous *Logik der Forschung* (1937), is that of induction.

Zawirski's approach to Popper's concept of 'falsifiability' as the most important criterion of a theory's being scientific was rather critical. This applied likewise to Popper's rejection of induction as a proper method in science.

According to Popper there is a basic asymmetry between verification on the one hand and falsification on the other. Empirical verification (of the consequences) of a theory does not prove that the theory in point is a true one. On the contrary, the falsification of a theory is a final proof that the theory is false (*Modus tollendo ponens*).

Notwithstanding his great respect for Popper, Zawirski subjects Popper's stand to a detailed critical analysis. Here in brief are his arguments:

1. Induction and deduction are equally important in the natural sciences. Both are indispensable for science. Induction is more important at the stage when a theory or a law are formulated *in statu nascendi*, while deduction plays its part during the process of testing.
2. Falsification, like verification, is never of a final character, since one does not test one singular theory, but always a group of theories or laws. One never knows for certain on the basis of a negative outcome of an experiment which theory exactly has been falsified.
3. Falsification (as well as verification) is based on protocol sentences. But protocol sentences are not therefore at rock-bottom; they can be undermined as well. They depend on a number of factors: technical, theoretical and linguistic.
4. The development and changes of science are of an infinite nature. One never knows whether truth had been already achieved. Although, thanks to Tarski, we have a definition of a true sentence, we do lack a criterion of truth: '. . . the mere definition of a true sentence does not provide us with a criterion of truth' (*op. cit.* 30).

Zawirski concludes: '. . . a falsification by way of an experiment or an observation has the same relative character as verification; hence if we look more closely at the verification and falsification processes, we shall find that the apparent asymmetry does not exist' (*op. cit.* 29).

In his book published 25 years later, *Conjectures* and *Refutations* (Routledge and Kegan Paul, London, 1971 [1st ed. 1963], Popper answers Zawirski's objections (without mentioning his name) in a footnote (pp. 41–42):

In the *Logic of Scientific Discovery* I discussed, and replied to, some likely objections which afterwards were indeed raised, without reference to my replies. One of them is the contention that falsification of a natural law is just as impossible as its verification. The answer is that this objection mixes two entirely different levels of analysis . . . . On the first level, there is a logical asymmetry. A one singular statement – say, about the perihelion of Mercury – can formally falsify Kepler's laws; but these cannot be formally verified by any number of singular statements. The attempt to minimize this asymmetry can only lead to confusion. On another level, we may hesitate to accept any statement, even the simplest observation statement; and we may point out that every statement involves *interpretation in the light of theories*, and that it is therefore uncertain. This does not affect the fundamental asymmetry, but it is important: most dissectors of the heart before Harvey observed the wrong things – those which they expected to see. There can never be anything like complete safe observation, free from the dangers of misinterpretation. (This is one of the reasons why the theory of induction does not work.) The 'empirical basis' consists largely of a mixture of *theories* of lower degree of universality (or 'reproducible effects'). But the fact remains that, rclative to whatever basis the investigator may accept (at his peril), he can test his theory only by trying to refute it.

Popper is, of course, right when stating that there exists an asymmetry between falsification and verification. Zawirski's formulation, above, that the 'asymmetry does not exist' is not correct. However, this is solely a logical asymmetry. In natural sciences the situation is much more complicated, and some other factors, 'technical, theoretical, and linguistic ones' play a decisive role. In this respect there is no difference between verification and falsification. Popper's mistake here is a subtle one, it is an overestimation of the importance of the logical asymmetry and what follows, an anathematizing induction which could in effect cripple science, by depriving it of one of its very useful tools.

To the problem of the relation between a hypothesis and the empirical basis, Zawirski returned once more in a short paper 'Observations on the Method of Natural Sciences' (cf. this volume).

In this paper he analyses not only the problem of testing one hypothesis, but also tackles the very hot question put forward by Duhem, viz. the possibility of an *experimentum crucis*. According to Zawirski there are many similarities between these two problems.

Zawirski started by stressing that, in the case of a negative outcome of an experiment, a simple application of *modus tollendo tollens* would not be appropriate.

Nevertheless some reservations are indispensable here. The total refutation of the hypothesis takes place only when no change had been introduced into the ideas entering into the composition of the hypothesis and playing a role in the description of the experiment. The point is that, if one changes the meaning of the terms applied, even to a very small degree, the refuted hypothesis can be taken up again. Thus, for example, the corpuscular theory of light, which was rejected after Foucault's experiments, was again reintroduced into physics owing to the quantum theory when, along with various changes in the theory of light, the meaning of the word 'corpuscle' was changed. The same applies to the case of '*experimentum crucis*', and '*instantia crucis*', i.e. to the case when, out of the two competing hypotheses or theories equally suitable for being adopted, the one selected is that which predicts a newly discovered fact. Out of the two hypotheses equally suitable for being adopted on the basis of the same phenomenon discovered, one hypothesis will be verified, while the other one will be falsified. There exists an analogy between the testing of a singular hypothesis and the solution of a conflict between two competing hypotheses. In either case we have to choose one of the several possibilities which form an alternative. In the latter case the alternative is only more complex. In the former case one expects to obtain an affirmative or negative answer to one question only. In the latter case the positive answer with regard to one hypothesis is, at the same time, a negative answer with regard to the other hypothesis. Nevertheless, one can easily note an analogy between the two cases and notice the difference between them if one becomes aware of the content and the complexity of the hypothesis. Let us assume that the hypothesis is a logical conjunction of three propositions 'p', 'q', 'r' and that it is rejected in view of the falsity of its consequences. One obtains a negation of this conjunction. However, this negation '$\sim$(p,q,r)' can be transformed in accordance with the laws of logic in such a way as to yield a seven-member alternative:

$$\sim(p,q,r) = [(\sim p.q.r) \vee (p.\sim q.r) \vee (p.q.\sim r) \vee (\sim p.\sim q.r) \vee (\sim p.q.\sim r) \\ \vee (p.\sim q.\sim r) \vee (\sim p.\sim q.\sim r)]$$

Having rejected the conjunction 'p.q.r' we face the necessity of choosing one of the seven remaining possibilities. At the same time, the more complex the hypothesis the more possibilities we would have at our disposal. It is only in the cases of very simple hypotheses that the rejection of one hypothesis is equivalent to the verification of only one other hypothesis. However, such cases are rare. (This volume, pp. 187–188.)

Nevertheless, says Zawirski, the situation is not so hopeless in practice. In every law there are involved many definitions and conventions and sometimes it suffices to alter one of these. However, sometimes science comes across situations where the whole edifice of physics has to be changed. This is a revolutionary situation, and no small changes could solve the difficulties. There are likewise cases, says Zawirski, when two originally inconsistent theories are later found to be capable of co-existence:

A few years ago we suggested in this connection the use of the three-valued logic in quantum physics. However, prior to doing this it was necessary to examine thoroughly whether the ideas of wave and corpuscle are really contradictory. It was found that they are not. The wave and the corpuscular theory are, to a certain extent, mutually inconsistent, but one of them is not a negation of the other. That is why one has rather to adopt a wait-and-see attitude than to find a more specific answer, viz. that the *h* constant and, consequently, the law of Heisenberg create an ultimate limit for the testing of the nature of the world microstructure. (*ibid.*)

I think that Zawirski's presentation of the situation in quantum mechanics (QM) is a slight over-simplification. The two complementary aspects, the wave and the corpuscular one in the quantum theory, unfortunately cannot be treated in the same way as the old wave and corpuscular theories. They are, in QM, a part of a quite different theoretical system, and all the difficulties of the so-called 'incommensurability' problems would be involved here. However, Zawirski cannot be blamed for failing to notice the problems involved, since the discussion of incommensurability took place one generation after his publication.

I should like to mention here, rather sketchily, an interesting and quite comprehensive essay: 'The Axiomatic Method and Natural Sciences' ('Metoda Aksjomatyczna a Przyrodoznastwo'), *Kwartalnik Filozoficzny 1923*, Gebethner and Wolff, Kraków, 1923, 125 pages.

Here Zawirski subjects the new achievements in the analysis of deductive systems, due to the development of the axiomatic method, to a penetrating analysis. Even more interesting is the application of these achievements to physics. His final conclusion is that science can neither exist nor develop without adopting certain ('metaphysical') constitutive principles (like the admission of regularities in nature, the principle of correspondence, and so on). He proposes to interpret Kant's *a priori* not as a feature of the human mind but as an acceptance of such principles *a priori* when starting the scientific investigations. The content of the '*a priori*' cannot be petrified but should vary with the development of science. Besides, one will notice how far Zawirski departed from his youthful radical empiricism expressed in his doctoral thesis (*vide supra*).

Here is a brief summary of his essay.

As usual, the author devotes the first part to historical considerations related to the genesis and evolution of the axiomatic method. The genetical rationalism[7] admittedly has been related, at least partly, to the opinion prevailing for many centuries, that some geometrical axioms have an intuitively certain character, that they are innate and, what is more interesting, that they show an amazing agreement with reality. The discussion relating to Euclid's fifth postulate, and the emergence of non-Euclidean geometries which included an axiom contradicting the fifth postulate, had cast a new and unexpected light upon the character of axioms and the conditions which must be satisfied by the set of axioms in a deductive system.

It was found, first of all, that, contrary to the belief prevailing from ancient times, axioms may completely lack the character of obviousness. The same applies to symbols and terms occurring in the formulation of the axioms. Their meaning is constituted by the axioms which, in this capacity, ought to be treated as definitions of these terms. Thus one eliminates some intuitive elements from the system.

The intuitive obviousness of some sentences, and this should not be forgotten, writes Zawirski, has played an important role in the process of formation of our knowledge. However, the departure from this requirement is a new stage in the development of the deductive systems and extends widely the range of our knowledge. Perhaps the most spectacular expression of this case is found in the non-Euclidean geometries whose propositions are non-obvious, and even exceed the boundaries of our imagination.

The axiomatization of deductive theories plays an immense epistemological role. Let us adduce some examples. The process itself, of looking for sentences from which all other sentences could be derived, while not an easy one, is nevertheless highly instructive. The point is that it enables us to understand better and more profoundly the structure of the theory involved. The conditions to be satisfied by the set of axioms of a deductive system are as follows. The axioms should be:

(1) Complete. This means that they should comprise all necessary axioms, and not a single superfluous one.
(2) Consistent. They must not imply statements which are mutually inconsistent.
(3) Independent. No axiom can be a consequence of other axioms.

Notwithstanding the apparent simplicity and obviousness of conditions which should be satisfied by axioms, it was found that is is by no means easy to establish that these conditions have been fulfilled. For example, condition (2) of the consistency of axioms has been found to be extremely difficult to ascertain.

The point is that inconsistency can be hidden very deeply, and it could sometimes not be revealed during the analysis of the axioms, but only following an inference from statements which can be separated from the axioms by a number of logical steps. We cannot even foresee whether and when an inconsistency will be revealed. Sometimes it is so difficult to establish the consistency directly that one has to look for an indirect method. Thus, e.g., Hilbert established the consistency of geometrical axioms by their interpretation in terms of integers, assuming that a set of integers cannot be inconsistent. The problem of finding the proofs of consistency for a given system, without assuming the consistency of other systems, remains unsolved up to now.

Condition (3) on the independence of axioms is likewise not easy to establish, as can be seen from the history of mathematics.

Axiom A is considered to be independent if one obtains a consistent system after adding to the remaining axioms an axiom which contradicts A ($\sim$A).

We know that the independence of Euclid's fifth postulate from the remaining ones had been achieved only after many centuries and owing to the formulation of the non-Euclidean geometries. For a very long time, attempts had been made to deduce this postulate from the other ones. To no avail! Poincaré writes in connection with this subject:

What vast effort has been wasted in this chimeric hope is truly unimaginable. Finally, in the first quarter of the nineteenth century, and almost at the same time, a Hungarian and a Russian, Bolyai and Łobachevski, established irrefutably that this demonstration is impossible. (*Science and Hypothesis*, chapter III, pp. 55–56; see as well, Irena Szumilewicz: Poincaré, Wiedza Powszechna, 1978, p. 129).

The axiomatic method did open new splendid horizons before science. Not only did it contribute to the deeper understanding of the structure of deductive systems but it also enabled the deeply-hidden relationships between theories to be found. Thus, e.g., one can give to the same system of axioms various interpretations [models], and transfer the ready-made statements derived in one model to another model of a given theory. An example of this can be found in the duality of projective geometry.

One can adduce many more examples for which the system of axioms possesses more than one model. Thus, e.g., the axiomatization of the laws of logic, comprised by Couturat's algebraic logic, contains a series of formulae which can be interpreted as relating both to relationships between ideas and to relationships between statements.

In this way the axiomatization of a theory, the baring of its logical skeleton, enables us to find deeply-hidden similarities between apparently quite different theories.

The axiomatic method also gives rise to other openings. Out of the axiom of one theory one can form another system by introducing, instead of some axioms, other axioms contradicting the former ones. This is precisely the way in which the non-Euclidean geometries came into being. It should be mentioned that a correspondence does exist between the non-Euclidean geometries and the Euclidean one. The formulae of the non-Euclidean geometries pass, in some boundary cases, into the Euclidean formulae.

Another advantage to which Zawirski rightly draws our attention is the heuristic value of the axiomatic method. Instead of looking for a system of axioms for a given theory and for the related system of ideas, we can proceed in the opposite way. We formulate a system of axioms and only then look for the interpretation of terms defined by the axioms which play the role of the definitions of these terms.

It is evident that the axiomatic method is related to abstraction from the intuitive sense of expressions in which the axioms are formulated. Now one has to differentiate between the degrees of this abstraction. A simple division into 'pure' and 'applied' theories does not suffice in this case. We can distinguish the degrees of abstraction. For the $n$-th degree of abstraction, $n - 1$ is the applied theory. The first degree of abstraction occurs when one abstracts from the intuitive meaning of the terms or symbols of the given special theory, while preserving the sense of logical terms occurring in it. The next step consists in rejecting the intuitive meaning of all terms, including the logical ones. This is connected with an absolutely pure theory in which the combination of signs becomes similar to the movements of chessmen. An example of such a pure axiomatization is Hilbert's attempt at the axiomatization of arithmetic and logic. Of course, at this last degree of abstraction the problem of truth and falsehood in its classical meaning loses any sense.

Zawirski believed that, even in the case of extreme abstraction in a most formalized theory, some intuitive factors cannot be completely eliminated. They include the postulate of existence.

The next chapter of his essay is devoted to the *axiomatization of physics*.

There exists a fundamental difference, says Zawirski, between the methods used in natural sciences on the one hand and mathematics on the other.

> . . . all empirical sciences, all sciences dealing with facts, use a research method which presents the reverse of the deductive method used in mathematics . . . in the latter we reach new propositions which are deduced as consequences from the axioms or from propositions which had been deduced previously from the axioms. Here we proceed in the opposite direction, viz. we explain various phenomena in such a way that we look for some reasons from which such propositions regarding the facts would follow as their consequences (16–17).

Descartes was, strictly speaking, the creator of the program of the axiomatization of physics. He had formulated the program of 'universal mathematics', by assuming that our cognition of the external world ought to be purely mathematical, and that physics ought to be apprehended in a system like geometry. Descartes tried to carry out his idea, yet he had at his disposal only scanty means, and physics was then at too low a degree of development. Nevertheless this intention can be found even today to be a program which has successively been attempted by Newton, and in the 18th century by Lagrange. Hilbert was an ardent adherent of the application of the axiomatic method in physics ('Axiomatisches Denken' *Mathematische Annalen 78*, 1918). He writes:

I am convinced that everything that can be ever a subject of scientific thinking, provided it is mature enough for the generation of a theory, is subjected to the axiomatic method, hence indirectly to mathematics. While inquiring deeper and deeper into the strata of axioms we acquire an ever more profound understanding of the essence of scientific thinking and we realize better and better the unity of our knowledge.

At the beginning of the 19th century and following the example of Newton and Lagrange, scientists began to axiomatize other sections of physics. Fourier set up his analytical theory of heat, whereas Fresnel, on the basis of several hypotheses, managed to explain a whole series of optical phenomena.

In the mid-nineteenth century, the electric and magnetic phenomena could be set into a series of axioms. They expressed Coulomb's law as a basis of electrostatics and an analogous Coulomb's law for magnetostatics . . . the entire theory of electric and magnetic phenomena could be reduced to eight basic equations.

In the second half of the 19th century further research in this direction was crowned by two highly important results: the kinetic theory of matter had shown the possibility of reducing heat to mechanics, while Maxwell's electromagnetic theory of light allowed the laws of optics to be reduced to the axioms of electrodynamics. In this way one could reduce the apparently mutually independent axioms of various sections of physics to two groups: the basic equations of mechanics and the basic equations of the electromagnetic field. However, phenomena began to increase in number which indicated that even these two groups were not independent of one another. The theory of relativity cast new light onto the relation between the laws of mechanics and the laws of electricity. The theory of relativity demonstrated that the electromagnetic field can be completely defined by means of a four-dimensional vector potential. Zawirski writes: ' . . . one could reduce the entire system of physics to fourteen differential equations, ten gravitational and four electromagnetic equations' (*Ibid.* p. 24).

On the other hand, Hilbert drew attention to the fact that the fourteen basic equations of physics can be deduced from one variational principle, from Hamilton's principle, which can be considered to be the most general and the most basic principle of physics. Hamilton's principle has been known in physics for a long time, and it did appear beside a few other equivalent principles, such as the principle of d'Alembert, the principle of Gauss, etc. All of these principles are different formulations of the same proposition, the principle of Hamilton being the most general of them, since, in one equation, it comprises the relationship of such magnitudes as have a significance in all sections of physics, viz. space, time and potential.

Compared with the principle of conservation of energy, the principle of Hamilton has a distinct superiority, since it enables one to determine the direction of processes, while the principles of conservation yield only the energy balance at various stages of the process without, however, determining its direction. The content of Hamilton's principle with regard to purely mechanical systems is as follows:

> . . . the movements of a system of material points subjected to certain forces, when passing from a certain initial state to a final state, are always carried out in such a way that a certain magnitude determined by the position, velocity of material points and forces acting upon them assumes a lesser value than in the case of any other transition from the same initial state to the same final state (p. 30).

By selecting a suitable function of the universe, Hilbert deduces from Hamilton's principle all of the fourteen basic differential equations of physics, the ten gravitational equations and Maxwell's four electromagnetic equations, hence everything which, according to him, is required for the elucidation of all, even the most hidden, natural processes. The mathematical analysis of the results achieved suggests to Hilbert the daring assumption that electromagnetic phenomena are due to gravitation. It is precisely this relation between electricity and gravitation that aroused in Hilbert the hope to achieve the solution of all physical problems by a purely mathematical method.

This view of the relation between physics and mathematics has been expressed very clearly by Haas:

> . . . in the final analysis the so-called axioms of physics will be nothing more than a kind of dictionary which is necessary for translating the purely mathematical properties of Minkowski's universe, which emerge from geometric axioms, into a language used by experimental physics based on sensual experience. ['The Foundation of Physics' ('Grundlagen der Physik'), *Götting. Nachr. Math.-phys.*, 1915, Heft 3, p. 407]

In the end of his review, Zawirski stresses that the introduction of axiomatics in physical sciences is an area where mathematicians and physicists have the primary interest. The role of philosophy is to examine the philosophical significance of their work. Although, Zawirski writes, the above-mentioned problems are controversial and debatable, this is no obstacle for

> . . . subjecting . . . even at the present state of affairs, the philosophical aspects of this research to a fundamental examination, now that the splendid perspectives suggested by the new state of affairs exert some irresistible and odd fascination which forces a thinking man to engage in more profound philosophical reflexions. (37).

It is to these reflexions that the last third part of the essay is devoted. Its title is: 'The Mathematical Axioms of Physics and the Intuitive Principles of Cognition'.

The application of the axiomatic method in natural sciences has resulted in serious changes, both in mathematics and in natural sciences. From many quarters statements were put forward, either that physics had become something similar to geometry (Hilbert, Haas, Weyl) or (an only apparently different statement) that geometry had become a physics-like science.

According to Zawirski, the term 'geometry' is not univocal. It is used in at least three different meanings:

(a) Investigation of spatial properties of bodies appearing in the real universe,
(b) Investigation of the properties of the ideal, empty, space (without bodies),
(c) Investigation of the diversities, i.e. of the set of ideal objects linked and determined by axioms.

Now the proposition of the change relates only to geometry in the first meaning (a). The proposition according to which geometry became a physics-like science means merely that the laws of geometry are so chosen as to express real relations in the existing world. On the other hand, the proposition that physics had become a geometry-like science means that its laws, despite their empirical genesis, present an axiomatic deductive theory.

A further problem, which became the subject of lively discussion, consisted in how one ought to evaluate Kant's synthetic *a priori* propositions in view of the fact that, in the case of the geometric axioms, one does not require them to be intuitively obvious.

We find here a few possibilities. The first, and the most superficial, consists in asserting that Kant had been wrong, and in rejecting his propositions. However, such a simple rejection would be tantamount to letting something really valuable in Kant's considerations slip away.

A judgement can be interpreted not as a necessity connected with the structure of our mind, but as a proposition introduced in an arbitrary way. However, such an interpretation preserves rather little of what is valuable in Kant's theory. Now Zawirski proposes that the necessity and universality of some principles should be recognized as a characteristic feature of science which, for its existence and development, requires certain constitutive principles to be adopted. Such principles are, for example, the admission of the existence of certain regularities, or the requirement of correspondence between laws or theories. The principle of correspondence is the only rational explanation of the fact that the laws which had been formulated earlier, could serve as a step toward the knowledge and discoveries of later laws. The point is that, without assuming regularity in nature, science would be unable to formulate laws or theories.

Now these requirements undergo variations along with the development of science. In this sense we cannot agree with Kant that they constitute an invariable feature of the human mind and are intuitively obvious. The constitutive principles undergo changes along with the development of science and, what is more important, they must be intuitively obvious.

Kant's error consisted in the petrification of a definite stage of the development of science.

One ought to separate the meaning of the '*a priori*' factors from the meaning of the constitutive elements of science. The fact that some constitutive notions and principles of sciences must exist will remain true even when their *a priori* character is fully rejected and disregarded (p. 66).

The *a priori* notion is neither a psychological nor a metaphysical one (Hermann Cohen, Alois Riehl), but only a logical and methodological one. What is here at stake is a certain assumption which is indispensable for science.

Zawirski writes: are these assumptions indispensable in view of the character of the human mind (in which case they are subjective), or else in view of the development of science (and thus they are objective); this remains an open question.

One thing is certain, viz. that the constitutive principles of science are neither an unchanged, firm basis of science nor an arbitrary private agreement.

In view of the facts that the constitutive principles of science undergo change along with the development of science and, in particular, since they lack an obvious character, the content of one of the basic functions of science, i.e. the explanation, must undergo a change. "The principle that explanation should consist in reducing the unknown things to the known ones will not be true then" (p. 90). The range of the known things is too narrow to suffice for explaining the tremendous variety of things in the universe. "It must not and it ought not surprise us if new theories do reduce the phenomena which, true, are known to us, to laws which are very general and relatively simple, but which nevertheless look rather enigmatic". Both quantum theory and relativity physics do not look exactly simple and known; they are inconsistent with current intuition, and yet they explain the world and enable us to predict many phenomena and processes better and more accurately than the physics which appeared to be close to intuition.

## 5.3 TIME

Zygmunt Zawirski's lifelong passion undoubtedly was time. His most important achievements were in this field. In addition to many papers dealing with time, he also wrote two books which we shall discuss. The first, *The Eternal Returns of the Universe* (*Wieczne Powroty Światów*) was published in Cracow in 1927. Zawirski first devoted more than ten years to this problem, and only later decided to publish a book. Work on the subject of eternal return, involving a rather odd theory, awoke in his mind, he admits, a profound interest in the problem of time.

The time enigma had been a focus of investigation almost since, human curiosity had been awakened at the earliest stage of culture. No wonder! We have only to think of the mystery connected with the existence and the flow of time. The past has ceased to exist, the future has not yet arrived, and the present is only a point of contact between the past and the future. We cannot grasp any part of time, and yet its relentless flow is so closely connected with human fate, with the irreversibility of its direction, with the pilgrimage of mankind from childhood to old age and the unavoidable ultimate end. The irreversibility of time, the threat of death which, like the sword of Damocles, hangs over everyone's head, for ages attracted the attention of philosophers and scientists. We find this problem, again and again, in the ancient books, even in the Bible.

The most distinguished thinkers, from the Greek philosophers to the modern times, were fascinated by the time. We have only to cite some names: Anaximander, Heraclitus, Plato, Aristotle, St. Augustine, St. Thomas Aquinas, Pascal, Leibniz, Bergson.

At the turn of the 19th century, the problem of time could be found at the center of scientific and philosophical discussions.

In his book *The Eternal Returns of the Universe* Zawirski deals with the idea of cyclic time. According to this theory the universe exists eternally, but this is by no means tantamount to saying that no changes take place in it. On the contrary, the universe is subjected to variations, both of evolutionary and revolutionary natures, but, after a relatively long period, it returns to the initial point, in order to begin a new history which is identical with the previous one.

Zawirski's book consists of two parts. The first is of a historical character; the other is a critical analysis of the theory of cycles, in the light of contemporary science.

The historical part is highly instructive. It presents the first, systematic exposition of the idea of returns chronologically and objectively arranged. The reader is often struck by the fact that, throughout history, the idea of eternal returns had been brought forward frequently and from many standpoints. One cause is psychological. The idea presents a remedy for the profound human fear of death. In this approach death has only a temporal character. The theory presents also a coherent combination of the variability of events and processes with the duration of an unchangeable world.

Zawirski starts with Hellenic times. We follow here, in Greek views, the Babylonian influence of Chaldean astronomers and of the popular idea of the great astronomical year. In the course of this 'astronomical year', various changes (fires, floods, etc.) take place, and, finally, everything returns to the starting-point. Indeed Pythagoreans, or at least some of them, proclaimed the idea of returns, which they linked to belief in metempsychosis. It appears that astronomical observations, especially of the cyclic nature of changes, had an effect upon the founding of their belief in the eternal returns. Indeed, Aristotle's disciple, Eudemos, while explaining the essence of time to his disciples is reputed to have uttered the following sentence:

If we believe the Pythagoreans, that the same things do return fully, then I myself will address you one day, holding the same stick in my hand, and you will sit in front of me just as now, and similarly the case will be with everything else (Zeller, *Die Philosophie der Griechen*, vol. I, pp. 410–411, ed. 4).

The idea of eternal return appears in a fully mature form among the Stoics. They combined metempsychosis with Heraclitus's teaching of the destruction and re-building of the worlds (the spread of fire and its extinction, 'the way up and down'). Heraclitus compares the periodical conflagration and creation of worlds with the playing of a child who builds hills of sand in order to pull them down. An interesting element within the Stoics' idea is the 'determinist' conviction that cause-effect chains appear regularly. The repetition of one link of a chain was supposed to bring about the inevitable succession of the entire chain of events.

The Greek atomists: Leucippus, Democritus, Epicurus (and the Roman, Lucretius) also were adherents of the theory of eternal returns. Some difficulty in the reconciliation of the idea of returns with atomism was found in the infinite number of atoms in the universe. So the atomists accepted an amendment. True, there are infinitely numerous worlds and atoms, but in the structure of a single world only a finite number of atoms take part. The number of combinations between atoms is, to be sure, very great, but still finite. If so, repetitions must take place within a sufficiently long time.

If the matter remains eternally the same, . . . why, after a certain long period of time, identical things could not be generated, from the same matter and due to the same causes? (Zeller, *Die Philosophie der Griechen*, vol. III, p. 156, ed. 3).

During the Middle Ages two tendencies appeared: one towards reconciliation between Christianity and the idea of returns, the other toward a decisive rejection of this idea. We find in the Scriptures distinct influences of Pythagoreans and Stoics. We should also mention the Bible. Thus the *Book of Ecclesiastes* contains the passage:

That which hath been is that which shall be, and that which hath been done is that which shall be done, and there is nothing new under the sun. Is there a thing whereof it is said: 'see, this is new?' It hath been already in the ages, which were before us. (*Ecclesiastes*, Chapter 1, verses 9 and 10).

As far as criticism from the point of view of Christian philosophy, the main obstacle was the problem of free will and the responsibility of every man for his deeds. Thus, Origen, the Church writer belonging to the Alexandrian School of Stoic and Neo-Platonic influence, emphatically rejects the succession of identical cycles because it is contrary to free-will. 'The new Socrates', he writes, must not necessarily marry the same Xanthippe and be impeached by the same people. Some small changes are admissible.'

Two Catholic philosophers, who had enjoyed probably the greatest authority proceeded even farther: St. Augustine and St. Thomas Aquinas. They reject the idea of eternal returns as alien, and even contrary, to Catholic philosophy.

St. Augustine discusses this subject in his main work *De Civitate Dei*. According to him the return of the world would sentence man to eternal adversity. His point is that the human soul would never be able to be saved from the eternal roundabouts, and would be always forced to return to the 'false happiness' and 'true misery'. The soul would never be able to enjoy true happiness since he would never be sure of his durability.

St. Thomas Aquinas advanced more serious arguments: in his work *Contra Gentiles* he writes that we cannot exclude the returns of things which are similar, but we cannot admit the return of the same, identical things. Thus, human life cannot be restored by any cosmic changes nor by motions of celestial bodies, because 'such a resuscitation of a dead body transgresses the ability of the action of nature'. Besides, the soul being something distinct from the body and possessing no corporeal nature is not subject to the laws governing the motions of celestial bodies and to astronomical cycles.

In modern times, we observe a revival of the idea of returns. The fundamental difference consists in the justification of this idea in the laws of natural sciences. Here the situation was particularly favorable since the laws of mechanics were symmetrical with respect to the direction of time. According to these laws, all phenomena may run both from past to future and from future to past.

Kant was a master scholar of the mechanical laws. He was an adherent of the theory of returns, especially during the period when he was writing his *Allgemeine Naturgeschichte and Theorie des Himmels* (1755). Having presented his theory of the cosmic nebula from which the solar system was formed, Kant gives serious thought to the end of the solar system. According to him, the velocity of the planets will decrease gradually. Due to the force of gravitation, the planets will spiral onto the sun. The heat thus generated will transform matter again to the state of cosmic nebula. Is it not to be assumed, says Kant, that 'nature which had been able to pass from the state of chaos into a regular system would be also able to develop with the same ease and to renew the regular combination?'. Kant's answer to this question is in the affirmative. It is in gravitation that Kant sees the main force that is to bring about the repeated formation of the solar system. This is all the more interesting since at present both the idea of black holes and the theory of stellar evolution invoke gravitation as the main force.

One ought to underline that Kant applied his theory not only to our solar system, but also to other systems. According to him the stars are solar systems which resemble our system: 'as the same number of suns are the centres of similar systems in which everything can be arranged in a manner as grand and orderly as in our system' (*op. cit.*, [Philosophische Bibliothek, Band 48], pp. 113–114].

Just as Kant, so also the physicist Helmholtz sees in gravitation the main force which winds the spring of the world.

Another adherent of the idea of returns was Auguste Blanqui (1805–1885) who had written in prison his discourse *L'Eternité par les Astres*.

Friedrich Nietzsche presents the idea of eternal returns in a very suggestive and beautiful form. Nietzsche writes:

> Zarathustra . . . thou art the teacher of the eternal return this is thy destiny! . . . Look, we know what is thy teaching: that all of the great things do return eternally, and we do the same together with them, and that we have been here for all times, and all things along with ourselves . . . The souls are just as mortal as the bodies are.
>
> But there will return this node of causes into which I myself am also involved it will create me again! For I myself do belong to the causes of the eternal turn.
>
> I shall be repeated in turn along with this sun, this earth, this eagle and this serpent not for a new nor for a better life, nor for a similar life: I shall return eternally to always the same life, both in the great and the small things . . . [F. Nietzsche, (*Also sprach Zarathustra*) *Thus spake Zarathustra* 1883–4].

A kind of support for the idea of eternal returns, as mentioned before, was in the symmetrical nature of all known laws of nature (with respect to the direction of time). This was felt to be against human intuition. Such was the situation prevailing in science almost until the mid-19th century. But a radical change took place in the second half of the 19th century. In 1824, the French engineer Sadi Carnot, when examining the mechanism of the heat engine, discovered the Second Law of Thermodynamics. However, in his interpretation, that law was not only at variance with the first law, the conservation of energy, but it also fails to stress with sufficient sharpness the irreversible character of the transformations of thermal energy into mechanical work. Ultimately we owe the so-called classical formulation of this principle to Clausius who, in the 'seventies of the 19th century, presented it as the so-called law of the increase of entropy.

According to Clausius, in every isolated system which is not in a state of thermodynamic equilibrium, all processes which occur spontaneously lead to an entropy increase, to degradation of energy.

During this period, the tremendous significance of the second law consisted mainly in that, for the first time, a law was discovered and formulated which was asymmetric with respect to the direction of time. The past of a system not in equilibrium differs from its future objectively by having a lower value of entropy.

Clausius applied the second law to the entire universe which, according to him, tends, just as all other systems do, toward the state of maximum entropy, to thermal death. The entropy will reach a maximum value. The second law of thermodynamics seems, as we see, to supply serious arguments against the theory of eternal return.

Zawirski, who clearly leans toward the theory of returns, subjects the arguments against the irreversible character of changes in the universe to extensive analysis.

The theory of universal thermal death is a conclusion drawn from the extrapolation of the second law to the entire universe. Let us therefore begin with arguments which undermine the validity of such extrapolation:
(1) The first to argue against the extrapolation was surely Ernst Mach. Being a radical empiricist he wrote:

. . . phrases about 'the energy of the *universe*', 'the entropy of the *universe*', and so on, have no meaning. For such phrases contain applications of metrical concepts to an object which cannot be measured. [*Principles of the Theory of Heat*, English translation, ed. B. McGuinness, Reidel/Kluwer 1986, p. 439, note 20].

(2) It appears to be questionable whether conclusions drawn from our terrestrial observations, which refer to relatively small space-time ranges, can be applied to the entire universe about which we know relatively little. We do not know even whether the universe is temporally and spatially finite or infinite.
(3) Even if the second law of thermodynamics could be applied to the entire universe, the entropy increase could have taken place asymptotically, and consequently thermal death would not occur.
(4) Henri Poincaré argued against the hypothesis of thermal death in his work *Le Mécanisme et l'Expérience*, (1893). He wrote that our observations, when not sufficiently universal, can appear to be incompatible with mechanics, and yet this incompatibility may be only apparent. Poincaré formulates the differences between irreversible and reversible phenomena: the division into these two classes of phenomena is not sharp, as both are quasi-periodic. Phenomena which we call 'reversible' have a relatively short period of return to the preceding state, while in the 'irreversible' phenomena, the time of return is relatively long.

Conclusions about the thermal death of the universe can simply be proof of short-sightedness. The time of observation is too short. The first flowers of the spring, says Poincaré, would be able to say similarly that the world gets warmer and warmer, since they did not witness the reverse change.
(5) The universe cannot be treated as a 'thermodynamically isolated system'. The point is that every thermodynamically isolated system has an environment. Conclusions leading to generalizations which relate to thermodynamically isolated systems formulate a transition to a limit, where the energy exchange with the environment tends to zero. However, it is not possible to carry out such reasoning in the case of the entire universe, i.e. a system which, *ex definitione*, does not possess an environment.

(6) Unquestionably the strongest argument against the theory of the thermal death of the universe was suggested by Ludwig Boltzmann. In lieu of the Clausius formulation of the second law, Boltzmann suggests a statistical formulation. The statistical apparatus suggested is based on the so-called ergodic presupposition, i.e., stated roughly, on the presupposition of the equal probability of all possible micro-states. According to Boltzmann's interpretation, an increase of entropy in a thermodynamically isolated system, under the influence of spontaneous processes is highly probable, and yet there also exists a small probability of a decrease of entropy in such a system.

According to Boltzmann a determinate, irreversible direction of thermodynamic phenomena by no means implies the temporal finiteness of the universe, whether in the sense of temporal beginning or end. According to the fluctuation hypothesis, the universe finds itself in a state of thermodynamic equilibrium. Individual small segments of the universe may depart from equilibrium; they are the fluctuations:

> One can imagine the universe as a mechanical system consisting of a colossal number of components and possessing a gigantic period of existence, such that the period of existence of our stars seems in comparison to be insignificantly small . . . It is in this universe that there ought to exist, here and there, relatively small spaces . . . which, during a relatively small period of time, deviate from thermal equilibrium, there existing in equal degree such parts in which the probability of the state increases, as well as parts in which it decreases. Hence both directions of time are indistinguishable for the universe, just as there is no down and up in 'space'. [L. Boltzmann, *Vorlesungen über Gastheorie,* 1898. See also Irena Szumilewicz, *Teoria śmierci cieplnej wszechświata* (The theory of thermodynamical death of the Universe), PWN, Warsaw, 1961, p. 29].

It can be seen that the idea of fluctuation connects in a consistent manner the *reversibility* of the phenomena in the entire universe with the *irreversibility* of processes within the fluctuation range.

(7) Poincaré's theorem and the fluctuation hypothesis of Boltzmann had directed the attention of scientists to look for anti-entropy processes. Before long, an experimental confirmation of the statistical interpretation of the second law of thermodynamics was found within micro-systems. It is here, where the number of particles is relatively small, that the anti-entropy processes can be most easily observed. That was precisely the judgement of the famous Polish physicist, Marian Smoluchowski. The Brownian movement, the phenomena of opalescence and of the fluctuation of concentrations have demonstrated that, in a system which is in thermodynamic equilibrium, fluctuations occur around the medium state. The statistical character of the second law had

been beautifully and in all aspects confirmed experimentally by Svedberg, Perrin, Chaudesaigues, Dąbrowski, Seddig, Zangger and Böhi.

Zawirski examines these arguments and refers also to the opinion of Smoluchowski, which had been stated *expressis verbis* in 1913, in the assembly hall of the old Lwów University, viz. that 'the eternal return of all things is possible' (*Wieczne Powroty Światów* – [*The Eternal Returns of the Worlds*] p. 74).

As an important argument speaking for the theory of returns, Zawirski refers also to the theory of relativity: in accordance with selectivity considerations, Zawirski argues the world is finite:

Most far-reaching is doubtless the detail relating to the finiteness of the world. Of course this is merely a consequence of the general theory of relativity, which, as yet, has not been verified, but which can be verified indirectly . . . the assumption of the finiteness of the world, of the finite number of atoms, has been the basis of all reasonings relating to the necessity of some combinations being repeated, were it only at extremely long intervals (*ibid.* 79).

It is especially on the basis of Poincaré's phase theorem that Zawirski adopts the conception of quasi-returns, i.e. of returns not to identical states, but to states which are very close.

Had it turned out, says Zawirski, that the phase theorem cannot be maintained, there remains, to base the theory of returns on facts, which 'demonstrate exceptions from the second law of thermodynamics' (*ibid.* 97).

Finally the author cites Einstein on this subject: 'The eternal return cannot be denied with complete certainty.' (*ibid.* p. 98).

Zawirski ends by stating that

. . . the scientific odds for the theory are, on the whole, not unfavourable, in the case both of inductive, and deductive argumentation . . . If the Universe is an unending melody which is played by Divine Eternity, then our existence seems to be those tones which are repeated now and then in this melody.

It appears to us that the conclusions drawn by Zawirski when appraising the probability of the theory of returns are doubtful. To accept a theory only because it 'cannot be denied with complete certainty' seems to be risky and against scientific standards. Zawirski's approach can be explained only by his emotional attachment to the idea of returns.

Zawirski's *opus magnum* was, no doubt, 'The Evolution of the Notion of Time', Cracow (in French), 1936. In an international competition under the auspices of the Italian bimonthly *Scientia* he was awarded the First Prize for it. At that time this Prize was very respected and of the highest order.[8] Since a translation of two-thirds of this book is included in the present volume, we will confine ourselves to a very short review.

No other publication known at that time was so up-to-date as Zawirski's. Duhem's *Le temps selon les philosophes hellènes* was devoted solely to the ancient period; Baumann's *Die Lehren von Raum, Zeit und Mathematik* ends with Hume; and Werner Gent's *Die Philosophie des Raumes und der Zeit* begins with Aristotle and ends at 1768. Hence, one great merit of Zawirski's book consists in that it carries the development of the idea of time down to the latest period, and takes into account problems related to relativistic physics as well as to quantum mechanics. The book is written in a clear and lucid style. Zawirski introduced a convincing development classification, with a division into historical periods and isolation of the main problems to facilitate orientation in this vast material. We recall that conceptive clarity and stylistic lucidity are characteristic of the mental formation to which Zawirski belongs as the faithful disciple of Twardowski.

The book consists of two parts. The first contains the history of theories of time beginning with the Pythagorean period and going up to Zawirski's own times. The second is critical, devoted to the analysis of some '*aporias*' related to the notion of time, and to the author's ideas of how to solve them.

From among the ancient doctrines, the author bestows most attention upon Aristotle whose conception he believes to be the closest to his own. From the pre-Aristotelian period, he takes into account mainly the doctrines of the Pythagoreans, Eleatics and Plato; and from the thinkers after Aristotle, he takes the theories of the Stoics, atomists, sceptics, Plotinus and his successors, such as Proclus and Simplicius.

From the Christian philosophy of the medieval period Zawirski chooses only the most important items, i.e. the theories of St. Augustine and of St. Thomas Aquinas.

From the standpoint of the evolution of the time concepts, Zawirski divides modern history into four periods. The first one, which extends to the time of Newton, remains still under the influence of scholastic doctrines, which is particularly exemplified in the case of Descartes and Spinoza. The second period begins with Newton and ends with Kant. During this period the scholastics are replaced as the subject of criticism by Newton and his school. Kant opens up the third period which spans almost the entire 19th century. At this stage psychological studies develop and the critique of the bases of Newton's conception deepens. This results in opening the road to new conceptions of time in the fourth period which lasts through the 20th century. These new conceptions are connected with Bergson's theory and with phenomenological investigations of Husserl's school. The main set of problems is related here in the first place to the theory of relativity, which is discussed at length with the use of formulae and analysis. Zawirski devotes a lot of time to the conventionalism of Poincaré and to the conception of Enriques who fights against radical conventionalism as well as against difficulties connected with the quantum theory.

In order to enable the readers to get a better orientation, Zawirski divides this set of problems into groups. According to him, the basic problems are: 1) the reality of time and 2) the psychological genesis of the idea of time.

From the point of view of the reality of time all of the theories are divided into realistic and idealistic ones. The latter comprise theories which either degrade time to the role of delusion (as in the case of Eleatics), with which reality has nothing in common, or else to the *a priori* form of our sensuality (as in Kant). The realistic views are, in turn, divided into extreme (substantial) and moderate realism. The extreme realism occurred in antiquity (some neo-Platonists), and, in modern times, in Newton. This view attributes to time a reality which is independent of external objects. Time flows 'by virtue of its own nature', and it would exist even if the world would cease to exist.

The extreme realism is opposed by moderate realism, according to which time, while not being an *ens per se*, still denotes something real, that co-exists with the changes and would disappear already if the world would, so to say, set and become motionless, just as in the fable of the sleeping princess (after the princess had been pricked with a needle). The moderate realism, which the author himself supports, can assume various forms. Time accompanies everything occurring in the world; it is a 'feature of a feature', according to the Epicureans; it can be likewise, as Leibniz said, the 'order of inconstant possibilities' or an 'order of sequences'. This group comprises the view of Kotarbinski to whom time is an apparent name (such are to him the names of all properties, classes and relations); also the view of Nicod and some others of kindred nature, for which the term 'time' is a collective name of the 'class of concrete durations'. Moderate realism proclaims the thesis that time does not exist, but this sentence has for them a different meaning than for Kant or the Eleatics. 'Time does not exist' because there is no such substance or thing with which this name could be co-ordinated, but events do exist and the temporal order in nature would exist even in the absence of consciousness and of the observer. On the contrary, for Kant time is related to consciousness, and would disappear only together with the observer. Idealism and extreme realism would occur only sporadically. Throughout history the dominating opinion was that of moderate realism, whose adherents included Aristotle, the scholastics, Leibniz, and what is more important it owed its scientific support to relativistic physics.

Another basic problem, apart from the reality of time, is the problem of the genesis of the idea of time. Zawirski distinguishes here two possible answers: the source of the idea of time is found either in the facts of mental life or in those of the physical world. In addition to these two stands there exists also an intermediate stand which takes into account both sources. Depending on the answer to the questions relating to the genesis of the idea of time, the above-mentioned stand of moderate realism can assume the form of either psycho-logical or naturalistic or even a universal doctrine. The psychological theories stress the qualitative properties of time, while the naturalistic ones stress the metric properties.

The contrast between these trends runs across the entire history of philosophy. The followers of Plato, as well as St. Augustine, and Locke, form so to say the stages of the evolution which has led us to Bergson. The stands of Aristotle and of St. Thomas Aquinas may be associated with modern physics.

Zawirski thinks that, since time has both qualitative and quantitative properties, one has to reconcile both aforementioned tendencies.

In the critical part of his book the author subjects to analysis the suggested solutions of problems which have appeared in the course of history. These are, successively: the presupposition taken over by scientists, the correctness of reasonings, and terminology. Zawirski concludes that, much as the relation between generalizations and the experimental data leaves much to be desired, on the whole the reasonings are correct. As far as terminology is concerned, the situation is much worse: there appear to be many sins related to the ambiguity of terms, faults in their definitions, etc. Zawirski thinks that most of the doubts can be applied to the presuppositions accepted as the starting point. It is best in this case to apply Ockham's razor *Entia non sunt multiplicanda praeter necessitatem*. 'So the presuppositions ought to be both as few as possible, as well as obvious and intuitively understandable and sure'. Ockham's razor is, however, found to be insufficiently sharp. Zawirski quotes examples of moderate and extreme realism, in order to demonstrate that intuition and obviousness do not provide a sufficiently acute criterion. It follows from the examples that, sometimes, contradictory presuppositions may appear to be equally obvious.

The problem of the verifiability of presuppositions produces insuperable difficulties. '. . . none of the three possible theories is verifiable'. For instance, we cannot verify what would happen with time if the world ceased to exist, or if it would suddenly become motionless and changes of all types would cease to exist. In this situation Zawirski chooses, as the 'least adventurous', the stand of moderate realism. The word 'time', as denoting no object (just as the names of all relations and classes) ought to be considered to be an 'apparent name'. One has to explain here how to understand the formula 'what really exists'. There are two possibilities, the first of which is the stand of Leibniz, viz. that only substances do really exist, and that every true sentence which apparently speaks about something else, can be transformed into a sentence dealing with substantial beings. On the other hand, the second possibility corresponds to the approach of Bergson and Whitehead and assumes that, in the basic sense, events do exist. According to Whitehead 'the object is a feature of the event' or 'an event is a situation of the object'. Zawirski explains at length the difference between his stand and Kant's conception. According to Zawirski the proposition that time is an apparent name means that it is neither a substance nor a thing. In the understanding of Kant, time is only a form of sensuality, and its existence is related to the existence of consciousness. Much as Zawirski underlines very carefully the difference, we have to state that the selection of the term 'apparent name' for time does not seem to be correct. We attribute the term 'apparent name' to such words as *nymphs*, *fauns* etc. The latter are merely the creations of our imagination, and therefore are even more weakly anchored in reality than Kantian time. It is worth mentioning here that Zawirski's criticism levelled at Kant does not seem to be convincing in another respect as well.

Zawirski's objection to Kant is that he does not take into account the anisotropic character of time. He writes: 'Kant failed to take into account all properties of time . . . no one-dimensional continuum constitutes time by itself, as long as the difference between present, past and future is not associated with the relevant ideas. It is precisely this fundamental property of time that Kant passes over in silence in all of his works' (Zawirski, p. 46). Now here Zawirski is wrong, for Kant discusses anisotropy of time. Here are the relevant quotations:

Second Analogy, Principle of Succession in Time in Accordance with the Law of Causality. All alterations take place in conformity with the law of the connection of cause and effect' Immanuel Kant's, *Critique of pure reason*, Macmillan and Co. Limited, 1933, p. 218.

But the concept which carries with it a necessity of synthetic unity can only be a pure concept that lies in the understanding, not in perception; and in this case it is the concept of the *relation of cause and effect*, the former of which determines the latter in time, as its consequence, not as in a sequence that may occur solely in the imagination (or that may not be perceived at all). Experience itself – in other words, empirical knowledge of appearances – is thus possible only in so far as we subject the succession of appearances, and therefore all alteration, to the law of causality . . . (*ibid.* 219).

Kant explains that, when saying 'the necessity of the causal relation' he understands that the succession of the causally connected events cannot be reversed. On the contrary, the succession of events which are not connected causally depends on our will. He illustrates this idea by two examples: When we see a house, the succession of our perceptions depends solely on ourselves; we may begin with the roof and end with the gate, but we can reverse the order. The situation is quite different when:

For instance, I see a ship move down stream. My perception of its lower position follows upon the perception of its position higher up in the stream . . . The order in which the perceptions succeed one another in apprehension is in this instance determined, and to this order is bound down . . . But in the perception of an event there is always a rule that makes the order in which the perceptions (in the apprehension of this appearance) follow upon one another a *necessary* order' (*ibid.* 221).

I think that the quotations adduced suffice to quash the objection relating to Kant's allegedly not taking into account the anisotropic character of time.

As regards the other main problem, viz. that of the origin of the idea of time, Zawirski does reject both apriorism in the Kantian sense and the epistemological apriorism. Bergson's conception (with his psychological concept of time) is also subjected to criticism. Zawirski thinks that the main difficulty, when trying to solve this question, is connected with the insufficient standard of psychological investigations. Our author lays upon Bergson the blame for exaggerating the role of intuition and memory in the creation of the idea of time. The same refers to the qualitative characteristic of time, which had exaggerated to the detriment of the metric properties of time. Time can possess metric properties without being measured by any observer; this does not change the fact that time is measurable.

The last question to be discussed is the oldest and most general of various ideas of time. Zawirski thinks that this is the notion of the time interval. Kant distinguishes among three kinds of time relations: simultaneity, succession and duration. Duration is a term involving the greatest difficulties for historical reasons. According to Kant duration means constancy and immutability of the given existence as opposed to existences which undergo changes. The Scholastics considered that the permanent duration of the immutable being was something different from time, since in such a duration there is no succession. Descartes was first to protest against such a radical difference. The last link in this development is Bergson, who proceeded even farther. According to him there is not only no difference between permanent duration and the time understood as a succession of states, but there does not exist any permanent substance. Everything undergoes changes in a 'creative evolution', i.e. always something new appears.

The third and the most important group of aporias deals with the structure of the notion of time. In this matter Zawirski refers to the book of Reichenbach, 'one of the most eminent philosophers among the relativists' as he calls him. Zawirski accepts Reichenbach's[9] division of the time properties into the metric (quantitative) and topological (qualitative). The qualitative notions include the relations of succession, simultaneity and the relation 'in between'. The metric notions include in the first place the equality of two time intervals.

On the whole Zawirski accepts favourably Reichenbach's suggestions; he also accepts the proposed definitions. In the basic epistemological field he begins the appropriate discussion with Reichenbach's statement that the definitions involved are allegedly of conventional character and do not express any cognition.[10]

Zawirski underlines quite rightly, when starting a discussion with Reichenbach, that, as far as the definitions of objects and real relations are concerned, the scientist's choice is limited by reality itself. He gives as an example the history of the abolition of the classical definition of simultaneity. He writes:

For all that we claim that the relativity of simultaneity expresses cognition and not pure convention, and yet no *circulus vitiosus* is found . . . One presupposes the validity of the (classical) intuitive notion of simultaneity even for distant events. One measures the velocity of the signal (light) on the basis of such presupposition (this is borne out by the history of physics). Next one introduces a postulate, indispensable to science, viz. that each definition should be tested by experiment (no physicist can rightly give up this postulate), and one subjects a principle suggested by experience, in particular by the investigations of the Dutch astronomer de Sitters, viz. that the velocity of light should be considered to be constant for all systems admissible in STR. One finds then that the initial presupposition which permitted the validity of the intuitive notion everywhere and always, cannot be maintained without limitation, wherefrom one attains the final conclusion that it is false. (*ibid.* p. 53).

At that we end our account, and refer the reader to the major texts of Zawirski's Time book, included in the present anthology. I hope that the reader will enjoy this book which can be much better evaluated today than by his contemporaries – although he had been granted the Eugenio Rignano Prize. It was one of the best books written on the subject of Time, and is by no means an obsolescent one. The majority of the issues Zawirski took into account are still being discussed today and can be found at the very heart of current investigations. And what is more his critical judgements are far in advance of his own period. From all other fragments included in this anthology one may see both the clarity of presentation and the large set of problems Zawirski was able to cover during his scientific activity. He took an active part in the scientific life and reacted promptly and with deep insight to the problems of his period. Being an eminent representative of the Lwów-Warsaw School, he was one of the most impressive and original Polish philosophers and scientists.

## NOTES

[1] Suffice it to adduce several examples:

* General Kazimierz Pulaski (1747–1779) fought for the independence of the United States of America as Commander of the Pulaski Cavalry Legion. He was killed in 1779 at Savannah.
* General Tadeusz Kościuszko (1746–1817), fought in America in 1776–1783; built the fortifications at Saratoga and West Point.
* General Jozef Bem (1794–1850), a gunner, fought against Russia during the Polish uprising of 1830–31; later in 1849 General Office Commanding the Hungarian Army against Austria and Russia during the Nationalist rising.

Again, in the novel *Rudin* (Heinemann, London, 1894, translated by Constance Garnett) by the famous Russian novelist Ivan Turgenev, one finds near the end:

'Tiens'! said one of the escaping revolutionists to another 'on vient de tuer un Polonais'. This 'Polonais' was Dmitri Rudin – a Russian.

Indeed, any barricade fighter in 1830 or 1848 in Western Europe was presumed to be a Pole.

[2] H. Skolimowski, *Polish Analytical Philosophy* (London, Routledge and Kegan Paul, 1967), pp. 69–70; translated from J. Lukasiewicz 'O metodę w Filozofii' ('On Method in Philosophy'), *Przegląd Filoz. 31* (1928), pp. 4–5.

One ought to note that, after a few years, Lukasiewicz changed his opinions about Aristotle dramatically. After a profound analysis of modal notions in Aristotle, he put forward one of the most fruitful ideas of three-valued logic and of many-valued logics in general.

[3] (1) 'Jan Lukasiewicz's three-valued logic'/'Logika trójwartościowa Jana Lukasiewicza'/ *Proceedings of the Poznán Society of the Friends of Science*, nr. 2–4, 1931.

(2) 'Relationship between logic and mathematics', printed in the Jagiellonian University, Cracow, 1927

(3) 'The Meaning of Many-Valued Logic', *Przegląd Filozoficzny*, 1934

(4) 'Relation of Many-Valued Logic to the Calculus of Probability', (The Society of the Friends of Science), T.P.N. Poznán, 1934

(5) 'Les rapports de la logique polyvalente avec le calcul des probabilités', *Actes du Congres International de Philosophie Scientifique*, Sorbonne, Paris, 1935.

[4] True enough, Brouwer rejects the absolute truth of the principle of the excluded middle, but he rejects its being false, since he admits the law

$$--(p \cup -p)$$

The law of double negation is the combination of two simpler principles

$$(p \rightarrow --p) \text{ and } (--p \rightarrow p)$$

From among two simpler principles Brouwer does reject only that which has the negation on the left-hand side. Yet he maintains the validity of the former of the two principles.

[5] $v(p)$ signifies the value of the proposition $p$.

6 Two definitions of the measure of probability are known: (a) *Classical definition*, where by the probability (*P*) of the event (*E*), i.e. by *P*(*E*), we understand the ratio of the number (k) of events favorable to the given event (i.e. such from which the event (*E*) does arise, to the number (*n*) of all events of the basic set, *P*(*E*) = k/n. [Under the condition that all of the events are equally probable.]
(b) *Frequency definition*. Probability as the limit of relative frequency. Let the event *E* consist in that the result of the measurement of a certain physical magnitude is comprised in the interval [a, b]. Let us denote by *n* the number of effected measurements, by *k* the number of such measurements as belong to the interval [a, b] and by *P* the probability. Then the probability *P*(*E*) of the event equals *P*(*E*) = lim ~ k/n.

7 The problem of the genesis of our ideas is a *psychological* one. *Genetical rationalism* is a view that there are inborn ideas, *ideae innatae*, which all people share, and which are intuitively certain. According to the opposite view, *genetical empiricism*, the human mind is a *tabula rasa* and only experience writes signs on it *nihil est in intellectu, quod non prius fuerit antea in sensu.*

The problem how to obtain an adequate knowledge is a *methodological* one. Here we have two opposite views: *methodological rationalism* and *methodological empiricism.*

8 Zawirski's prize as mentioned in a Polish newspaper of 9.1.1934.

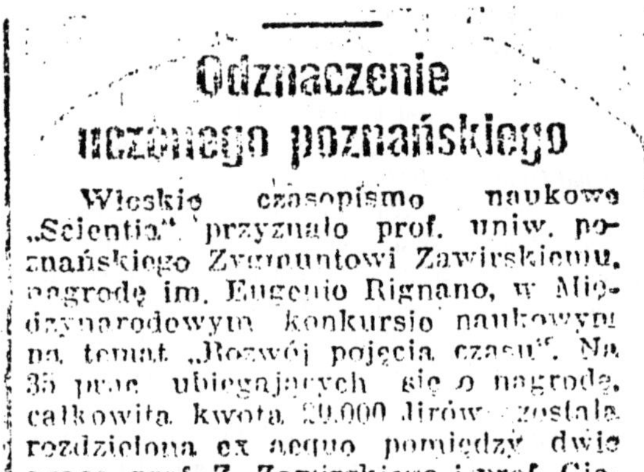
Odznaczenie uczonego poznańskiego

Włoskie czasopismo naukowe „Scientia" przyznało prof. uniw. poznańskiego Zygmuntowi Zawirskiemu, nagrodę im. Eugenio Rignano, w Międzynarodowym konkursie naukowym na temat „Rozwój pojęcia czasu". Na 35 prac ubiegających się o nagrodę, całkowita kwota 20.000 lirów została rozdzielona ex aequo pomiędzy dwie prace, prof. Z. Zawirskiego i prof. Giovanini Giorgi z Palermo.

*Kazier Poznanski*, 2d.9.I.34. nr. 9.

**A Poznan scientist gets a prize**

The Italian science periodical *Scientia* has awarded Zygmunt Zawirski, a professor at the University of Poznán, the Eugenio Rignano Prize in an International Science Competition for his work on the 'Evolution of the Idea of Time'. From among 35 essays competing for the prize, the total amount of Lire 20,000 was divided equally between two essays: that of professor Z. Zawirski and that of professor Giovannini Giorgi of Palermo.

9 Wieczne Powroty Swicotów, *Kwartalnik Filozuficzny*, Kraków; 1927.

10 *The Philosophy of Space and Time* by Hans Reichenbach, Dover Publications, Inc, Constable & Co Ltd, London, 1958.

# PART II

# ZYGMUNT ZAWIRSKI: SELECTED WRITINGS ON TIME, LOGIC AND THE METHODOLOGY OF SCIENCE

CHAPTER 1

# REPORTS

## 1.1. AN ACCOUNT OF THE EIGHTH INTERNATIONAL PHILOSOPHICAL CONGRESS

Between September 2nd and 7th, 1934, there took place in Prague the Eighth International Philosophical Congress which differed considerably from all preceding Congresses in that it was devoted primarily to social or even political problems of the current period, and to the trends which transformed the current collective life. This specific tenor of the deliberations of the Congress was doubtless due to the fact that the Congress took place in the capital of a young republic whose President Thomas Masaryk was a former professor of philosophy and sociology. Just as many others, so also the Czech philosophers not only include sociology within the philosophical sciences, but also consider it obviously to be the most important philosophical science, demand a turn towards practical philosophy and stress the practical obligations of a philosopher with regard to cultural life. This explains the fact that such a subject as 'the crisis of contemporary democracy' was the object of deliberations in one of the principal meetings of the Congress. This philosophical attitude of the Congress organizers was also followed by a lively interest in the problem of the philosophy-religion relationship, to which further impulse was given by then current anti-religious slogans, e.g. the well-known Bolshevik definition of religion as opium for the masses, by the crisis affecting Christianity owing to the ruin of the Eastern Church, and by the new slogans in the Protestant world.

Perhaps a more important phenomenon, likewise significant for the Congress, was the appearance of a novel philosophical trend vigorously represented in part by the Vienna Circle and in part by most of the Polish philosophers who attended the Congress. The traditions of Mach's positivism merged in the Vienna Circle with contemporary trends of mathematical logic toward the revision of philosophical problems in such a manner both that the Polish logicians came closer to that group of philosophers and that, in the section of the Congress on 'logical analysis of cognition' they played the dominating role, along with the representatives of the Vienna Circle. The problem of the 'importance of the logical analysis of cognition' was also to be one of the subjects discussed in the plenary sessions, and the principal speakers on this subject were to be Professor Łukasiewicz from Warsaw and Professor J. Joergenson from Copenhagen. However, later, for unknown reasons, this subject had been relegated to the sessions of the first section, where almost all of the Polish lectures were concentrated. In view of this change, no Pole, despite the original promises of the Czechs, spoke in the plenary sessions (although at the previous Congress held at Oxford as many as three Poles did gain this privilege).

The following problems were put forward as the subjects of the plenary sessions: (1) limits of cognition in natural science, (2) the descriptive and normative standpoints in the social sciences, (3) religion and philosophy, (4) the crisis of democracy, (5) the mission of philosophy in our time. On each of these subjects there were two main speakers who mostly represented two opposite camps, after which a lively discussion began which, of necessity, had to be limited. The subjects of the plenary sessions corresponded to the analogous subjects of sectional sessions, the difference being that the first section, devoted to the importance of the logical analysis of cognition, did not have, as mentioned already, a counterpart in the plenary sessions. In spite of the previous tradition there was no section on the history of philosophy.

We shall describe below an account of the plenary sessions and, further, of lectures in the section on the logical analysis of cognition, since we continually attended only that division in its full presentations.

At the first plenary session Professor Bachelard from Dijon and Professor Driesch from Leipzig spoke on the limits of cognition in natural science. Bachelard defended the thesis that there are no absolute limits to cognition, that every insoluble problem is, as a matter of fact, a wrongly posed problem, the formulation of which itself suggests an incorrect method of solution. For a scientific mind, the delimiting of a provisional limit is already, to a certain extent, its transgression, since we have to do here with a sphere in which the human mind has an opportunity for an extraordinary expansion and creative ideas, while the limit, which a metaphysician wants to demonstrate to the scientist, is, for the latter, a completely irrelevant limit; the point is that it indicates the realm of fictitious, completely sterile and fruitless problems. However, later on, the speaker did not display a completely hostile position with respect to metaphysics, as he did admit that science itself achieves transcendency to the extent that it assumes a reality which differs from what is originally given, and that a purely sensualistic attitude cannot be maintained in science. The world which is hidden behind the phenomenon is even more clear and understandable to us than the world of phenomena. However, sometimes we have to break through primordial intuitions and through forms of traditional philosophy, which are apparently not definitive, but should undergo transformations as need may be.

Driesch's lecture entitled 'Natural Science and Philosophy' distinguished between two philosophical disciplines; one of them, a 'science of order' (*Ordnungslehre*) remains in a close relationship to natural science, while the other, 'metaphysics', utilizes the results of natural science, to be sure, but is not interested more closely in the investigation and theoretical activity of the natural sciences. The lecture dealt with the science of order only. The starting-point of the latter are certain primordial meanings (*Urbedeutungen*) which are the source of some statements about order, which are valid for all objects of experience. In this way the science of order, which is at the same time a science of meanings, sets certain prescriptions for scientific theories. Driesch turned against those philosophers who uncritically take note of some scientific theories, providing them only with their visas, just as this happens now so frequently with regard to current physical theories. Next he turned both against the logistic theorists and against the 'phenomenologists', although with regard to the latter he himself approached them quite strongly in his opinions (though at the previous Congress in Oxford he came out with a sharp philippic against Husserl's school). His objection to the phenomenologists was that, in their grasp of nature, they lean too much upon certain matters of course, and upon self-evident things (*Selbstverständlichkeiten*), while nature is, if anything, something peculiar and incomprehensible, and can be determined only in its logical, i.e. purely ordinal, structure. When formulating next his guiding principles for scientific theories, he declared himself for the absolute nature of motion, time and space, for the Euclidean nature of the last, and declared himself against all attempts to revise the idea of causality, viz. both the older ones, which tried to replace the idea of causality with that of the functional relationship, and the newer ones, based on Heisenberg's formula. With regard to the biological sciences, he sided with vitalism and the idea of entelechy, which, according to him, is not at all a metaphysical idea smacking of mysticism, but a natural-science idea belonging to the stock of the indispensable ideas of the science of order in nature.

The main discussion took place after Driesch's lecture though

Bachelard's lecture also included points which could be disputed. Driesch was rightly charged with an excessively dogmatic opposition with regard to modern physical theories. Driesch, who up to now had been working mainly in the field of the philosophy of the biological sciences, wanted to express, in the present paper, an opinion concerning the whole of contemporary philosophical problems, i.e. also such problems which, up to now, had not been the subject of his close interest. However, we have serious doubts as to whether his pronouncements in this direction had been really the result of more profound considerations.

The second plenary session dealt with the descriptive and normative stands in the social sciences. The speakers on this subject were Willy Hellpach from Heidelberg and T.V. Smith from Chicago. The former represented, in sociology, the attitude of Nazi Germany, the latter, the democratic ideas of the United States of America. According to Hellpach, the central object of sociology is the people as a fact of nature, as a certain mental form and, at the same time, as a creation of volition. As a fact of nature it grows up from the drive to collective life and is built on the basis of the tribe and race, while, owing to the formation of language and of juridical ideas, it assumes a certain mental form. The creative will finds its expression in the leading individuals, in the organization of the state, and in the readiness of all to subject themselves to the leading individuals. The state is not something different from the 'face' of the spiritual culture of the people, and that is why every genuine culture is intolerant. Any cultural imperialism, which goes beyond the life of the people among which it originated, carries in itself the germ of the decline of that people. Hellenism was the end of the Greek culture, and Christianity, to be creative, had to be divided quite early into the Eastern and Western part; the entry of the Germanic world into the circle of Christian culture of the West carried the germ of the future secession in the Reformation.

In the defence of democracy, as presented by Smith, the fact was striking that he attributed a great role in the improvement of the collective life to the creative imagination which, according to him, often undergoes various phobias and manias.

The discussion comprised, in the first place, attacks on Hellpach because of his statements that every culture is intolerant, that Hellenism was the symptom of the decline of Greek culture, as well as because he accepted, as the central object of sociology, the ethnic unit, and not a multiplicity of such units. As far as the first objection was concerned, Hellpach's defence consisted in saying that even cultures tending towards expansion, such as Christianity and liberalism, were intolerant. But Hellpach did originally present the matter in such a way that exclusivity, and not a tendency towards expansion, were the indispensable condition for cultural creativity.

During the third plenary session, devoted to the relationship between religion and philosophy, there spoke, on the one hand, the learned Jesuit from Munich, Erich Przywara (a German of Silesian stock, hence his Polish-sounding surname), and on the other, Professor Léon Brunschvicg from the Sorbonne. The former presented the similarities and differences between religion and philosophy from the point of view of the subject, method, and standpoint. When the subject is in question, then both philosophy and religion deal with world and God, religion being more interested in God, while philosophy more in the world itself, and the equivalent of the idea of Deity in philosophy is often merely the abstract of the 'absolute' or of something which is to be 'first and final'. With respect to method, the difference is marked in that a philosopher starts most frequently from certain mental necessities, while in religion it is belief that is decisive, as an expression of absolute obedience and devotion. Finally, as far as the differences and similarities in attitude are concerned, religion looks upon the world as a place for the actions of divine mercy and for the incessant miracles of God. The philosopher, on the other hand, sees only a continuous becoming, and looks for the creative elements in nature. To the theologian this creativity comprised in creation is merely an incorrect grasp of what he calls the presence of the God-Creator in creation.

We can see from this sketch that Father Przywara did not play at polemics or discussions but, opposing philosophy to theology, he had in mind the traditional metaphysics which is a kind of lay theology. The lecture was dull, and the author made use of a kind of peculiar terminology, which was far from clear.

Brunschvicg's lecture is likewise difficult to summarize. The quotations borrowed from Spinoza and Kant implied that these two great thinkers grasped most profoundly the relation of religion to philosophy. The speaker referred, however, to the time of the Stoics, since it was they who first introduced the symbolic and allegorical interpretation of religious creeds, and were the creators of the first theodicy. Much as the religious myths often had an illogical and non-moral content, yet one could always, by using the method of symbolic interpretation, find in them some more profound *ratio physica*. The same Stoic method was later used by Philo

Judaeus for interpreting the Old Testament, pointing to the figurative style of the Bible, which ought not to be understood verbatim. Modern philosophy went much further in this sense, applying the same method to the New Testament and to the Christian dogma. (It is known that for Kant, both the teaching of Christ's divine nature, and the teaching about mercy are not taken *verbatim*, but in a certain reasoned interpretation.) However, while with the Stoics and their imitators the allegorical interpretation is conducted for the purpose of defending a certain definite dogmatic metaphysics, Kant's merit consists in that he had pointed to the possibility of a 'third way' between dogmatism and skepticism, by which the old teaching of the divine nature of the Word residing in our heart can be seen in a new light. He had also demonstrated that the problem of religion cannot be solved in a speculative way with the aid of a theory based on physics or biology. From the finding that causality does exist in the world, it does not follow at all that the 'causality of the world' exists. In his conclusion the speaker repeated the idea of Kant, viz. that there exist, it is true, many creeds, but only one religion, and that this one and only religion which lies at the basis of all historical religions, ought to have a uniquely spiritual content. He also repeated Spinoza's statement that Christ should be considered to be a *summus philosophus*, and added that a tragic moment in the life of Christendom would occur if one had to take, with regard to the New Testament, the same stand as that with which Christ reproached the Jews, as far as their attitude with respect to the Old Testament was concerned.

Right at the outset Brunschvicg stressed that he did not wish his lecture to be treated as opposed to the attitude of the former speaker; nevertheless, the lecture did not appeal to many theologians. (Some of them came to Prague directly from the International Neo-Thomist Congress which took place a few days earlier at Poznań.) One of them stressed that, if the philosophical defence of religion is to be based on the renewal of the hazy pantheism of Spinoza, then the game is not worth the candle. It seems however that this reproach was based only on the purely superficial side of the lecture, on the use of quotations taken from Spinoza, as the attitude of the speaker was rather more in agreement with Kant than with Spinoza.

The lectures in the fourth group, dealing with the crisis of contemporary democracy, were delivered by the Italian, Bodrero, and the American, Montague. In turn, they moved the deliberations of the Congress to the field of socio-political problems. Emilio Bodrero, a

Professor from Padua, a representative of Italian Fascism, presented a criticism of democracy, stressing that liberal democracy has no remedy for the evils of capitalism, and that only a corporate state is able to remove, economically and politically, the negative sides of capitalism. Professor Montague from New York admitted that democracy in its present form cannot last, and that Fascism and Communism will really triumph unless democracy removes its shortcomings. The reforms ought to relate to both parliamentarism and the capitalist economy. Parliamentarism ought to be supplemented by a kind of dictatorship, derived from elections (elective dictatorship), while the drawbacks of economic life should be removed by introducing a twin-track economy, an individualistic beside a collective one. Unemployed people should be united into communist communes which everybody would be able to join for a period of three years, after which one would be able to return to a private life. These communes should form a sort of islands in the society, i.e. the goods produced by these communes ought to be consumed exclusively by the communes, in order to prevent any sudden changes in the prices of products in the surrounding capitalist world.

As defenders of democracy in the discussion there appeared in the first place French speakers, Professor Victor Basch (Sorbonne) and the Member of the Institute Joseph Barthélemy; raising their oratorical talent to a summit, they fought against adversaries, using logic, wit and vicious scoffs. They considered the system of fascist corporations to be a relic of the Middle Ages, which France had conquered long ago. The deliberations were not supposed to be of a propaganda nature; opposite trends were to be considered objectively; yet temperament could be expected to carry people away with their feelings. During the session devoted to the crisis of democracy, Barthélemy also delivered one of the best lectures on this subject, admitting that democracy not always had the best men in the forefront and that the main difficulty consisted in combining the principle of liberty with that of order. We believe that it was a good thing that these questions were touched upon at the Philosophical Congress but, when we take into account that sociologists also have their specialist congresses, we must admit that, for a Philosophical Congress, too much space was devoted to them to the prejudice of other philosophical subjects and problems.

The last plenary session was devoted to the problem of the future of philosophy and its mission in the present period. The Italian, Orestano, Secretary of the Royal Academy in Rome, characterized the present times as a period of the hypertrophy of criticism, which is to be followed by a new period of construction. According to him, the task of philosophy consists in the collaboration of criticism and synthesis, and its constant aim ought to be the elevation of human dignity. The other speaker, Professor Emil Utitz, from Prague, stressed, it is true, quite strongly that philosophy ought to have a scientific character, but that this does not yet suffice, and besides, it should be something more, performing, as it does, an exceptional role in the spiritual life of mankind. When occupied wholly with philosophizing, man attains the full development of his *ens*, his being, and neglecting philosophy would be tantamount to a sinful decline of mankind. Philosophy is not only the exponent of human reason, but also of the entire human personality.

Let us now pass to an account of the sessions of doubtless the most important part of the Congress, the first section, which was devoted to the significance of logical analysis for cognition. The first to speak was Professor Łukasiewicz from Warsaw, whose lecture on this subject was originally, as already mentioned, to be delivered at the plenary session. Being a specialist in mathematical logic, he presented the most important results of this science, selecting in the first place such points as would be best accessible to philosophers without a sufficient orientation in logic. At the outset he stressed the wrong opinion of some philosophers, viz. that logistic or mathematical logic is merely a certain trend which exists in logic along with others having equal rights, whereas it is really an autonomous science which represents contemporary scientific formal logic; and it is impossible to admit, alongside of it, another logical trend that would merit the name of scientific logic. From the point of view of history, it is the higher stage of the development of ancient formal logic. However, it is not the immediate continuation of Aristotle's logic since the latter presents only a small fragment of the contemporary logic of names, beside which there exists the logic of sentences, and it is precisely the logic of sentences that is to be considered as logically prior to the logic of names. Together with the logic of relations, the latter logic can be grasped as logic of predicates.

The speaker stressed later the difference between a logical thesis and an inference rule which entitles one to draw out new theses from those already acknowledged. Due to this distinction, which likewise had been overlooked by traditional logic, it is possible to have both logical disciplines grasped axiomatically in the form of deductive systems. These systems are strictly formalized, i.e. the correctness of their deductions can be verified without the meanings of the symbols used being taken into account. They also can become an object of 'metalogical' investigations which comprise the problems of consistency, independence and completeness. The speaker also mentioned the multi-valued logics, of which the three-valued logic had already been axiomatized. He finished his lecture by presenting the following postulates: (1) The algebra of sentences must hereafter be considered, both in the school syllabi and in scientific research, as the basic component of formal logic. (2) On the other hand, Aristotle's syllogism cannot be understood as representative of that logic, and neither should it be used as the basis of logico-philosophical speculations. (3) The history of logic ought to be rewritten, viz. by an historian who had profoundly mastered mathematical logic. (4) If we want to create a scientific philosophy, we must admit that the logistic education of the future researchers in the field of philosophy is a necessary and indispensable postulate.

It is impossible to cite here the content of each of the lectures read in this section. We have already mentioned that they can be divided, as a whole, into two groups (except a few outside lectures): the lectures of the representatives of the Vienna Circle and those of the Polish logicians and philosophers. The interest in logical problems among Mach's disciples and adherents in Vienna began with the publication of Wittgenstein's book *Tractatus logico-philosophicus* (London, 1922); Wittgenstein was closely connected with Vienna, yet he was at the same time Russell's disciple. It was he who contributed to the propagation of the thesis that all sentences of logic and mathematics are tautologies; at the same time he held the radical opinion that the analysis of cognition cannot result in sensible sentences. This opinion is not shared by the contemporary representatives of the Vienna Circle, such as Schlick, Carnap and others. Carnap presented the method of logical analysis in the following way (which is developed further in his latest work *Logische Syntax der Sprache*). In the first place one ought to distinguish logical analysis from the psychological analysis and metaphysical analysis. For Carnap, metaphysical problems are not acknowledged at all, whereas psychological problems, being of an empirical nature, belong to genuine sciences. Therefore there remain, as scientific non-empirical problems, only logical questions. Often the task of logical analysis is characterized in such a way that it ought to set forth the sense of certain sentences and ideas. This characteristic is a right one; however, to avoid any misunderstandings, it ought to be supplemented with remarks which would explain more closely this analysis of meaning. The point is that the question relating to the sense of an expression can be understood purely psychologically, viz. 'what presentations or thoughts are usually linked to the given presentation'. It is true that this question is sensible, but an answer to it can be obtained simply by way of empirical investigations, and that this is not the matter in the case of logical analysis. Carnap likewise rejects the phenomenological approach to this question, adopted by adherents of the school of Husserl who, besides the empirical and formal (analytical) sentences, admits the existence of synthetic sentences. According to Carnap, such a stand results from the confusion of psychological problems with logical ones.

Hence the final answer to the question 'what does the logical analysis of an expression consist in' is: it consists in constructing a certain language system and in fitting the expression in question into this system. The construction of the system is effected by giving the rules which determine the idea of 'a sentence in a given system', and by giving the transformation rules by means of which the idea of the 'sequence in the given system' is determined. The last idea is the foundation of logical analysis, and is the fundamental idea of the formal theory of a language system, which we call logical syntax. The speaker explained this definition by means of examples. Thus whoever wants to give an answer to the question relating to the sense of an expression in accordance with the above-mentioned understanding, has to establish: (1) In sentences of what form does the relevant expression occur? (2) From what sentences are these sentences deduced and what sentences result from them? This method leads to the exclusion of metaphysical problems as being senseless, while it admits, it is true, the reasonableness of empirical psychological questions, though demarcating them strictly from logical ones. Thus logic becomes identical with logical syntax, the language is mathematics, the 'language' being understood as a system of rules of a certain kind.

Apart from Carnap the following representatives and sympathizers of the Vienna Circle delivered lectures: Schlick, Neurath and Reichenbach. The lecture by Schlick 'The Idea of the Whole' was aimed against the deduction of metaphysical consequences from the ideas of the whole, i.e. a form which opposes the combinations that produce only a simple sum, i.e. the additive combinations. As criteria for the form or whole, one usually adduces that: (1) the whole possesses properties which cannot be deduced from the properties of the parts (sometimes it is added that they cannot be deduced causally as effects), and in addition (2) that the whole does not undergo a change, even if we transpose the elements entering into its composition. An example can be a melody, or every living organism which preserves its character as a whole, even as the components undergo a change. The wholes thus defined can occur, according to some, both in the inorganic and the organic world (Köhler), whereas others (Driesch), maintain that the wholes occur only in the organic world, because only in this case not only do the parts condition the wholes, but also the wholes condition the parts, and this is, for him, the proof for the existence of entelechies. Schlick pointed to the lability of this distinction, stressing that the statement that the whole determines the behavior of the parts means solely that, if parts and all relations between them are given, the whole is given likewise, and the other way round.

Otto Neurath defended the thesis, known from his articles in *Erkenntnis*, viz. the thesis of universal physicalism, of the unification of all natural and humanistic sciences into one science, with the aid of the language of physics as the universal language. Later, Reichenbach spoke on 'The Significance of the Idea of Probability for Cognition'. The primary source of the idea of probability lies in the problem of the laws of nature. The idea that each 'becoming' in nature is subjected to strict rules is found to be a fiction which cannot be applied to natural phenomena. The application of a strict law, usually expressed in the form of an implication 'If *A* then *B*' presupposes that the initial conditions related to *A* are accurately known, and this is impossible. The agreement between the real state of affairs with the idealized *A* can be established only within certain limits. Causality is expressed by the sentence that, as *A* is defined more and more precisely, the probability for *B* tends to a limit = 1. However, as early as 1925, Reichenbach expressed the view that nature can be created in such a way that the rise of this probability up to the limit 1 can be excluded, and this had been precisely confirmed by the further development of quantum theory, viz. by Heisenberg's formula.

Another source of the idea of probability is the problem of induction. If, on the basis of *n* repetitions of a certain phenomenon I infer further possible repetitions, I deal with a probabilistic conclusion, the probability of which increases along with *n*. A positive result is never tantamount to a verification with certainty, but merely to an increase of probability. Hence in the case of inductive inference, we deal only with sentences with which are coordinated, as logical values, not truth and falsehood, but a continuous scale of probabilities. Thus, in the case of Reichenbach, the theory of probability is associated with a certain type of many-valued logic. Reichenbach has created the axiomatics of the calculus of probability, from which we can deduce that, if one accepts the frequency definition of probability, all of the presuppositions of the calculus can be reduced to a single one, viz. to the principle of inductive inference. Sentences relating to the future ought to be considered, according to Reichenbach, as a new type of sentence, differing alike from the true ones and false ones. This is betting on the future, just as, during horse-racing, one bets on a horse. Among the possible stakes, an order of value does exist, and one of the stakes is the most advantageous one. The inductive inference distinguishes precisely the most advantageous betting. Scientific predictions are stakes according to the best principle which yields the maximum number of right cases.

Professor Reichenbach's lecture was closely linked to Professor Zawirski's (Poznań) lecture 'The Significance of Many-Valued Logic and Its Relation to the Probability Calculus'. The speaker first pointed to the genesis of many-valued logic which had been created, independently of one another, by Professor Łukasiewicz and by the American logician Emil Post. However, the logics created by them were not satisfactorily linked to the probability calculus, since the logical sums and products of the new values did not correspond to all possible cases of the probability calculus. Now Reichenbach obtained the desired concordance, to be sure, but only because he introduced, for the calculation of the logical sums and products, the formulae taken from the probability calculus. Now the speaker wanted to attain the desired value exclusively on the basis of relations between logical functors, without using the formulae of the probability calculus. The method of solution, as suggested by himself, was considered by him to be superior to that of Reichenbach since, when introducing, in accordance with the calculation, a larger number of values for the sum and product, he preserved at the same time only one functor of implication and equivalence, whereas in the case of Reichenbach implication and equivalence also became quite unnecessarily many-valued. In the discussion, Reichenbach defended himself by saying that the many-valuedness of implication and equivalence follows of necessity from the probability calculus. However, this necessity will take place in logic only, when one introduces into it, as a new logical functor, the formula for a relative probability, which the speaker had avoided precisely by constructing the many-valued logic upon the logical constants of implication and asymmetric negation (which had been introduced into logic by Post for the first time). Owing to its many-valuedness, the relative probability is not fit for the role of a logical functor. To be sure, we can risk something of this kind, but we would be loth to introduce such an innovation, inconsistent with the idea of the truth functor and with the previous tradition of logic.

Professor Łukasiewicz reserved for himself a closer examination of this problem in the future; Dr. Tarski, on the other hand, declared during the discussion that the connection between many-valued logic and the calculus of probability, as given by Reichenbach and Zawirski, does not appear to be indispensable. Of course, the need for this connection will remain an open question. A serious motive for this undertaking was, as stressed by both Reichenbach and Zawirski, the fact that the probability ('inductive') decisiveness in many cases should be considered to be the final solution of the problem. For instance, the sentence that something will be repeated an infinite number of times cannot ever become a definitive truth.

The idea of probability was also involved in the lecture of Miss Hosiasson from Warsaw, who understands probability as a property of conclusions based on an incomplete premise, hence upon partial premises, that conclusion being of a greater probability whose partial premises can be supplementary to the complete ones by using a lesser number of premises, or, generally, by means of weaker premises. The lectures of Professors Ajdukiewicz and Ingarden (from Lwów) deserve special attention. The former spoke about the applicability of pure logic to philosophical problems; in particular, his question was to find the relation between the psychophysical problem of identity and the logical thesis of extensionalism. It is known that some representatives of the Vienna Circle developed ideas which introduce a certain kind of scientific materialism, or some sort of monism. Now, some principles do exist in logic which justify to a certain extent and *prima facie* such identification of, e.g., physical processes with some physiological processes, and even the identity of all events in general. Such is the principle of the thesis of extensionalism which can be formulated in different ways; and one formulation, combined with Leibniz's definition of identity, leads to the consequence that might be paraphrased as follows: two properties of equal domains are identical. Thus the extensionality thesis leads in fact to a certain theory of identity, while most contemporary logicians adopt the extensionality thesis which had been developed precisely on the basis of mathematical logic. However one must not forget that the philosophical problems were formulated in everyday language, and if mathematical logic accepted the extensionalist thesis, it does not follow that this thesis is valid in the colloquial language in which the philosopher formulates his problems.

A still more far-reaching criticism of the logical positivism of the Vienna Circle was to be found in the lecture of Professor Ingarden from Lwow, 'A Logistic Attempt to Reconstruct Philosophy'. It is known that the Vienna Circle considers that science is only empirical natural science which presents only one science and whose results can be expressed in the language of physics. Mathematics and logic are not separate sciences, but, being systems of tautologies, they serve only to transform and connect scientific sentences. Philosophy does not possess its own cognitive realm or its own cognition sources; its task consists in explaining the scientific ideas and scientific sentences by means of logical analysis, exactly as represented by Carnap. Any other philosophy ought to be rejected as senseless metaphysics. To substantiate this standpoint one makes use of the 'metalogical' language theory and a series of statements which can be briefly couched as follows. Sentences are sensible when, and only when, they are either verifiable or their structure is in agreement with logical syntax. The sense of a sentence is its verifiability. Science is composed of 'intersubjective' sentences. Any other sentence does not belong to science. Only sentences which are concerned with something physical are intersubjective, while sentences concerning something non-physical, are either senseless or else they ought to be translated into the language of physics. The directly given things cannot be cognized, they can be only experienced. One can cognize only those formal relations which can be communicated. What cannot be communicated is noncognizable.

Ingarden brought forward the following critical remarks against the above argument. The sentences about sentences, the so-called 'metalogical' sentences would be, with respect to the above criteria, either senseless or metaphysical since they are themselves not something physical, and they are not something physical because their sense is their verifiability. Now verifiability is nothing else but the possibility of verification, and a possibility is nothing physical. Indeed if verifiability consists in the coordination of some data to the sentence, then this coordination is also nothing physical. We must therefore distinguish between the reasonableness of sentences and their verifiability, and since the sense of a sentence is not anything physical, we have to adopt a separate cognitive field for philosophy if philosophy is to be based on 'logical analysis'.

In general, one had the impression that the arguments of Polish speakers were of a higher precision than the lectures of the representatives of the Vienna Circle. There were also other good lectures outside these two circles. Thus the lecture of Professor Jörgenson from Copenhagen 'On Logical Foundations of Sciences' stood out. Charles Morris from Chicago spoke about 'The Idea of Meaning in Pragmatism and in Logical Positivism', stressing that both these trends ought to be complementary. In the same section, a lecture was delivered by F.C.S. Schiller from Oxford, the well-known creator of English humanism and pragmatism, his subject being 'How is Exactitude Possible?' He attacked the logicians for their claims to exactitude which is absent outside the definitions of mathematics because no real object will enter into our definitions without resistance. Plato asserted that God always thought mathematically, but he ought to have added that he was doing it fairly perfunctorily and inaccurately, as nowhere in nature would one find exact counterparts of our ideal notions. One ought to create new words for every new situation in order to eliminate the lability of expressions. But such a language, built from *hapax legomena*, would have been quite unintelligible. On the whole it was an attack on logistics, not very serious yet not without value.

From the Polish lectures delivered in this section (and in other sections), let us mention the lecture of Father Kobyłecki 'On the Relation as the Fundamental Idea of Mathematical Logic and Philosophy', that of Father Kowalski from Gniezno 'On the Foundations of Philosophy and on Its Mission in Our Time', that of Jasirowski from Wilno 'On the Relation of the History of Philosophy to Systematic Philosophy,' and of Father Siwek from Rome 'On the Consciousness of Free Will'. Two lectures submitted from Poland were not delivered because of the absence of the speakers: Henryk Mehlberg 'On Physical and Extra-Physical Time', and Edward Habermann from Poznań on 'Causality and Probability in the Field of Chemical Processes.'

During the session of the International Committee which took place during the Congress, it was decided to organize the next Congress in Paris in 1937, viz. to celebrate the 300th anniversary of the publication of the treatise *Discours de la Méthode*. A telegram with congratulations was also sent to Bergson who was elected an honorary president of the next Congress.

Another very important thing should be mentioned here. It is known that the Vienna Circle, which represents logical positivism, felt a certain spiritual unity with contemporary Polish logicians. Suffice it to mention

Carnap's visit to Warsaw to deliver lectures, and the invitation of Tarski to Vienna about two years earlier. In recent times the Vienna Circle proposed an idea of organizing an international congress devoted to the philosophy of science in Paris in 1935, and with that end in view they invited Polish logicians to a preliminary conference in Prague, which took place two days before the congress (August 31 and September 1). Quite a number of lectures were delivered at that conference for the purpose of the mutual exchange of ideas and of establishing a closer acquaintance. On the Polish side, lectures were delivered by Professors Ajdukiewicz, and Łukasiewicz and Dr. Tarski. Ajdukiewicz presented his philosophical position in his lecture 'On the Scientific Perspective of the World', and in another lecture 'On Positivism in Poland' he turned his attention, however, not so much to the older scientific workers, who in Poland remained under the influence of Avenarius, Mach etc., but rather to the circle of contemporary logicians and, in accordance with the present interests of the Vienna Circle. Professor Łukasiewicz delivered a highly interesting lecture 'On the History of the Calculus of Sentences Starting with Ancient Times', while Dr. Tarski had a highly specialized lecture on the 'Methodological Research into the Definability of Ideas'. In addition to Poles and representatives of the Circle, there were the Frenchman Rougier, the American Morris, the Dane Jörgenson and others. Our representation turned out to be excellent and we were able to observe that the richness of our achievements called forth general admiration and amazement among people who heretofore had not known us more closely.

## 1.2. IMPRESSIONS FROM THE FIRST INTERNATIONAL CONGRESS OF SCIENTIFIC PHILOSOPHY

In accordance with the decision adopted by the Vienna Circle and by the representatives of contemporary logic at the introductory conference held in Prague in 1934 the First International Congress of Scientific Philosophy took place in Paris between September 15th and 23rd, 1935. However, prior to discussing the deliberations of the Congress, we should like, first of all, to excuse ourselves as to why, instead of giving the proceedings, as usual, we only confine ourselves to impressions. The reason is quite simple: we are unable to give an exact account since, before the beginning of the Congress, printed texts of the lectures had not been prepared as would be normal, nor were abstracts available, although such abstracts had been sent by the authors to the organizing committee. That is why a certain disorder found its way into the deliberations of the Congress, and for the same reason the reviewer must necessarily confine himself to what either he remembered or had been recorded by him.[1] The Committee tried to excuse themselves for the unpleasant situation by stating that, instead of the promised subsidy of 10,000 franes, the French Government had provided only 5,000, while the second instalment which was precisely earmarked for covering the printing cost, had been cancelled as a result of budgetary reductions initiated by Laval's government.

The host of the Congress was Émile Rougier, who took an active part in the introductory conference at Prague, as a professor at the Besançon University and, at the same time, a professor at the newly-founded Cairo University. It is well-known that the Moslem East had begun, a few years earlier, to follow in the steps of Europe as far as the organization of scientific work is concerned; universities on the European model were established in Istanbul and Cairo. For the time belong, all this is being done rather as window-dressing than because of real need. A place was founded in Istanbul for the numerous professors dismissed by the German government; one of them is the eminent Berlin philosopher, Reichenbach, who took part in this Congress. From conversations with Reichenbach and Rougier we understood that their work in the new place is very difficult and thankless. In Istanbul all of the professors who came from Germany lecture in German; until now, this language is poorly known in the East. For this reason an interpreter sits beside every professor during lectures, whose duty it is to translate every sentence of the lecture into Turkish. Who can guarantee that his translation is correct? The professor, who does not know Turkish, is of course unable to check the correctness of the translation. Rougier complained of the intellectual apathy of Moslems, who do not think it worthwhile to acquire the knowledge of anything beyond the Quran. One of them boasted of having learned the entire text of the Quran in Persian by heart, although he did not understand that language. Now this is a perfect illustration of their intellectual ambitions and of the way in which these ambitions can be satisfied! Rougier was moved in his pessimism even so far as to maintain that all of the famous Arabic philosophy had not been created by the Arabs themselves, but by foreigners who composed the huge empire of the Caliphs. Peoples of the ancient Orient: Syrians, inhabitants of Egypt and Chaldea, who had remained for many centuries under the continuous influence of Hellenic culture, all of them, having later dissolved themselves in the sea of Moslems, had produced the Arabic philosophy and culture.

In his view, a native Arab has never been capable of using anything. It is quite possible that this statement of Rougier contains much exaggeration; yet there can be no doubt but that the Moslem East is still buried in a drowsy state, and that one will have to wait for decades before it will be awakened indeed to the intellectual life, and before the discussion of scientific and philosophical problems will have become its natural spiritual need.

After this digression let us revert to the subject proper. The Congress assembled about 160 participants, who represented about 20 nations, not only European but also American and Asian. The solemn inauguration of the Congress took place on Sunday, September 15th in the afternoon, at the Institut de Coopération Intellectuelle, where the allocution of Rougier was followed by the orations of the representatives of individual scientific groups. Most interesting was the appearance and allocution of Russell who used to avoid speaking before any philosophical congresses, only to honor this one with his presence. He spoke about being constrained to devote a part of his life to unlearning that which he learned from the works of metaphysicians, and about the necessity of devoting six days to mathematics in order to spend one day with advantage for philosophy. In his introductory speech Rougier characterized the efforts of the contemporary philosophy of science to become the logical analysis of scientific theories, with the elimination as being pseudo-problems of all problems for which one is unable to indicate what experiments could give a relevant affirmative or negative answer. In his capacity as the representative of the Italian group, Enriques expressed, in his address of welcome, the fear lest the formalism of contemporary logicians should be changed into a kind 'of scholasticism of symbols in themselves', without any concern as regards their meaning. When replying to this speech, the physicist Frank, a member of the Vienna Circle, demonstrated that, despite these apprehensions the Vienna Circle can boast of positive results by pointing out that in the sphere of cognition, there can exist only statements about facts and the purely tautological transformations of these statements. Morris spoke in the name of American pragmatists, while, deputizing for the absent Professor Łukasiewicz, the Congress was welcomed by Professor Ajdukiewicz from Lwów.

The proper deliberations of the Congress began the next day and took place in two halls of the Sorbonne, viz. in the Descartes and Guizot amphitheatres. The point is that the number of lectures was so considerable as to make it necessary to organize two parallel sections, since the time devoted for lectures and discussions was only one full week. In view of the lack of printed abstracts, new difficulties arose for people who wanted to listen to two lectures which were accidentally divided between different sections. Thus, e.g., we did not hear the lecture of M. Jaśkowski on Brouwer's logic, in spite of our ardent wish, as we were obliged to deliver our own lecture in another section. Exceptions as far as the division into sections concerned, happened during two mornings only, Monday and Wednesday, devoted to 'plenary' sessions; similarly, on the last day it was possible, despite the printed program, to combine everything in one hall thanks to the fact that some lectures scheduled to take place were cancelled due to the absence of the members.

Monday morning there spoke, among others, the chief representatives of the Vienna Circle, Carnap and Neurath. Carnap presented three phases which can be distinguished in the evolution of the V.C. programme. The first phase is characterized by the anti-metaphysical attitude inherited from Mach, i.e. the conviction that metaphysics, *qua* science, is impossible. In this phase, the theory of knowledge appeared to them to be the fundamental philosophical science. In the following phase, the representatives of the Circle become conscious of the fact that, not only can metaphysics not enter into the composition of the philosophy of science, but also the traditional theory of cognition must be excluded from this role, frequently based on metaphysical foundations as it has been, and too such a theory of *a priori* cognition as that created by Kant in his transcendental philosophy. The third phase is distinguished precisely by the transition from such a theory of cognition to the logic of the sciences (*Wissenschaftslogik*) as the proper aim of philosophy, differing from the psychological or sociological treatment of philosophical problems. In a philosophy thus understood, its role is reduced to the investigation of the formal structure of the languages of scientific theories.

A supplement to Carnap's lecture was the lecture of Neurath, who added that the new empiricism does not want to be a new system, but a certain encyclopaedic conception of science, related to the language of physics, which ought to play the role of the universal language. Neurath's finishing phrase was: 'We are the Encyclopaedists of the 20th century'.

Reichenbach – a representative of the former Berlin group of sympathizers of the Vienna Circle (we would like to recall that Josef Petzoldt, Avenarius's pupil, whose critical empiricism was very close to Mach's positivism, was active in Berlin) turned his attention to the successive transformation of the meaning of Kant's *a priori*, in which we should differentiate between a purely logical meaning and a psychological one. The lecture of Enriques, who was the first to lecture on that day, related to the problem of the meaning of the *a priori* factors of cognition, although his position did differ considerably from that of logical positivism. Enriques stuck to the point that a certain apriorism must be accepted as before, were it only in the sense of an idea with a variable range, since it is contained in the form of rational postulates of various scientific theories. However, the idea of this variable *a priori* was strongly disputed in discussion as being highly indefinable. As long as it has a constant meaning, one knows at least what one has to deal with; but as soon as it becomes variable, one does not know any longer what one has to fight against, and it becomes virtually valueless.

In the afternoon of the same day the sectional debates began; it is hardly possible however to acquaint the reader with the content of the lectures one after another. That is why we will satisfy ourselves with characterizing the individual national groups, as these coincided to a certain degree with the subdivision into groups having slightly different views. Let us start with the characterization of the French group. With the exception of Rougier, no other eminent French philosopher took part in the Congress. Brunschvicg was absent, and Lalande put in an appearance only as a guest during some lectures. Many young people could be seen, but no logicians of note were to be found among them; Lautmann and Chevalley, two young scholars (the former of whom spent a long time studying in Germany), and General Vouillemin, possessed perhaps the best orientation in logic. The first two criticized the anti-metaphysical attitude of the Vienna Circle representatives. A very pleasant impression was made by the disciples of the eminent physicist de Broglie, his assistant Destouches, who had been with de Broglie in Poland the previous year, and also Renaud and Pétiau. They broached various problems at the borders of physics; Destouches maintained that, in the case of the microcosm one has to content oneself with a kind of statistical theory of time and space. Besides these, France was also represented at the Congress by many eminent representatives of special sciences. Two philologues, Massignon and Masson-Oursel specializing in Oriental languages, demonstrated on the basis of these languages, in what way language structure is linked to the structure of the thought. Among the mathematicians taking part in the Congress, two delegates from the

University of Lausanne ought to be mentioned: Juvet and Reymond. Juvet pointed to the importance of the theory of sets for the problem of axiomatization, while Reymond broached the problem of the relation of negation to the principle of the excluded middle. Finally, among the biologists one should mention Count du Noüy who opposed the views of Frank, professor of physics and member of the Vienna Circle, as far as the relationship between biological sciences and physico-chemical sciences was concerned. Frank believes that the processes of life can be explained by means of physico-chemical laws, while Count du Noüy stressed that, even at the molecular scale, processes do take place which transgress the laws of modern physics. Jörgenson a philosopher from Copenhagen, took part in the discussion of this subject; he declared that, on the basis of long conversations with his colleague Niels Bohr, the creator of the modern quantum theory, he arrived at the conclusion that the attitude of the Vienna Circle as far as the appraisal of the achievements of quantum physics and their significance for biological sciences is concerned, is doubtless one-sided. Comprehensive debates on this subject were planned in Copenhagen, and Einstein had been invited to take part in them. To be sure, Einstein's attitude with regard to Bohr was rather odd; he advanced objections of a general nature against Bohr, just the same as had once been advanced against Einstein's own [relativity] theory. We mention all these details, since an idea did emerge during the Congress, viz. that of convening shortly a smaller conference in Copenhagen, composed of persons working on the relationship between the biological and the physico-chemical sciences, with the participation of Niels Bohr.

During one free afternoon, Count du Noüy invited the members of the Congress to his place. In his flat situated in the aristocratic part of Paris, near the Hôtel des Invalides, on the sixth floor, with a spacious balcony from which a glorious panorama of Paris could be seen – the members of the Congress could agreeably spend a few hours, enjoying various liquors and refreshments.

The Polish group of the Congress was not as large as one could expect in the beginning. Professor Łukasiewicz, the most representative person in this case, was missing. We felt his absence very keenly. Also, Kotarbiński failed to arrive, and Chwistek too, much as he had promised lectures and discussions on three different subjects. In addition to the present writer, the professors included only Ajdukiewicz from Lwow, and a few days before the end of the Congress, there also came Professor Jasinowski from Wilno. Fortunately Tarski was not absent; also a few young persons came from Warsaw, though fewer than expected: Miss Hosiasson, Jaśkowski, Lindenbaum. Lectures were also given by Father Kobylecki from Nowy Targ and Mrs. Lutman-Kokoszyńska, an alumna of the Lwow University, at present married and living in Katowice.

Professor Ajdukiewicz delivered a lecture on definition. He described a definition of a certain expression on the basis of a certain language as a sentence or system of sentences, which allows each sentence to be constructed from the defined term and from constants other than the defined one, to be translated on the basis of the directives of the previous language, extended to the defined term, into a certain sentence of the previous language. Next he gave the conditions of the consistency and non-creativity of the definition (he being interested in the so-called non-creative definitions), after which he carried out the differentiation of the mid-language and meta-language definitions; the former ones manipulate the defined expression itself, while the latter ones only the name of the defined expressions. Within the compass of the former ones, he differentiated between the factual and the verbal ones, and in the end he dealt with the problem of the truth of definitions, at the same time stressing the need for relating the term 'true' to a certain language. If we take into account this relativization, we will not find odd the statement of the author, viz. that nominal definitions are either true or false not on account of the primary language which becomes richer by an expression because of these definitions, but rather on account of the new language which arose due to the enrichment of the primary language by this expression, these definitions are already true.

Prior to discussing other lectures read by the Polish members of the Congress, we ought first to turn our attention to the further lectures of the Vienna Circle. The representatives of that Circle formed a rather homogeneous and coherent union; a certain opposition (a 'legal' one according to Neurath) was represented within the Circle by the young Viennese Popper who, in agreement with the content of his recent book *Logik der Forschung*, denied any logical character to the so-called inductive conclusion-inference ('die ganze induktionslogische Problematik ist eliminierbar, ohne dass dadurch neue Schwierigkeiten entstehen'). One special day was devoted to the problem of induction and to the logic of probability, the chief spokesman of the Vienna Circle on this subject being Reichenbach. We shall refrain from summarizing his allocution which repeated the fundamental ideas expressed by him on this subject during the Prague Congress and during the introductory conference which preceded that Congress. The lecture of the present writer, Professor Zawirski, was linked closely with the content of that lecture and so was that of Miss Hosiasson from Warsaw. Zawirski wanted to draw the attention of the audience to some differences between his position and that of Professor Reichenbach in matters dealing with many-valued logic. Reichenbach interpreted certain expressions obtained directly from the calculus of probability as a kind of generalized logic, while we obtained a certain parallelism between the formulae of the calculus of probability and those of Łukasiewicz's and Post's many-valued logic, after having first carried out some transformations in that logic. Thus we achieved a certain formal agreement between the many-valued logic proper (the 'topological' or 'modal logic' according to Reichenbach's terminology) and the calculus of probability (the 'metric' or 'extensional logic' of Reichenbach). We also came to believe that a formal concordance took place here not of some two systems which had nothing in common with one another, but of two systems one of which provides the empirical basis for the other. The formulae of the calculus of probability, based undoubtedly on the two-valued logic, create an empirical foundation for new unitary sentences relating to the future, which themselves already belong to the many-valued modal, or topological, logic. Reichenbach, on the contrary, originally considered this topological many-valued logic to be completely valueless (he believes that it is impossible reasonably to attribute, in a verifiable manner, more than two logical values to the unitary sentences), and therefore he looked for the source of generalized logic in the metric side proper, i.e. in the formulae of the calculus of probability.

However, since the Prague Conference, Reichenbach gradually ceased to consider this topological, or qualitative interpretation to be utterly valueless. In our case a change has likewise occurred, viz. while considering more closely the value of various definitions of the measure of probability, we assumed that the only empirically admissible definition is the frequency definition. We did not study this problem before, but in recent times, mainly under the influence of Metallmann's work, we accepted as the only valid one the position of the frequency theory which Reichenbach, too, considered to be the only one admissible. It appeared to us then that our adversaries who reject any links between the calculus of probability and many-valued logic most often simplify the problem when sticking to the classical definition of the measure of probability; it is only then that one can be satisfied with the plain answer that probability concerns the names of the sentences about which we pronounce the sentences of the two-valued logic. However, if we assume the position of the frequency definition, then the matter will not be as simple as that, because the probability sentence also will be found to be only probable, and so *ad infinitum*, and therefore we shall never reach the sentence of the two-valued logic.

Miss Hosiasson, on the other hand, defended very valiantly the traditional position, turning mainly against Reichenbach, while also attacking some points of our stand. She differentiated three possible meanings in which one could consider the calculus of probability to be a kind of generalized logic, and tried to prove that, in each of these three meanings, such an assertion is false. The lecture was on a high level, but precisely for this reason it was not easily accessible to the unprepared listeners; it seems that Reichenbach did not lie when he admitted not to have understood anything from the argument of the lecturer. One of her highly important arguments – as we could establish during a conversation following the lecture, was that truth and falsity is a functor of sentences as arguments, while probability is not a functor of the sentence, but a functor of the sentence-forming functor. Yet we are forced to admit that we cannot see any reason to adopt such an approach to the subject.

The attempt to defend Reichenbach's ideas, which we put forward by indicating that, according to his stand, a probability sentence itself is only probable, did not satisfy Miss Hosiasson; she remarked that she did not see why the iteration of probability should present difficulties in the case of two-valued logic. And so we admit that there would be no difficulty were iteration taking place only a finite number of times; however, if this goes on to infinity, as is the case with Reichenbach, then it is impossible to obtain a sentence of two-valued logic referring to probability. On the whole we must admit that Miss Hosiasson is a very serious adversary, who has been dealing for a long time with problems at the boundary between logic and the calculus of probability; she bears the imprint of the high logical culture of the Warsaw Center, and one should also bear in mind that she is an exponent of ideas which are not only her own. We cherish the hope that further discussions on this common topic will doubtless result in lifting all ambiguities and misunderstandings.

It was found the next day that the culmination point of the Congress was not the lectures on many-valued logic and probability, but the papers dealing with semantics, among which the most outstanding was the paper by Tarski, which was generally admitted to be the most important event of the Congress. It dealt with the idea of truth as the fundamental idea of semantics, and presented briefly the results of Tarski's recent work *Pojęcie prawdy w językach nauk dedukcyjnych* (The Idea of Truth in the Languages of Deductive Science) (Warsaw, 1933), which is likely to appear in German translation. It is known that Carnap, in the book *Logische Syntax der Sprache*, while limiting the task of philosophy to research into the formal structure of the languages of scientific theories, wanted to eliminate completely the idea of truth from the task of philosophy thus understood (p. 164 of the book quoted: 'wahr und falsch sind keine echten syntaktischen Begriffe'). Now Tarski has shown that, for formalized languages of finite order, this notion can be defined, even quite correctly, viz. with the aid of purely morphological terms, and that, notwithstanding, such a definition realizes those intuitions which are comprised in the so-called classical understanding of truth, according to which 'true' means in agreeement with reality.

We would like to draw the attention of the Reader to two details only which cast some light on that construction of the substantially accurate and formally correct definition of the term 'true sentence'. In the first place, we have to distinguish the sentence from a singular name of that sentence, which can be for instance expressed with the aid of inverted commas. When we say that the sentence 'it snows' is true if, and only if, it snows, we use the inverted-commas expression 'it snows' as a certain singular name for the respective sentence. However, such a definition of truth is not correct and, as can be proved, it leads to antinomies, when treated on the basis of everyday language. But it is not only on the basis of everyday language that the above definition will lead to contradictions, but probably we shall find contradictions involved in every language which, in addition to its names and expressions, will also contain the names of such sentences and expressions as well as such semantical expressions as a 'true sentence', 'name', 'denotes' etc. In short, difficulties will arise on the ground of each such language which, while cultivating the universalistic tendency of the everyday language, will not refrain from talking about itself and will contain its own semantics. The difficulties will disappear only when we shall begin to distinguish the language about which we speak from the language we use for speaking as a *sui generis* meta-language, a language of a higher order. It is only in that language that it will be possible to express the definition of a true sentence belonging to the language about which we are speaking and it is only then that we will be able to explain in an accurate way from the intuitionistic point of view the meaning of the phrase '*x* is a true sentence when and only when $p$', where $x$ must represent the name of a sentence $y$ of the language about which we speak, while the name represented by $x$ belongs to the language in which we speak, and $p$ also represents only the translation of the sentence $y$ into the language we use for speaking. Then antinomies will arise no longer, and the suggested definition will be concordant with the intuitions relating to the word 'truth'.

Obviously the definition of the true sentence must be conducted separately for each of the languages of deductive systems, and it is found that it can only be constructed in the case of poorer languages, i.e. languages containing expressions of finite order (where all variables belong to the same semantic category, or in which the number of semantic categories is greater than 1, but is finite, or when, though we have to do with an infinitely great number of semantic categories which are qualitatively different but in a finite number of orders; the difference of an order is not co-terminous with a qualitative difference of categories, e.g. there can exist classes of the second order, i.e. the classes of individuals, or two-member relations of the second order, i.e. relations between individuals, three-member relations of the second order providing they take place also only between individuals, etc.). On the other hand, in the case of formalized languages of infinite order, such definition cannot be constructed, though we can establish for them a method of operating accurately with this notion, treating it as a primitive notion, while in the everyday language not only the definition but even a correct use of this notion is impossible. The results obtained for the notion of truth can be also extended to other semantic notions.

To the lecture of Tarski was linked the directly following lecture of Mrs. Lutman-Kokoszyńska who told the Vienna Circle point-blank what Tarski failed to add. The lecturer, a very good-looking alumna of Lwow University, referred to the notion of the syntax, i.e. of the logical syntax of the language within the meaning of Carnap, first distinguishing, along with Carnap, pure syntax (which uses in these investigations solely logical and arithmetical terms) from the descriptive syntax which, in the investigations into a language, also uses some descriptive terms. Within the compass of the latter, we ought according to the lecturer to distinguish between the syntax proper which admits only the specifically syntactic terms (the names of expressions and of the structural relations between the expressions), and the extended (non-specific) syntax which makes use of all of the descriptive terms of the language being examined. Now the adherents of logical positivism wanted to construct the science of a language by means of the syntax proper and, since it was found that some semantic notions, as e.g. the notion of truth, could not be crammed into the frame of their syntax, they proclaimed them to be non-scientific. However, the investigations of Tarski into the notion of truth have shown that it is possible to define the classical notion of truth quite accurately on the basis of the extended syntax (since the extended syntax of a language corresponds approximately to what Tarski calls meta-science), according to which a sentence is true if it is exactly so as the sentence affirms. This is in a way a defence of so-called 'absolute' truth, as it was understood by the Scholastics and even by Aristotle. Much as the notion of the true sentence is here relativized to a certain language, yet such a relativization is understood by itself and does not do any harm to the sense of truth as concordance with reality, as it only makes the truth of the sentence dependent on its meaning, and this meaning depends strictly on the language to which the sentence belongs.

As mentioned already, the opposition to the stand of the Vienna Circle came also from certain French philosophers; especially Lautmann and Chevalley criticized the anti-metaphysical attitude of the Circle. However this criticism did not, and could not, make an impression on the representatives of the Circle such as Tarski's criticism, based on methods admitted precisely by the Circle. In the discussion Neurath and Carnap took the floor on the part of the Circle. Neurath deplored the unnecessary mentioning of the word 'absolute', which always gave rise to so many misunderstandings, and expressed doubts whether the considerations presented by Tarski can be of service for the special sciences. Now Carnap's relation to the content of the paper was more factual; as a scientist, he expressed his readiness to accept all corrections and complements to the lacunae in his system, provided they were duly motivated. Though usually silent, Russell spoke up too, pointing to the fact that the work of Tarski in a way carries into effect his previsions which he expressed in the foreword to the work by Wittgenstein, *Tractatus Logico-Philosophicus* of 1922. The point is that, while Wittgenstein was skeptical with regard to the scientific nature of sentences of the scientific language, Russell, in the foreword to the work of his disciple, had already noticed then that the difficulties could be removed by creating a language of the higher order for the preceding language (in the way we ourselves note that Wittgenstein's skepticism turned not only against a semantics such as is accepted by Polish logicians, but even against such syntactic studies as are conducted at present by the Vienna Circle). At the same time Russell paid compliments to the Vienna Circle, stating that he liked very much especially the physicalism of the Circle, and that he was even of the opinion that logic should also be treated as a part of physics in a way more widely grasped, since the language symbols are physical bodies.

The lectures of the last day were devoted to problems from the history of logic. This group included the lecture of Professor Jasinowski who, among others, pointed to the fact that the logical attitude of Stoics and their opposition with respect to the logic of Aristotle will become easier to understand if we bring home that Stoics were nominalists who assumed general notions to be merely the products of our mind. Professor Scholz from Münster showed to the members of the Congress a part of Frege's manuscripts and his photograph, informing them at the same time of the planned, publication of Frege's work in the field of logic. At this juncture Russell delivered a speech stressing Frege's enormous merits in the field of logic. The commemoration of Frege's memory was connected with an ovation offered to Russell, and since Padoa mentioned at this juncture praise for the works of the recently deceased Peano in the field of logic, homage was also paid to his memory. Miss Antoinette Raymond, the daughter of a Swiss mathematician, spoke about the tangent points between the logic of the Stoics and Russell's logic. Alas poor thing, she ignored everything about Professor's Łukasiewicz's work on this subject; basing herself on the older French papers relating to the logic of the Stoics, she reached, to be sure, the point that Stoics called in the first place for examining the logical relations of sentences treated as a whole (*comme un tout*) and that they placed their syllogism above that of Aristotle; however, when she wanted to write this Stoic syllogism, she wrote it incorrectly. The point is that she presented it in the following form: 'If there is *a*, there is *b*; *a* is, hence *b* is, too.' Anybody who is acquainted with the theory of sentences will see immediately that the word 'is', when put in unnecessarily, changes the entire meaning. Now 'is' is a sentence-forming functor of the name argument, and since as the lecturer herself declared *a* and *b* are to represent the sentences as a whole, the addition of this 'is' only spoils the sense. The only correct way of reading is as follows: if *a*, then *b*; now *a*; hence *b*. The ancient texts do not give this incorrectly inserted 'is'; the difference between the contemporary way of expressing this syllogism of the Stoics and their own consists only in that we represent the sentence variables by means of letters, while the Stoics used the ordinal numerals for this purpose: *si primum, secundum; atqui primum; secundum igitur*. In his lectures, Professor Lukasiewicz points explicitly to this, and other similar erroneous ways of reading the formulae of the sentential calculus, which often occur, especially with beginners (*Elementy logiki matematycznej*, 'The elements of mathematical logic', p. 41). Miss Antoinette heartily thanked Professors Ajdukiewicz and Scholz for having called her attention to Professor Łukasiewicz's paper.

On the last day of the Congress, various committees were elected: one devoted to the publication of the encyclopaedia, the other to the establishment of terminology in the field of mathematical logic. The latter one is undoubtedly highly necessary, when one takes into account that the Warsaw logicians have introduced quite a number of new terms, unknown in the West, which might have been wrongly construed. The Vienna Circle often makes use of a slightly different terminology, which makes mutual understanding difficult.

As suggested by the French members, it was decided to hold the next congress of the philosophy of science again in Paris in 1937, during the Great Exhibition, so as to precede by a few days the usual international philosophical congress scheduled to take place in the same year, just as the introductory conference at Prague coincided in 1934 with the general Philosophical Congress. It is possible however that, in 1936, a smaller meeting of philosophers interested in the progress of the physicochemical and biological sciences will be convened in Copenhagen.[2]

CHAPTER 2

# LOGIC

## 2.1. THE RELATIONS BETWEEN MULTI-VALUED LOGIC AND PROBABILITY CALCULUS

In our communication presented at the International Congress of Philosophy in Prague, we proposed, along with Reichenbach, a connection between the probability calculus and a kind of multi-valued logic (I take the liberty of using the neologism introduced by Cavaillès). However, the attempt to bring the theory of probability into contact with this logic has aroused some objections. Although Reichenbach replied immediately to these objections, we desire to return to this question, since our point of view is not completely identical with that of Reichenbach. The first objection indicated that this combination was impossible *a priori*, since, in the logic of propositions, the values of the sum and of the product are univocal functions of their propositional arguments, while the probabilities of the sum and of the product are not univocal functions of their arguments. Different probability values can correspond to the same values of the argument, hence the former values cannot be considered as new logical values, in addition to true and false. The second objection can be summarized as follows: why should one construct a new logic if the theory of probability can be accommodated into the frame of ordinary (two-valued) logic?

As far as the former objection is concerned, it can be directed solely against Reichenbach's point of view, but not against ours, since we have increased the number of logical operators of the sum and of the product in such a way that only one logical value corresponds to each of these factors. Let us take, for example, the logic of five values symbolized by the series of numbers 0, 1/4, 2/4, 3/4, 1. It is true that, in the logic of Łukasiewicz and in that of Post, the logical sum acquires only one value; for the values of the arguments $v(p)^3 = 2/4$ and $v(q) = 2/4$, the sum obtains the value 2/4, while the calculus allows us, with the same values, to obtain the value 2/4 as well as 3/4 or even 1. However, we have assumed for each of these cases another definition of the sum, and thus we obtained three different formulae, and the values of each of them became univocal functions of their arguments.

This innovation was not whimsical: we were authorized to adopt it for very important reasons. Two possible definitions of the logical sum (and of the product) p v q = (p ⊃ q) ⊃ q, or pvq. = ~ pvq, which are equivalent in ordinary logic, are not equivalent in the multi-valued logics. For $v(p)$ = 2/4 and $v(q)$ = 2/4, the former sum has the value 2, while the second yields 1. It was therefore necessary to introduce these two formulae into the new logic. It is true that this was not yet sufficient; by using these formulae we only rendered an account of the minimum value and the maximum value of the sum. The intermediate values remained unexplained. In order to express them by means of a logical operation we introduced the symbol of an asymmetrical negation, used for the first time by Post. By means of this new logical constant, we can express all of the values of the sum and the product of propositions in the new logic, in conformity with the probability calculus. However, we ought to admit that these formulae, in which the asymmetrical negation occurs, are extremely complicated, and it would be very desirable if they could be expressed in a simpler way by using new logical constants. Perhaps we could for this purpose use a generalization of the functor 't' introduced by Słupecki in three-valued logic.

However, it is also possible to reach our result by using another method, viz. by considering each proposition of the n-valued logic according to Post's postulate as one class of (n – 1) two-valued propositions. In our example of the logic with five values, the different formulae of the sum are justified by the difference in the order in which the true propositions follow the false propositions in these classes. If v(p) = 2/4 and v(q) = 2/4, the corresponding classes, containing two true propositions for four true or false ones, can have the form (the true proposition being denoted by 1 and the false proposition by 0): p = (0,0,1,1) q = (0,0,1,1) or else p = (1,1,0,0), q = (0,0,1,1), or again p = (0,0,1,1), q = (1,0,1,0). Let us now interpret the logical sum p v q as a number of two-valued logical sums formed by joining the first member of the first series with the first member of the second series, the second member of the first with the second member of the second series, etc., and we shall obtain in the former case p v q = (0 v 0, 0 v 0, 1 v 1, 1 v 1) = (0,0,1,1), the value of the sum will be 2/4; in the second case pvq = (1 v 0, 1 v 0, 0 v 1, 0 v 1) =(1,1,1,1) it will be 1, the case of the maximum value for the sum; finally, in the last case pvq = (0 v 1, 0 v 0, 1 v 1, 1 v 0) = (1,0,1,1) the value of the sum will be 3/4. In Łukasiewicz' logic and in that of Post, the values of the sum correspond to the former arrangement, but, since all of the arrangements are equally realizable, one has to admit all of the possible results of the sum. Now, Reichenbach indicated the possibility of calculating the various values of the sum as functions of all possible arrangements for two arguments, by introducing, as a third argument, the variable 'u' which expresses the degree of coupling of the values of two propositional variables. Thus he has exposed himself to the objection that, in his logic, the sum and the product are not univocal functions of propositional variables. However, he indicated very rightly that, much as this innovation is without a precedent, nothing prohibits us from the generalization of the logic in such a way that the sum depends on a third parameter which expresses the degree of coupling of the two former ones. 'This generalized logic does remain extensional, since the contents of the propositions does not play any role here, and the degree of coupling is likewise a formal quantity just as the values of other variables.'

However, do we not return to the field of ordinary logic while only an apparent illusion of a multi-valued logic is created, if we are obliged to substitute, for each multi-valued proposition, a class of two-valued propositions? The metrical interpretation, in the framework of ordinary logic, is the only one that has a real value, while the qualitative interpretation, in the framework of multi-valued logic, is merely a quite superfluous decoration. We find ourselves faced with the second objection mentioned before. At the same time we are shown a much simpler solution of the problem. The probability is related to propositions, hence it is not the propositions *in suppositione formali* that here form the arguments, but rather the names of these propositions, propositions *in suppositione materiali*. We set forth the propositions about the probability of the mentioned propositions, so to say the proposition in inverted commas. If the names of the propositions ought to be divided into more than two classes, nothing follows for the propositions themselves, and nothing entitles us to look for more than two values of the latter ones.

To all of this we give an answer which might not be identical with that of Reichenbach. By increasing the number of logical functors of the sum and of the product, we have obtained only a parallelism between the multi-valued logic, considered solely from the qualitative point of view ('topological logic' of Reichenbach) and the calculus of probability which possesses a quantitative nature ('metric logic'). But, we did not want to deny the fact that the metric side had been constructed by means of two-valued logic. However, if we examine the relation between two-valued logic and the probability calculus, we cannot forget that there exist two possible definitions of the measure of probability. The solution proposed by our opponents is mostly founded on the classical definition of the measure of probability, but it becomes untenable, according to Reichenbach's opinion, if one considers it from the point of view of the empirical theory which defines probability as a limit of the relative frequency, since "the proposition itself which lays down the probability becomes in turn contaminated by the same character, and so forth until infinity".[4]

In the paper we presented at Prague we did not want to be carried away to a controversy of discussing which of the two definitions is better. However, we are now convinced that we should accept the frequency theory, in spite of all of the objections raised against it. The classical theory is interesting as a purely mathematical theory, which has nothing in common with the possibility of application in experience. The beautiful defence of the classical theory, as given by Waismann in a paper of his appearing in the periodical *Erkenntnis* had been subjected to a rigorous critique in the Polish literature by Metallmann in his work: 'Le déterminisme des sciences naturelles'. It follows from this critique that there exist cases when the collectivist theory, founded on the concept of frequency, cannot be applied, and that there exist other cases where our trust in the precision of a mechanism makes this application superfluous for us. However, we then become responsible for a risk by blindly estimating the cases as equally probable. It is only then that the subjectivist theory which considers the probability calculus in the framework of ordinary logic can be justified. In all other cases we are forced to use the frequency theory which leads necessarily, according to Reichenbach, to a multi-valued logic.

We shall now demonstrate the new light which is cast by multi-valued logic onto the problem of induction. Post observes that it is possible to construct deductive systems in which the axioms and theorems can possess logical values inferior to truth. In such systems the inference rules, especially those of detachment (*Abtrennungsregel, modus ponens*) ought likewise to be modified, by permitting those with a value less than 1 to be added to the system. Now Post was the first to have formulated such a rule of detachment based on the special concept of implication. Post's definition of implication proves that, if the degree of the antecedent is lower than or equal to that of the consequent, the implication becomes true; if the degree of the antecedent is higher than that of the consequent, being at the same time equal to $\mu$ or higher than $\gamma$, the value of the implication should be identical with that of the consequent; but, if the value of the antecedent is lower than $\mu$, being equal, for instance, to $\gamma$, while the degree of the consequent has, for instance, the value $\gamma$, then the value of implication is equal to $1 - \mu + \gamma'$. It is only by accepting this definition that the value of the consequence being put aside (according to the detachment rule based on the concept of this implication) will never drop below the value $\mu$, if the value of the premises is equal to, or higher than, $\gamma$, which is very important if we want to construct a polyvalent deductive system in which the propositions with the degree $\gamma$ ought to appear.

We see immediately that the inference rule, thus formulated in the multi-valued deductive systems of degree $\Delta$ (where $0/\mu/1$), can be compared to the so-called rule of inductive inference. The choice of the degree $\mu$ depends on us, which means that it does depend on us as to what degree of probability will satisfy us in our system. In practice it is impossible to determine rigorously the limit of the relative frequencies; at best we can indicate the interval within which this limit is contained.

However, one thing ought to be changed in the considerations of Post. If we base ourselves firmly on the point of view of the frequency, just as Reichenbach wants, the classes of propositions of the two-valued logic, which correspond to multi-valued propositions, should contain a number of propositions, which has nothing to do with the degree of value. The same remark applies to the number of relative frequencies. The degree of value $\mu$ depends on the number of logical values, which is equal to the number of limits around which the frequency values are grouped. This number of limits can be revealed or suggested solely by experience.

The definition of probability as a property of conclusions, according to two-valued logic, is drawn from partial premises. It seeks to measure the probability according to the number or the force of complementary premises, but it does not tell us how this process can lead to univocal metric results. We should rather introduce, right from the start, the definition of probability as the limit of a series of frequencies. We can introduce the concept of the new implication into the definition of the relative frequency, and even into the consideration of the relation of relative frequencies at their limit, but this is not necessary. It suffices to introduce a new conception of implication at the moment when multi-valued propositions are being introduced, by making a series of frequencies with their limit set to correspond with each of these propositions.

## 2.2. THE GENESIS AND DEVELOPMENT OF INTUITIONISTIC LOGIC

When discussing intuitionistic logic we have in mind the trend originated by the Dutch mathematician L.E.J. Brouwer, and later developed by his disciple A. Heyting. Further, quite essential corrections were introduced into this system by the young Viennese scientist Kurt Gödel, and the final form was given to it by the Polish logician Stanisław Jaśkowski. Let us start with the presentation of Brouwer's ideas.

(1) To Brouwer, the basis of arithmetic and the theory of sets is the 'proto-intuition' of the natural series of numbers and mathematical induction, understood not as some general logical formula, but as a principle of operation used, step by step, in every section of mathematics. Brouwer does not agree with the thesis of logicism which would like to reduce mathematics to logic alone, nor with the thesis of the Hilbert school formalists who believe that mathematics is reduced merely to the language of mathematical symbols. Mathematics should be understood as something independent of the mathematical language, of the purely paper proofs which do not give rise to an intuitive conviction of truth. And since logic is often treated as something closely connected with the form of language, one ought also to get rid of the language form of the logic thus understood. Besides, it is not at all true that logic is to be the sole foundation of mathematics; logic is rather something that had been abstracted from the mathematics of finite sets and has been transferred without justification to the mathematics of infinite sets.

Among further postulates which are advanced by the new trend in relation to the previous ones and rather the most important is the one that connects the notion of existence with the possibility of construction. While pondering on the origin of the statements of mathematicians which do not arouse intuitive conviction, Brouwer arrives at the conclusion that they originate from the fact that one disregards the possibility of carrying out a construction, while attributing existence to cases which not only can never be carried out in practice, but in which the statement of the existence of 'existence theorems', loses its sense. An example of such a statement is the familiar Zermelo axiom which states that, if we have any number of non-empty and mutually-exclusive classes, then there exists at least one class which arises by picking one element out of each of these classes. This statement ceases to be obvious in the case when the elements of the classes from which we have to pick singly do not differ in any respect from one another. Thus the difficulty here lies not in the notion of infinity but in that, when the elements do not differ in any respect from one another, one does not know which elements are in question when we speak about the selected class. We gave above a rather incorrect formulation, viz. that the class was formed 'by picking'. Hence one could have the impression that the stress is laid on the practical impossibility of carrying out the selection. But this is not the question; of course, if there are an infinite number of elements in the class then one cannot carry out the process of selection. However, everybody can demand to be presented with a method which would decide the designation of the chosen element, in order to make it known which element is meant, as otherwise the elements of the selected class are not determined. Each statement about the selected class will then be a sentence without content.

Next, Brouwer reflects upon the origin, in modern mathematics, of empty sentences or, as he puts it '*Urteilsabstrakte.*' According to him, the source of such sentences is the incautious manner of using the word 'all' and of applying the law of the excluded middle to each class determined by using the word 'all'. It is likewise not a question of its being not allowed to use the word 'all' in the case of infinite sets – all right, this can be done if these are such infinite sets as, e.g., the natural series of numbers. The point it that, in this case, an order reigns, and one can make use of the principle of induction. But let us take a set such as the series of decimal places of a transcendental number, where the sequence in which figures succeed one another cannot be framed by any rule. Here, if we take into account all of these figures and if we want to apply to this set the law of the excluded middle, we can arrive at an empty sentence. The point is that the law of the excluded middle allows us to say: either at least one element is devoid of a certain property, or else all elements possess this property. It also permits us to say: or no element possesses this property. Now Brouwer says that such an excluding alternative is inadmissable, with respect to some sets, in so far as we are not supposed, in case the general component of that alternative proves false, to assume, on the basis of the law of the excluded middle as applied to such sets, that the singular statement must be true. The general sentences are not capable of being negated (*die allgemeinen Urteile sind nicht negationsfähig*). The negation of a general sentence would be a singular, existential statement. And of the existence of anything, one can be convinced by way of construction only. To exist, for Brouwer, means to be able to be constructed.

To illustrate this idea even still better, let us revert to the example of the number $\pi$. Let us assume that the question is whether, among the figures of this number there is such a figure of seven, after which the figure of seven would occur seven times. On the basis of the principle of the excluded middle along with the principle of inconsistency, everybody is entitled to state that at least one such case does take place or else that no such case does take place. And should it be found that, for some reasons, the general alternative would have to go, we would be entitled to state that at least one such figure of seven must exist. Now Brouwer affirms that, in this case, we must not assume the truth of such a singular statement as long as we have not really found such a figure in the expansion of the number $\pi$. It is also not a question of somebody carrying out the calculation. Brouwer does not require any checking, but only the construction of a formula that would indicate the position of such a number of seven.

Zermelo's axiom leads to the statement that each set can be well-ordered, i.e. to be arranged in such a series that each part would contain a first element. However, nobody can do it, i.e. nobody can point to a principle according to which this ordering should be carried out. It is obvious that, for an intuitionist, such a statement is not worth more than a theological dogma which would force us to believe in something that could not be checked by anybody. To introduce, into mathematics, such judgements about existence without a possibility of a construction is tantamount to introducing theological methods of demonstration into mathematics.

We now understand what it means, for Brouwer, that the law of the excluded middle is not generally valid, and why Brouwer claims this. The question is to defend the thesis that to exist in mathematics means to be able to be constructed. We also understand what it means that general sentences are not capable of negation. The negation of a general sentence is a singular statement expressing existence. The question is that a singular statement need not be considered to be true if a general sentence is found to be false. The law of contradiction is always true. But the law of the excluded middle must not be true always, i.e. from two contradictory statements one need not necessarily be true; in particular one must not, on the basis of this principle, assume the truth of a *singular statement* if the general alternative is found to be false. The point is that one must not do it when the general statement does not use the term 'all' for finite sets or for numerable sets, i.e. such infinite sets which are similar to the series of natural numbers where the elements can be numbered. The point is that wherever good ordering does exist, the principle of natural induction is valid. The infinite sets to which the induction principle cannot be applied, the non-numerable sets, the sets of the power of the continuum, the sets where the *'wahlfreie Ordnung'* governs: one must not use the law of excluded middle to such sets, if such an application would be used to demonstrate the existence of something that cannot be constructed.

Elsewhere classical logic is a must, i.e. it is always compulsory when we have to deal with finite sets or such infinite sets as are subject to mathematical induction. If the question were to apply logic to finite, and always small, sets, then with such sets logic would be in general unnecessary because in such cases every general statement and its negation can be verified. If I have in front of me seven pieces of chalk, and somebody asks a question about the truth of a disjunction in which I express the combination of the law of the excluded middle with the law of contradiction, viz. whether it is true that either all pieces of chalk are white or there exists at least one non-white piece of chalk then I do not need any logic for solving such a disjunction. The point is that I can check experimentally whether this disjunction is true, since the number of chalk pieces is finite.

Logic becomes really necessary only when I begin deliberating about an infinite, or a finite number of objects, which is, however, so large that experimental checking would not be worthwhile. However, when I begin to ponder on an infinite number of objects, does logic give any guarantee of truth? Such has been our assumption thus far, and that is why we have used the word 'all' without restriction. However, the negative results of this excessive self-confidence began to appear. In the theory of sets which adhered to the foundations of classical logic, viz. that its laws are valid for all finite sets, certain antinomies began to appear which recalled the antinomy of the liar, known in antiquity. It was found that these antinomies arise from the fact that we simply abuse the word 'all'. But it is not only the antinomies that have arisen therefrom. Brouwer says that some statements which do not lead to antinomies but which do not give him intuitive assurance also originate from abuse of the word 'all'. Perhaps they are true, but Brouwer does not want to consider them to be true.

Both the antinomies and these non-intuitive statements will disappear if we assume that the term 'all' should be applied only to those sets for whose elements the principle of mathematical induction is valid. For such sets the fundamental disjunction will be binding which expresses the combination of the law of contradiction with the law of the excluded middle. These are the denumerable sets.

But what should be done with the non-denumerable sets? Is the human mind to be powerless with regard to them? And besides, where is the explanation of the fact that the unlimited manipulation of the word 'all' leads, in one instance, to antinomies and, in another case, to the non-intuitive statements which exist only on paper, but do not exist in the mind? In order to answer this question, one has to go deeper into the genesis of Brouwer's standpoint.

(2) While looking for the genesis of Brouwer's position, we see that it is really the consequence, pushed to the ultimate limits, of the standpoint of Kant and Poincaré. In his inaugural lecture read at the University of Amsterdam in the year of Poincaré's death, Brouwer stresses[5] the fundamental contrasts between the two trends in mathematics, as regard the method of understanding the sources of precision, viz. the intuitionist and the formalist. The French mathematicians professed rather the former, while the German mathematicians, the latter. Brouwer considers that the trend of logicism also belongs to the formalist one. 'When asked, where does mathematical exactness exists, the two sides reply differently: The intuitionist says: in the mind, the formalist: on paper'.[6]

Much as Brouwer had numbered Germans among the formalists, he still admits that intuitionism did appear most strongly with Kant. But this is the old-fashioned intuitionism, completely given up at present; the point is that both time and space are considered here to be forms 'inherent in human reason'. Both the arithmetic and geometric statements are, for Kant, synthetic *a priori* ones, hence independent of experiment and 'not capable of analytical proof'. In view of the progress in science, Brouwer admits the position of that intuitionism became weaker and weaker, and the mortal blow was inflicted on it by non-Euclidean geometry.

However, if one rejects the apriorism of space, retaining only apriorism of time, the situation will undergo a radical change. The starting-point for

this neo-intuitionism considers the falling apart of moments of life into qualitatively different parts, to be reunited only while remaining separated by time as the fundamental phenomenon of the human intellect, passing by abstracting from its emotional content into the fundamental phenomenon of mathematical thinking, the intuition of the bare two-oneness.[7]

This two-oneness intuition, the basic intuition of mathematics, creates not *only the numbers one and two,* but also *all finite ordinal numbers* insofar as one of the two-oneness elements can be thought as a new two-oneness, and this process can be repeated without any limit; this yields further a basis for the least *ultra-finite ordinal number* $\omega$. Finally, this basic intuition of mathematics, which unifies what is connected and what is separated, the continuum and the discrete, yields directly the basis for the *intuition of the linear continuum*, i.e. the intuition of the meaning of 'between', which cannot be exhausted by interposing new units and which, therefore, *can never be understood as a simple collection of units*.

The time apriorism not only results in that the statements of arithmetic are *a priori* synthetic statements, but that likewise the propositions of geometry (much as space is not something *a priori*) also become synthetic *a priori* judgements since, as shown by Descartes by means of the calculus of co-ordinates, all geometries can be reduced to arithmetic, and thus not only the geometry of the one-dimension continuum but also the geometry of the two-dimension and three-dimension continua, and besides all non-Euclidean and *n*-dimensional geometries.

Upon this basic intution any mathematical sets of units can be developed, viz. solely by combining two operations, a finite number of times, 'to create a finite ordinal number' and 'to create an infinite ordinal number ω'. And this should be understood thus: any previously constructed set or any correctly performed operation of construction can be dealt with as a unit for further purposes. It is for that reason that intuitionism admits only the existence of numerable sets, i.e. of sets whose elements can be introduced into a one-to-one correspondence either with the elements of a finite ordinal number or with the elements of an infinite ordinal number ω. In the construction of such sets the everyday language or any symbolic language cannot play any other role than that of an auxiliary *non-mathematical agent*, of supporting mathematical memory and of making it easier for various persons to construct the same set.

It is for this reason that an intuitionist cannot ever be assured of the accuracy of a mathematical theory by such guarantees as the proof of inconsistency, as the possibility of defining ideas by means of a finite number of terms, or by such a guarantee as the practical certainty that a mathematical theory will never result in misunderstandings in human relations.

Brouwer considers next one of the systems of axioms, created by formalists for the world of mathematical objects which remain mutually in the 'set-to-element relation'. Such a system accepts the following six axioms: (1) a set is determined by its elements (2) for any two mathematical objects it is decided whether one of them is or is not enclosed in the other as its element; (3) to every set there belongs another set which contains, as its elements, only the subsets of the given set; (4) the selection axiom (its formulation is known from the preceding monographs); (5) the inclusion axiom: if, for any mathematical object, it is decided that a certain property does or does not occur in it, then there exists a set which contains only such objects for which this property does occur; (6) the composition axiom: 'the elements of all sets which belong to the set of sets form a new set'.

It is on the basis of this axiomatization that Brouwer subjects to criticism all statements of the theory of sets. Now, even in the case of finite sets, one notices a difference between the standpoint of intuitionism and formalism, but in this case the difference relates to the method only, not to the result. In the case of formalists the definition of a finite set denotes that it is a set whose elements cannot be introduced into one-to-one correspondance with the elements of one of its subsets. Next, using very complicated reasoning, one proves [Zermelo] the principle of complete induction as the fundamental property of these sets. For an intuitionist, this principle is obvious and does not require any proof; its obviousness for finite numbers results from the very construction of these numbers. This principle not only does not require any proof here, but the proof must lead to a vicious circle. Here Brouwer refers to Poincaré.

If we pass now to infinite sets, then the formalist, mainly using his inclusion axiom, introduces various ideas, absolutely nonsensical to intuitionists, such as 'a set whose elements are points of the space', 'a set whose elements are a continuous function of a certain variable', 'a set whose elements are discontinuous functions', etc.

The inclusion axiom in the end resulted in the Burali-Forti antinomy. On the basis of that axiom the set S is constructed, which contains, as its elements, all natural numbers ordered according to their magnitude; next it is proved on the one hand that S is a well-ordered set whose ordinal number must be greater than any ordinal number, and on the other hand it is possible to prove that 'since not all objects of mathematics are natural numbers, it is possible to create a natural number larger than S by adding a new element to S'.

What has been done in this case? The formalists altered the formulation of the inclusion axiom which later read as follows: 'If, for all elements of the set it has been decided whether, or not, a certain property does occur, then the set contains a subset which contains only elements for which this property occurs'. In this form the axiom allows us to introduce only such sets as are subsets of the previously introduced sets. If we want to operate with other sets, then their existence ought to be explicitly postulated. Since, in order to reason about something, we have to assume such axioms beforehand, we make use of this, provided such an assumption does not result in an inconsistency. Consequently the only modification that has been introduced by the discovery of paradoxes into the practice of the formalists, says Brouwer, is the fact that only those sets had been removed, which lead to paradoxes. On the other hand one operates without any hesitation with other sets introduced on the basis of the old inclusion axiom. The result of this is that, for a formalist, vast fields of investigation remain fully of interest, which are without any significance for the intuitionist.

We see here how Brouwer's criticism turns against the completely uninhibited introduction of the existence postulates and against the inclusion axiom which leads to antinomy. Brouwer believes that other antinomies, which did not grow out of the inclusion axiom, are less serious and less dangerous. Thus, e.g., when speaking of Richard's paradox, which developed from the question 'is there a least integer which cannot be determined by means of a sentence comprising at most twenty words.' Brouwer says that its source lies not in the axiom of inclusion, but in the changing meaning of the word 'defined' which occurs in the sentence quoted, owing to which we can define exactly by means of this sentence an unlimited number of cardinal integers.

Next, Brouwer passes in turn through various statements of the theory of sets to which the intuitionist cannot agree. The abyss which separates both sides is best revealed by the theory of powers. Two sets are said to be of the same size if we can establish between their respective elements a one-to-one co-ordination. We say that the size of the set A is greater than that of the set B if we can establish a one-to-one relationship between B and a part of the set A, while we cannot establish such a relationship between A and a part of B. A set which is of the same size as one of its subsets is called infinite; sets which do not possess this property are finite. Sets whose size is the same as the natural number $\omega$ are called denumerably infinite, and 'their size is called aleph-zero'. Now Brouwer declares, in accordance with what he said earlier, that the *aleph-zero* set-size is the only infinite quantity whose existence is admitted by intuitionists.

Let us now consider the idea of 'denumerably infinite natural number'. On the basis of the fact that this idea has a clearly and well defined meaning for both the formalist and the intuitionist, the formalist arrogates to himself the right to forming a set of all denumerably infinite natural numbers, whose power he denotes as *aleph-one*; the intuitionist *does not concede* this right. Further, the formalist and the intuitionist both agree that, first, the denumerably infinite sets of denumerably infinite natural numbers can be constructed in various manners, and that, next, for each such set, it is possible to indicate a denumerably infinite natural number which does not belong to this set. But again, from this fact which is conceded by both, the formalist deduces at once that aleph-one is greater than aleph-zero, which conclusion sounds nonsensical to the intuitionist.

Similarly, from the fact that to the satisfaction of both sides, the formalists and the intuitionists, it is impossible to 'construct' a set of denumerably infinite natural numbers that would have a power of less than aleph-zero or more than aleph-one, the formalist concludes that aleph-one is the second-smallest infinite number, a conclusion which is again a nonsensical sentence to the intuitionist.

The real number between 0 and 1 is, to a formalist, an elementary[8] series of numerals after the decimal point, while for an intuitionist, it denotes 'the right to construct an elementary series of decimal places after the decimal point by means of a finite number of operations . . '. However, when a formalist creates a 'set of all real numbers between 0 and 1', then these words are devoid of any meaning to an intuitionist, irrespective of whether one has in mind the formalist's real numbers defined by the elementary series of freely chosen numbers, or the intuitionist's real numbers defined by means of the finite laws of construction. Both sides agree that the denumerably infinite sets of real numbers between 0 and 1 can be constructed in different ways, and they also agree that for each of such sets we can indicate a real number between 0 and 1 which does not belong to this set. However, the formalist takes the liberty, on this basis, to advance the conclusion that the power of the continuum, i.e. the power of the set of real numbers between 0 and 1, is greater than aleph-zero, such a sentence being nonsensical to an intuitionist. A formalist puts a further question, whether there exists a set of real numbers between 0 and 1 whose power would be lower than the power of the continuum, but higher than aleph-zero; in other words whether the power of the continuum is the second-smallest infinite power; this question, unanswered up to now, is considered by the formalist to be one of the most difficult and most fundamental mathematical problems. On the other hand, to the intuitionist, putting the question in such a way is nonsensical, and the question easily will find an answer if we interpret it in such a way as to impart a sense to it.

Let us now put the following question, says Brouwer:

Is it possible to establish the one-to-one correspondence between the elements of the numerably infinite natural numbers, on the one hand, and the set of real numbers between 0 and 1, on the other, if we deal with both sets as with extensive ones owing to the construction of new elements, a construction of such kind as not to disturb the correspondence by the continuation of the construction of both sets?

The question thus posed will be answered by the intuitionist in the affirmative since 'the extent of both sets can be divided into phases in such a way that, during each phase, one can add a numerable infinite number of elements.'

Let us however pose the question in the following form: Is it possible to construct a law which would assign a denumerably infinite ordinal number to each elementary sequence of numbers and *which would give an a priori certainty* that two different elementary sequences would never have the same denumerably infinite ordinal number corresponding to them? An intuitionist must give to this question a negative answer. The point is that this law of correspondence must, in a certain manner, assign the construction of some denumerably infinite ordinal numbers in each of the consecutive places of the elementary sequence; hence there exists, for each place $c_\nu$ a well defined denumerably infinite ordinal number $a\omega$ whose construction is indicated by this special place; thus there exists also a well-defined denumerably infinite ordinal number $a\omega$ which is greater than all $a\gamma$ and which, therefore, cannot be exceeded by any ordinal number comprised in the law of construction; hence the power of this set cannot surpass aleph-zero.

To achieve any higher power the formalist defines, for the set with power $\mu$, a set of all other ways by means of which we can carry out the selection from the set with power $\mu$, and he demonstrates that the power of this set is higher than $\mu$.

It is true that the intuitionist agrees likewise with the formalist, that one can, using various methods, construct laws according to which different functions of the real variable can be coordinated to all elementary series of numerals, but that, on the other hand, we cannot construct a law according to which an elementary series of numerals would correspond to every function of a real variable, and on the basis of which an *a priori certainty* would appear that two different functions will never have the same elementary sequence as the corresponding one. However, from this fact the formalist draws the unjustified conclusion that the power $c'$ of the set of all functions of the real variable is greater than the power of the continuum – this conclusion being completely without sense for the intuitionist. On the other hand, the formalist, using the same path along which he had passed from $c$ to $c'$, passes from $c'$ to a still higher power $c''$.

The formalist has also another method of attaining higher and higher magnitudes starting from the set of magnitudes of ordinal numbers; namely he proves that the set of all ordinal numbers which correspond to the sets with the magnitude $\mu$ is itself greater than $\mu$. In particular he denotes by aleph-two the power of the set of all ordinal numbers of the power aleph-one, and proves that aleph-two is greater than aleph-one, and that, as far as the magnitude is concerned, it follows immediately after aleph-one. It would be possible to interpret this result in a manner which would have a sense for an intuitionist also, but this interpretation would not be as simple in this case as in the previous ones.

Finally Brouwer subjects to criticism the proof of Bernstein's theorem, viz. that 'if the set A has the same size as the subset of the set B, and the set B has the same size as the subset of the set A, then A and B are of the same size', or else the equivalent form 'if the set $A = A_1 + B_1 + C_1$ is of the same size as the set $A_1$, then it is of the same size as the set $A_1 + B_1$'. Brouwer considers this statement to be evident for denumerable sets. However, if it is, for an intuitionist, to have any sense for sets of a higher power, it is to be interpreted as follows: If it is possible, firstly, to construct a law which defines the one-to-one correspondence between mathematical objects of the type A and objects of the type $A_1$, and, secondly, if one can construct a law which defines the one-to-one correspondence between the mathematical objects of A type and objects of types $A_1$, $B_1$, $C_1$, then it is possible to define, on the basis of those two laws, by means of a finite number of operations, a third law which defines one-to-one correspondence between mathematical objects of the A type and objects of the $A_1$ and $B_1$ type.

In order to demonstrate the validity of this interpretation, Brouwer presents the proof:

From the division of A into $A_1 + B_1 + C_1$ we deduce, on the basis of the correspondence $\gamma_1$ between A and $A_1$, the division of $A_1$ into $A_2 + B_2 + C_2$ and, at the same time, the correspondence $\gamma_2$ between A and $A_1$, the division of $A_1$ into $A_2 + B_2 + C_2$ and, at the same time, the correspondence $\gamma_2$ between $A_1$ and $A_2$. From the division of $A_1$ into $A_2 + B_2 + C_2$ we deduce, on the basis of the correspondence between $A_1$ and $A_2$, the division of $A_2$ into $A_3 + B_3 + C_3$ and, at the same time, the one-to-one correspondence $\gamma_3$ between $A_2$ and $A_3$. The infinite series of repetitions of this procedure will divide the set A into an elementary series of subsets $C_1$, $C_2$, $C_3$ . . . the elementary series of subsets $B_1$, $B_2$, $B_3$ . . . and, as remainder, the set D. The correspondence $\gamma_c$ between A and $A_1 + B_1$, which we are looking for, will be obtained by co-ordinating a corresponding element $C_{\gamma + 1}$ to every element $C_\gamma$, and by co-ordinating each element of A to itself.

To verify this proof with an example, let us take, as A, the set of all real numbers between 0 and 1, represented by means of the infinite decimal fractions, and, as $A_1$, let us take the set of these decimal fractions, in which the number in the place $(2_n - 1)$ equals the number in the place $2_n$. Next, the decimal fraction which does not belong to $A_1$ will be numbered among $B_1$ or $C_1$, depending on whether the adduced equality of numbers occurs an infinite or a finite number of times. By substituting in turn for each number of any element of A a couple of numbers equal to that number, we determine at the same time the law which assigns the one-to-one correspondence $\gamma_1$ between A and $A_1$. For the element $A_1$ which corresponds to any well-determined element A, such as, e.g. $\pi - 3$, we can designate successively as many numbers as we like, hence it ought to be considered a well-defined one.

In order to design an element which corresponds to the number $\pi - 3$ in accordance with the correspondence $\gamma_c$ it is necessary first to decide whether, in the expansion of $\pi - 3$ into a decimal fraction, there occurs, an infinite or a finite number of times, the case in which the number in an odd place equals the number in the following even place; for this purpose we ought to find a process of constructing the elementary sequence with such pairs of equal numbers, or else we ought to deduce a contradiction from the assumption of the existence of such an elementary series. However, there exists no basis for believing that some of these problems will be soluble. Such a conviction could arise only if we appealed to the *principium exclusi tertii*, i.e. to the axiom of the existence of a 'set of all mathematical properties', an axiom which is even more far-reaching than the aforementioned axiom of inclusion.[9]

Hence, says Brouwer, it became obvious that Bernstein's theorem, and along with it the positive part of the theory of powers likewise does not admit of an intuitionist interpretation. The article ends in a remark that it is practically excluded to achieve a chance of reconciliation between the representatives of the two contending camps within the compass of finite time! He repeats Poincaré's words: 'Les hommes ne s'entendent pas, parce qu'ils ne parlent pas la même langue et qu'il y a des langues qui ne s'apprennent pas'.[10]

Several years later Brouwer started systematic work on the intuitionist theory of sets, and the result of this work was publication of 'Begründung der Mengenlehre unabhängig vom logischen Satz vom ausgeschlossenen Dritten', the two first parts of which appeared in 1917–1918. Brouwer gives information about the results of this work in the article 'Intuitionistische

Mengenlehre', published in 1919 in the *Jahresbericht der deutschen Mathematiken-Vereinigung*. He states there that from 1907 he has been defending the following two theses: (1) that the comprehensive axiom (also in the limited form which was later given to it by Zermelo), on the basis of which all objects which possess a certain property are united into a set, is inadmissible or useless in the theory of sets, that one has to adopt the constructive definition of sets as its basis in mathematics, (2) the axiom of the solubility of every problem, formulated in 1900 by Hilbert, is equivalent to the logical law of the excluded middle, and since the aforementioned axiom lacks a sufficient basis and "logic is based on mathematics, not the other way round, the logical law of the excluded middle is not an allowed means of proof in mathematics." We can merely attribute to it a scholastic and heuristic value, and statements for which the proof cannot be carried out without this means are without any mathematical content.

Thus the intuitionist understanding of mathematics is concentrated in these two theses. However, Brouwer does admit that until quite lately he had drawn from this standpoint only fragmentary consequences, and that he has been using in his mathematical researches only the old method, endeavouring, true enough, to reach only such results as, according to him, would have preserved their value even after carrying out the systematic construction of the intuitionist theory of sets, although perhaps in a modified form. Having carried out these systematic reconstructions, he now points to the following changes of content, not only formal changes in the classical theory of sets.

The basic constructive definition of sets is as follows:

The set is a law on the basis of which, if we choose continually anew an optional complex of numerals from the series 1, 2, 3, 4, . . . , each of these selections will, or will not, produce a definite sign, or else it will result in a stopping of the process and in the definitive destruction of its result, with the possibility of adding, for each $n > 1$ after each unstopped series $(n - 1)$ choices, at least one complex of numerals which, when chosen as the $n$-th complex of figures, does not involve the stopping of the process. Each sequence of signs thus produced by the unlimited series of choices is called the element of the sequence (hence, in general, such a sequence of signs cannot be represented as a ready sequence. The common method of the formation of elements of the set M is also briefly denoted as the set M.

On this notion of a set is later based the definition of a 'mathematical species' (der mathematischen Spezies) which, as a special case, comprises the notion of a set.'[11] Disregarding this definition, Brouwer reviews the changes in the theory of cardinal numbers, in the theory of ordered sets and of well-ordered sets, and then he discusses also some sections of the applied theory of sets, such as the theory of point sets on a plane, the 'inner boundary sets' and the classical definition of mensurability, thus:

(1) In the theory of cardinal numbers we find, according to Brouwer, in the first place the distribution of the notion of equal power. According to the classical theory, two sets of equal magnitude can be, for the intuitionist theory, of equal or semi-equal magnitude, or equivalent, of equal domain, equal extent or equal weight. It is only this distinction of notions, says Brouwer, that has enabled him to extend this notion of magnitude, which previously he had only retained for some sets, to all mathematical species, thus restoring, in a way, the comprehensive character of the species. In this connection one adopts at present, for the given denumerable sets, some distinctions between the various 'Mengen bezw. Spezies' (Cf. 'Intuitionistische Mengenlehre', 1919, *Jahresbericht der Deutschen Mathematiker Vereinigung*).

There remain the classical cardinal numbers $a$ (for the denumerably infinite sets) and $c$ (for the set of the continuous power), but one rejects the example – adduced in the classical theory – of the cardinal number greater than $c$ (the set of all functions of one variable).

(2) In the theory of ordered sets one leaves out the classical characteristic of ordinal numbers $\vartheta$ (the ordinal type of the set of all real numbers) and $\zeta$; the former is replaced by the following sentence of a complex content:

One find the ordinal number $\vartheta$ in every ordered species $P$ which contains such a numerably infinite, in the stricter sense an everywhere dense subspecies M, that, between every two elements $P$ there lie elements $M$, *that* the species of elements $M$ which lie before any element $p$ of the species $P$ is such a separable subspecies of the species $M$ (*eine abtrennbare Teilspezies von M*), which is either without an element or else possesses at least one determinable element, and *that*, for every fundamental series of the relation 'after' and 'not after' with respect to elements $M$, which satisfies the ordinal property, it is possible to construct an element of the species $P$, which fulfils these relations.

(3) In the theory of well-ordered sets both main properties of these sets fall victim to Brouwer's reform, viz. that every two well-ordered sets can be compared with one another, and that every subset of a well-ordered set possesses a first element. Since both these properties had been the most important means of proof of the classical theory, hence the new theory has almost no resemblance to its predecessor. The sentence giving the second of the above-mentioned properties of the classical theory is replaced by the following sentence:

The law which determines an element in a well-ordered species, and which, to every already determined element, either co-ordinates the stopping of the process or a preceding element, such law designates with certainty an element to which the stopping of the process is co-ordinated.

(4) The basis of the theory of sets of points in a plane is the set *a* of such sources whose every apex, in view of the rectangular co-ordinate system, possesses co-ordinates $a/2^r$ and $b/2^n$, while the lengths of sides parallel to the axis are either $1/2^n$ or $1/2^{n-1}$. Next, a point in the plane is understood as the infinitely continued series of squares *a*, each of which is inside the immediately preceding one.

Upon this basis there no longer comes into consideration, from the sentences of classical theory, the sentence of Cantor, and only its negative part remains:

There cannot exist any closed ordered point set whose power would be higher than the numerably infinite one, and whose every point would, on the one hand, point out the nearest, following point and, on the other hand, would be of a numerable order or would possess a finite interval from the species of the following points.

Now this sentence must be proved in a way that is completely different from the usual way based on the law of the excluded middle.

Instead of Cantor's positive theorem there occurs, in the intuitive theory of sets, a detailed characteristic of these sets of points and species of points which possess the relevant property.

We disregard the theory of measurability in which the intuitionist theory of sets introduces only insignificant changes; we also leave aside a new edition of the considerably enlarged present study, which was published in *Mathematische Annalen* (1925, 1926, 1927) under the title 'Zur Begründung der intuitionistischen Mathematik'; let us instead turn to the smaller papers by Brouwer, which contain more fundamental notes of a philosophical nature. Such is precisely the paper 'Über die Bedeutung des Satzes vom ausgeschlosenen Dritten in der Mathematik, insbesondere in der Funktionentheorie', which was published in 1925 in the *Journal für die reine und angewandte Mathematik*.

It is known that the mathematical function expresses a certain relation, in the simplest case between the elements of two sets, this relation being, on the whole, one-to-many, since to every 'value' of the argument there corresponds a certain, though not necessarily always a different value of the 'dependent variable'. If, to every two different elements of one set, two different elements of the other set are always coordinated, then the relation is a one-to-one coordination, and we say that the function is univocally reversible. If the elements of both sets are the points of a certain spatial extent, we also say that one set is the representation of the other set. Now Brouwer concedes that, within the range of finite systems, we can test the properties of the systems, i.e. the projections of systems upon other systems, by means of the attributed correspondences (i.e. one can either prove or disprove them). The projection indicated by the relevant property has a finite number of possibilities of execution, each of which can be taken up and either completed or stopped. Brouwer adds likewise in parenthesis that the principle of mathematical induction provides us with a means for carrying out such testing without analysing each case separately and individually, and consequently the verification can be carried out fast even in the case of systems with a very great number of elements.

Consequently, for the properties of finite systems, it is the *law of the excluded middle* that is valid, viz. that each property for every system is either possible or impossible; also the principle of the reciprocity of the complementary species (*von der Reziprozität der Komplementärspezies*) i.e. the principle that 'for each system it follows, from the impossibility of the impossibility of a certain property, that the property is correct'.

If, e.g., the position S(p,q) of two mathematical species *p* and *q* contains at least 11 elements, then, on the basis of the law of the excluded middle (which comes here into play is the principle of 'disjunction'), it follows that either *p* or *q* possesses at least 6 elements. Similarly, if it has been proved in elementary mathematics that, provided none of the positive integers $a_1, a_2, a_3 \ldots a_n$ is divisible by the prime number *c*, then their product $a_1.a_2.a_3 \ldots a_n$ is not divisible by *c* – it follows on the basis of the principle of the reciprocity of complementarity, that if the product $a_1.a_2.a_3 \ldots a_n$ is divisible by the prime number *c* then at least one of the factors is divisible by *c*.

We have to do with a manifestation of nature that numerous objects and mechanisms of the sensual world can be mastered with respect to the vast complexes of facts and events, provided they are considered as (possibly partly unknown) finite discrete systems bound, as far as some known parts are concerned, by definite laws of temporal relations. It is to these objects and mechanisms, in view of the relevant complexes of facts and events, that one can apply the laws of theoretical logic, including the law of the excluded middle, although a completely empirical testing of the conclusions obtained here, mostly *a priori*, is materially excluded, and in the case of some temporal (juridical and other *ex-post*) inferences, it is out of the question, to test, even partly. One should attribute to this incomplete verification of conclusions which are nevertheless considered as infallibly correct, as well as to the partial ignorance of the finite systems and to the circumstances that theoretical logic was more often applied to these material objects rather than to the mathematical ones – one should, I repeat, attribute probably to all of these, factors that one had attributed, to the laws of theoretical logic, including the law of the excluded middle, the *a priori* character and that one lost sight of the conditions of its applicability, inherent in the projection of a finite discrete system onto the relevant objects. In this way we have achieved the result that, for a completely primary and autonomous mental activity which is presented by mathematics of finite systems, one began to look for a deeper justification in logical laws. In accordance with this, during the logical treatment of the sensual world, any appearance of contradictions could not entail a doubt about the immutability of logical laws but, at most, could result in a modification and completion of mathematical fragments projected onto the sensual world.

The attribution of an *a priori* character to the laws of theoretical logic entailed the consequence that, until quite lately these laws, along with the law of the excluded middle, had also been applied in the mathematics of infinite systems without any reservations, nor would one become discouraged by the circumstance that the results obtained in this manner do not lend themselves, in general, either theoretically or practically, to empirical testing. On this basis, extensive wrong theories are built, especially in the latter half of the century, contradictions which had frequently been encountered when doing this gave rise to *formalistic criticism*, which, as a matter of fact, is tantamount to the fact that the *language accompanying the mathematical mental activity* became included in the mathematical operations. From this point of view the laws of theoretical logic look like operators applied to basic formulae or axioms, and one broaches the aim of such a transformation of the axioms as to render the verbal action of these operators (which are, themselves, left without change) no longer disturbed by the appearance of a verbal figure of contradiction. We are not obliged to call in question the achievement of this aim but this does not yield any mathematical value: a false theory which cannot be stopped by any annihilating contradiction is just as fallacious as a criminal policy that cannot be halted by any juridical repression, and yet remains criminal.

After these general remarks Brouwer demonstrates that the theory of real functions includes two theses of a fundamental importance, which, however, ought to be considered wrong, as being based on the law of the excluded middle, viz. that

(1) the points of the continuum form an ordered class of points,
(2) every mathematical class is either finite or infinite.

The falsity of the former property is demonstrated by the following example:

Let $d$ be the 8th figure after the decimal point in the expansion of $\pi$ into a decimal fraction, and $m = k_v$; if, in the expansion of the number $\pi$ it happens at $d_m$ for the $n$-th time that a part $d_m, d_{m+1}, \ldots d_{m+g}$ of this expansion will form the sequence 0123456789. Next let $c_\gamma = (-\tfrac{1}{2})^{k}/_1$, if $v \geqslant k_1$, otherwise let $c = (-\tfrac{1}{2})^{\gamma})$, then the infinite series $c_1, c_2, \ldots c_\gamma \ldots$ will define the real number $r$ for which neither the case $r = 0$ nor the case $r > 0$ nor $r <$ will take place.

Brouwer adds that, for the definition of $r$ one can use any other property, provided that for every positive integer one can deduce either the existence of such a property or its impossibility; however at the same time one cannot designate a number that would possess this property, nor is it possible to prove the impossibility of the property $\varkappa$ for all positive integers.

Along with this property there falls also the usefulness of the idea of Lebesgue's integral created by the Paris school, since it is associated with the idea of the 'measurable function', while, according to what was said before, even a constant function does not satisfy the conditions of 'measurability'. The point is that, for a function $f(x) = r$, where r is the number defined above, the values of x, for which $f(x)>0$ do not form any measurable species of points.

The fact that the other basic property is likewise wrong, can also be seen from the species $k_n$ of the positive integers, supplied by the above mentioned example. The point is that their value will depend on the circumstance, where, i.e. in what place, did and how many times in the expansion of $\pi$ the sequence of natural numbers from 0 to 9 will appear. Along with this second property the so-called 'enlarged principle of disjunction' also falls, according to which, if the aggregate S(p,q) of two mathematical species p and q contains an infinite fundamental sequence of elements, or if p or q contains this infinite series, and, along with this principle, there also falls the proof of the Bolzano-Weierstrass theorem based upon it, according to which every limited species of points possesses a boundary point. Less basic, but still indispensable for the construction of a logical theory of functions are likewise the two following theorems whose falsity is demonstrated by Brouwer again by means of examples:

(1) Every function f(x), continuous everywhere in the closed interval 'i' has a maximum, i.e. such a value of the abscissa $x_1$ in such environment – *a* that, for every *x* comprised in this surrounding *a* and likewise in the interval 'i'

$$f(x_1) \geqslant f(x)$$

(2) The Heine-Borel theorem asserting the possibility of a finite and closed set being covered by means of a finite number of intervals.

Since these fundamental theorems had been rejected it is understandable that an enormous part of the theories based on these statements also falls. Brouwer chooses, as an example, the statement that the monotonic and everywhere continuous function is 'almost everywhere' differentiable, and he demolishes this statement by using the example of a function in a closed unit interval, which is everywhere continuous and monotonic, yet nowhere differentiated.

The rejection of the law of the excluded middle also undermines some older parts of the mathematics of the infinite, e.g. the criterion of the convergence of infinite series according to Kummer, as this criterion is also based on the Bolzano-Weierstrass theorem.

In the same year 1925 Brouwer published, in the *Jahresbericht der deutschen Mathematiker-Vereinigung 33,* a brief and very important paper 'Intuitionistische Zerlegung mathematischer Grundbegriffe', in which he points to some consequences of the thesis that the law of the excluded middle is valid only for those sections of mathematics "which take place within a determined finite mathematical system", and consequently "only for such sections of natural science to which a certain finite mathematical system can be applied". He adds to the last-mentioned idea a note that the belief in the possibility of applying the law of the excluded middle when studying the laws of nature consequently implies a belief in the finiteness and the atomistic structure of the world. However, this does not mean that, for a physicist who holds this belief, an intuitive criticism would be valueless, since the methods of calculation he uses, even when studying a nature which is finite or assumed to be atomistic, are based on the mathematics of the continuum, i.e. on the mathematics of the infinite. He explains the intuitive thesis with the following example:

Let us call the real number $g$ *a rational one*, if one can assign two integers $p$ and $q$, such that $g = p/q$, and let us call it an *irrational one* if the assumption of rationality for $g$ can lead to absurdity; then, on the basis of the law of the excluded middle, every real number ought to be rational or irrational.

However, we shall define real number precisely in the same manner as this was done by the author in the previous discourse, when he wanted to demonstrate the falsity of the sentence according to which the points of the continuum would form an ordered species. According to this example the number $r$ understood as a limit of the infinite series $c_1, c_2, c_3, \ldots$, where $c_\gamma = (-\frac{1}{2})^{k_1}$ if $r \geqslant k_1$, and in the opposite case $c_\gamma = (-\frac{1}{2})^\gamma$ is neither rational nor irrational. This result can be also expressed in a different way.

Let us call the real number $g$ *comparable with zero* if $g > 0$, the assumption that $g$ is comparable with zero leads to an absurdity; then the number $r$ is neither comparable with zero nor non-comparable with it.

Consequently the example points at the same time to the invalidity of the principle of the reciprocity of the complementary species, which has already been mentioned by Brouwer in the foregoing article, since, as he adds now, it merely presents a certain corollary of the principle of the excluded middle and endeavors to "deduce the truth of a certain property from the absurdity of that property". The point is that the number *r* is not rational, although its irrationality is an absurdity. When talking of the '*Reziprozität der Komplementärspezies*' Brouwer has in mind the logical principle

$$\sim(\sim a) . \supset a$$

which reminds one of the principle of double negation. The principle of double negation, which expresses the equivalence $\sim(\sim a) \equiv a$ does not exist, for Brouwer as we shall see later. Brouwer admits only the principle $a \supset . \sim(\sim a)$ which is only one component of the equivalence contained in the law of double negation, while the other component which expresses an implication in the opposite direction, i.e. this '*Reziprozität der Komplementärspezies*' $\sim(\sim) \supset a$ is rejected by Brouwer. Otherwise he would likewise have to admit the law of the excluded middle, and it was Bernays who drew his attention to it, as mentioned by Brouwer in an appendix to the work discussed at present. Besides Brouwer also admits the principle of the absurdity of the absurdity of excluded middle, which is expressed symbolically as follows:

$$\sim\sim(a \gamma \sim a)$$

which he had already mentioned in one of his Dutch-language papers of 1908, as well as in a brief note (which we had omitted) to the previously discussed paper published in *Journal für die reine und angewandte Mathematik*. Now it would suffice to substitute, in the principle 'of the reciprocity of the complementary species', $\sim\sim(a v - a \supset . \supset a \gamma - a$ for 'a' in order to obtain the very law of the excluded middle by having recourse to the principle of 'the absurdity of the law of excluded middle and to the directives of detachment.

Next Brouwer discusses further properties of his logical system. Thus, an intuitionist, says Brouwer, has to reject the postulate of the *alternative of absurdity or of the absurdity of absurdity*, of any property for any system, which, besides, says more than the principle of excluded middle.[12] Thus the principle $\sim a \gamma \sim\sim a$ becomes superfluous.

For instance, one cannot decide, with respect to the above-mentioned positive integer $k_1$ on the basis of its property, "neither about its absurdity nor the absurdity of absurdity".

As regards the series of *n* denotations of absurdity: 'the absurdity of absurdity . . . of absurdity', one can, in accordance with the classical method of understanding delete, on the basis of the principle of the reciprocity of the complementary species, as many times as desired, each time two consecutive statements of the series, in order to arrive, finally, at the 'absurdity or truth'. It would appear *prima facie* that something of this kind is excluded in the intuitive understanding. Now this is not the case; on the contrary, the relevant process of cancellation is likewise admissible from the standpoint of intuitionism, provided *the last statement of absurdity in the series is excluded from this cancelling*. Thus it will be possible to prove the statement: *The absurdity of the absurdity of absurdity is equivalent to absurdity*. Since this equivalence can be expressed in the form of two implications, Brouwer's proof consists of two parts one of which shows that

$$\sim\sim\sim a \supset \sim a$$

while the other, that

$$\sim a \supset \sim\sim\sim a$$

The proof of the former statement is based on the law of contraposition

$$a \supset b. \supset .b\sim \supset \sim a$$

and on the law

$$a \supset \sim\sim a$$

The point is that, by inserting in the first formula '∼∼a' instead of 'b', we obtain

$$a \supset \sim\sim a. \supset .\sim\sim\sim a \supset \sim a.$$

Hence, by having recourse to the adopted formula $a \supset \sim\sim a$ and to the directive of detachment one obtains the desired theorem. The proof of the second theorem is obtained directly by inserting '∼a' instead of 'a' in formula $a \supset \sim\sim a$.

Brouwer takes us next to the highly interesting discussion of the conditions of 'fusion' between two points (*Verschmelnungsbeziehungen von zwei Punkten*). As basic notions he chooses the idea of coincidence (*Zusammenfallen*) and of distance (*Entfernung*). Two points $P_1$ and $P_2$ coincide, *fall upon one another*, if the square $P_2$ is contained in each square $P_1$, and if the square $P_1$ is contained in each square $P_2$.

On the other hand two points $P_1$ and $P_2$ are *distant from one another* (or they are *locally different*) if one can indicate such squares $P_1$ and square

$P_2$, that both would lie outside one another (*ausserhalbeinander*). Now if we apply to the ideas thus defined the laws of intuitive logic, we shall find that one has to adopt a third relation other than that of coincidence and of distance, viz. the relation of *deviation*.

Let us apply to the idea of coincidence the predicates of absurdity and of absurdity of absurdity; only in the former case will a new relation arise, which we shall call the relation of deviation, while in the latter case we shall find again the relation of coincidence.

Let us apply to the idea of distance the predicate of absurdity and of absurdity of absurdity, then, in the former case, we shall obtain coincidence, and in the latter deviation.

After having introduced the signs a⟷b (to express the contradiction of a and b), and a→b (i.e. that b follows from a), and bearing in mind that, for these relations, the properties do occur that, if a⟷b and b⟷c, then a⟷c, and that from a→b and b→c follows a→c, we shall be able to present as follows the logical connection between the three obtained relations of the fusion of two points:

Coincidence⟷deviation←distance.

Let us now pass to the relation of a point being included by a species of points. Point P is *included* in the species of points Q if it coincides with one of the Q-points. Point P lies at a distance from the species of points if it is distant from every point Q.

Let us now apply to the relation of inclusion (*Einhüllungsbeziehung*) the predicates of absurdity and absurdity of absurdity; we shall arrive then at two new relations which we shall call the relation of *deviation* and the relation of *connection*.

By applying to the idea of distance the predicates of absurdity and of absurdity of absurdity we shall obtain two new relations: of *adhesion* and *separation*.

The logical connection between the six relations of inclusion of a point in the species of points thus obtained can be presented as follows:

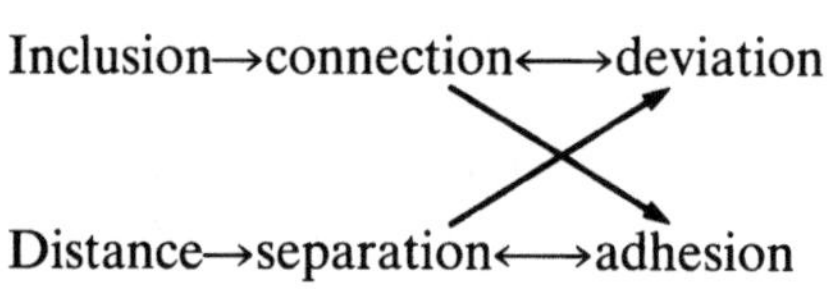


It can be seen that a real number can be 'rational' in three ways, depending on whether it is included in the set of rational numbers or is connected with them or adheres to them: likewise it can be 'irrational' in three ways, depending on whether it is distant from the set of rational numbers or separated or deflected.

Finally we have to analyse the relation of the fusion of two point species. The basic notions arising here are *coincidence, deviation* and *distance*.

Two species Q and R of points coincide if every point of the species Q coincides with a certain point of the species R and if every point of the species R coincides with a certain point of the species Q.

Two species of points Q and R deviate from each other if one of these species contains a point which deviates from others.

Two species of points Q and R are distant from one another if one of them contains a point which is distant from others.

Let us apply to the relation of coincidence the theorems of absurdity and the absurdity of absurdity; we obtain two new relations which we shall call in turn the relation of *unwinding* (*Loswindung*) and of *interlacing* (*Verflechtung*).

Let us apply to the relation of deviation the theorems of absurdity and of absurdity of absurdity, and we obtain two new relations which we shall call in turn the relation of (local) *congruence* and of *detachment* (*Loslösung*).

Let us finally apply to the relation of distance the theorems of absurdity and of the absurdity of absurdity, and we will obtain two new relations which we will call in turn the relation of the (local) *agreement* and *separation*.

The logical connection between these nine relations of the fusion can be represented by the following scheme:

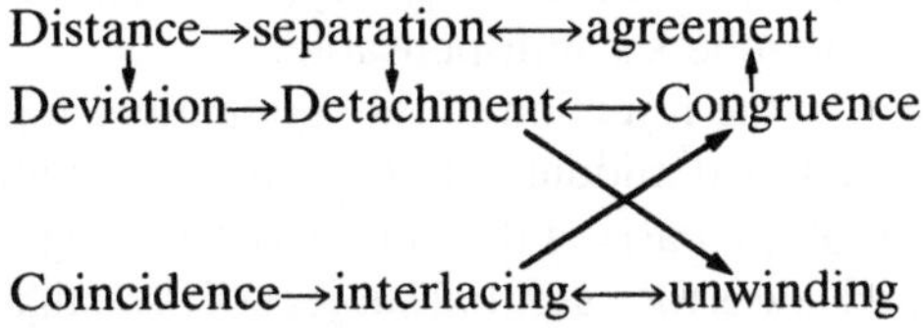


For the sake of elucidation Brouwer gives examples. Let A be the set of rational numbers, B – the set of real numbers ‘r + s’ where ‘r’ has the same meaning as in the example adduced at the beginning, while ‘s’ is a variable rational number: let C be the species of real numbers; D – the species of negatively irrational numbers (i.e. deviating from A); let E be a species of positively irrational numbers (i.e. distant from A); let F be the coincidence of A and D; let G be the coincidence of A and E. Then A and B are interlaced, but do not coincide; C and F are coinciding (congruence) but are at the same time unwound from one another (*voneinander*

*los gewunden*), hence are not interlaced; C and G display agreement without congruence.

The final summary of the results of his standpoint in the struggle against the formalist trend in mathematics was presented by Brouwer in a short paper 'Intuitionistische Betrachtungen über den Formalismus', which was read in 1928 (February 16) by Bieberbach in a session of the Prussian Academy of Sciences. Brouwer reduces his standpoint to four basic points, and declares that the difference between the trends of the intuitive and formalist reconstruction of mathematics will be removed, or at least the choice between them will 'be reduced to the matter of taste' (*Geschmacksangelegenheit*), when these four points, formulated for the first time by intuitionists and addressed to formalists, will gain general acceptance. Brouwer believes that this acceptance is merely a question of time since the case here concerns the results of pure reflection, which do not contain 'any controversial element', and which must be accepted by anybody once they are understood. Of these four points two have already gained understanding, and the formalist literature concedes this explicitly. The understanding of the two other points will undoubtedly signify an end to the contention about the foundations of mathematics.

Now these are the points in question:

*Point one*: it is indispensable to distinguish between the formalist efforts in the construction of the mathematical stock of formulae (*Formelbestand*), i.e. the formalist image of mathematics, and the intuitive theory of laws of that structure; (dealing with the content); it is likewise indispensable to understand that, for the latter theory, the intuitive mathematics of the set of natural numbers is indispensable.

*Point two* boils down to the rejection of the thoughtless application of the logical law of the excluded middle and to gaining the understanding that, *first*, examination of the basis of the authority and range of validity of the above-mentioned law is the essential subject of investigation in the foundations of mathematics, and, *second* that this range of validity in the intuitive mathematics (material mathematics) includes only finite systems.

*Point three* consists in the understanding of the identity of the law of the excluded middle with the principle of the solubility of every mathematical problem.

*Point four* consists in understanding that the justification (relating to the content) of formalistic mathematics by way of proving its consistency comprises a vicious circle, since such justification is based on the (content) correctness of the pronouncement that the truth of a sentence follows from its consistency, i.e. it is based on the (content) truth of the principle of the excluded middle.

As regards point one, the formalists did initially not understand it. Brouwer believes quite rightly that Hilbert is the main representative of formalism among the contemporary mathematicians, and, less rightly, he believes that the inaugural moment of his work in this direction was his lecture 'Mathematische Probleme' at the first International Congress of Mathematicians in 1900. Our point is that, as a matter of fact, this trend is not only a product of the development of all of 19th-century mathematics, but also the awareness of this trend in the case of Hilbert appeared much earlier. Quite rightly Brouwer stressed the achievements of Poincaré as far as distinguishing between formalist and intuitive mathematics is concerned, and speaking about himself he states that, as early as in his first work (written in Dutch in 1907), he had in mind this difference when opposing 'mathematical language' to 'mathematics of the second order'. Now the need for this distinction was finally accepted by the formalists, when introducing the term 'meta-mathematics' to denote that which Brouwer called 'mathematics of the second order', thinking precisely of mathematics prescinding from the intuitive contents of the laws. Now Hilbert accepted only in 1922, in his work 'Die Neubegründung der Mathematik', the need to distinguish mathematics as a purely formal science from meta-mathematics.

On the other hand, one cannot take seriously the claims of the formalist school which, by using this distinction, borrowed from intuitionism, wants to reduce intuitionism to absurdity.

Point two, which is turned against the thoughtless application of the law of the excluded middle, is raised continually by Brouwer, starting with his second work (written in Dutch in 1908). Now the formalists who even, initially, considered the law of the excluded middle to be equivalent to the law of contradiction (according to Brouwer), began only slowly to realize the difference in the twofold application of the principle of the excluded middle (in the case of Hilbert this only occurs in his works of 1922 and 1926).

As regards point three, of the law of the excluded middle being identical with the principle of the solubility of every mathematical problem, here also one can see a certain evolution of the formalists' view, this evolution having not been completed.

In 1900 Hilbert propounded a principle of the solubility of every mathematical problem as an axiom; it was, according to Brouwer, the period of the thoughtless application of the principle of the excluded middle. However, already in his work of 1918 'Mathematisches Denken' Hilbert gives to this axiom a twofold form, speaking once about the 'fundamental' solubility, and another time about 'the possibility of solution by means of a finite number of operations', and he admits that the axiom is the object of problems waiting to be solved. However this problem is treated by the formalists as being independent of the establishing of the foundations of mathematics, to which there belongs only the consistency of the principle of the excluded middle. The formalists discuss only the problem of the consistency of the axiom of solubility of any mathematical problem as a problem of fundamental character, which falls under the category of mathematical thinking which one should not previously set oneself to do.

Similarly Brouwer finds a complete lack, in the literature of the formalists, of understanding of point four, viz. that the (content) justification of the formalist mathematics by means of the proof of consistency forms a vicious circle.

The second part of the article by Brouwer we are discussing contains a proof of a highly interesting theorem. The intuitive consistency of the law of the excluded middle could doubtless pass for a circumstance encouraging the formalists in their efforts to prove the consistency of formalist mathematics. However, says Brouwer, the consistency of the single law of the excluded middle (i.e. of this principle as applied to a single mathematical property) does not yet visualize the combined consistency of the law of the excluded middle, i.e. of this principle as applied to any mathematical properties either in a finite number or for any species of mathematical properties. This combined law of the excluded middle is called by Brouwer the *multiple principle of the first kind* when the question is of its validity for a finite number of properties, but a *multiple principle of the second kind* if we apply it to any species of mathematical properties. The question arises now whether, indeed, the multiple law of the excluded middle is consistent. Brouwer gives a proof that the multiple principle of the excluded middle of the first kind is indeed consistent. The proof was carried out by making use of complete induction. However, for the multiple principle of the excluded middle of the second kind Brouwer proves that it does not possess an inconsistency!

(3) Brouwer's stand began to arouse a keen interest not only among mathematicians but also among logicians. They set about the task of exactly working out Brouwer's logic with the aid of symbols which had

been avoided by Brouwer to the detriment of the matter when formulating his logical principles. In Poland, it was Professor Łukasiewicz who came to be interested in this logic; as early as 1920 he constructed a ready-made system of three-valued logic which, in addition to truth and falsity, adopted a third logical value, viz. that of 'possibility'. These ideas had a completely different source, since the foundation of the new logic was the indeterminist view of the world; yet it displayed some points of contact with Brouwer's logic since both the law of the excluded middle and the law of contradiction cease to be true in this logic – nor are they false from its standpoint, but merely possible. Well now, Brouwer adopts something similar, but only for the law of the excluded middle, and with the difference that the law of the excluded middle is not false, but that it is impossible to prove its truth. Professor Łukasiewicz held a lecture on Brouwer's logic during the Second International Congress of Mathematicians in 1928 (in Bologna). Both in the same and in the following year Gliwenko's papers appeared dealing with Brouwer's logic, in the *Bulletin de l'Academie de Belgique*. It is in the last of these papers that Gliwenko shows that, if a formula is true in the propositional calculus of classical logic, it 'cannot be false' in Brouwer's system. Things behave differently in Łukasiewicz's logic. Some formulae which are true in classical logic are, in Łukasiewicz only possible, while others are simply false. Consequently Brouwer's logic is nearer to intuition than Łukasiewicz's logic.

Finally, we find the best study of Brouwer's logic in A. Heyting's paper 'Die formalen Regeln der intuitionistischen Logik' published in 1930 in the *Jahresberichte der Preussischen Akademie der Wissenschaften Berlin*.

The foundation of the propositional theory in this logic consists of the following eleven axioms:

(1) $a \supset aa$
(2) $ab \supset ba$
(3) $a \supset b. \supset .ac \supset bc$
(4) $a \supset b.b \supset c: \supset :a \supset c$
(5) $b \supset .a \supset b$
(6) $a.a \supset b: \supset :b$
(7) $a \supset a \vee b$
(8) $a \vee b \supset b \vee a$
(9) $a \supset c.b \supset c: \supset :a \vee b \supset c$
(10) $\sim a. \supset .a \supset b$
(11) $a \supset b.a \supset \sim b: \supset :\sim a$

All of these laws are well known in traditional logic. The first one expresses the law of tautology; the second, the commutative law of the product; the third formula expresses the so-called law of the factor, which indicates that both sides of the implication can be multiplied by any factor; The fourth formula is the well-known law of the syllogism; the fifth one is often called the law of Duns Scotus as it corresponds to his thesis that 'truth follows from everything', in other words that one can add to a true proposition any assumption, and the resulting implication will be always true, as it follows after all from its definition; The sixth formula corresponds to the *modus ponens* law of classical logic; the seventh one is the law of simplification for the sum; the eighth one expresses the commutative law for a sum; the ninth one corresponds to the law of composition; the last but one (tenth) relates to the second law of Duns Scotus, viz. that 'anything results from a falsehood', which is here indicated by the fourth that, if we have a pair of contradictory propositions, then any proposition can appear as the consequent of the implication, this being likewise in agreement with the meaning of implication established in mathematical logic. The last law expresses the idea that, if a couple of contradictory propositions follow from a certain proposition, then that proposition is false.

We are struck in the first place by the large number of these axioms. The laws of classical logic can be derived from four, three axioms, even from a single axiom. This depends partly on the selection of basic ideas, i.e. of the original 'terms'. If we choose as the basis of logic the so-called Sheffer symbol, we can, just as Nicod had already tried to do, reduce all logic to one axiom only. If the primary terms chosen are the symbols of negation and implication, then, as Professor Łukasiewicz had shown, one has to adopt three axioms, whereas, if the primary terms chosen are negation and the symbol of the alternative, as one finds in *Principia Mathematica*, four axioms are required. Of course, it is indispensable for the axioms to be independent of one another; in such a case they are also called the basis of the system. Now, the number of propositions which are the basis of the system – as shown by investigations – can be arbitrary, since everything depends on which propositions will be chosen as the basis of the system. It has been proved that, whatever integer $m$ is chosen, one will be able to find for logic a basis which will contain $m$ elements. The logic of Brouwer-Heyting, however, is not an ordinary two-valued logic in which every proposition can have two values, truth and falsity. As mentioned already, Brouwer rejects the law of the excluded middle because it is impossible to prove its absolute truth, although it is likewise not false, since as early as 1908 Brouwer did claim that the proposition that it is false is likewise false, which just expresses the law which is called by Brouwer the law of the 'absurdity of the absurdity of the excluded middle', and which is symbolized by the formula

$$\sim\sim(a \vee \sim a)$$

Thus we can say with some sense that the law of the excluded middle apparently possesses a third logical value, in other words that we have here to deal with a three-valued logic. The question arises how do the logical relations of the implications of the sum and the product of two propositions stand if we adopt the view that propositions can assume three logical values, i.e. truth, falsity and the third one which, following Professor Łukasiewicz, we will call 'possibility', although in Brouwer we ought to treat it somewhat differently than in Łukasiewicz. In the first place the question obtrudes itself, what logical value should be attributed to the negation of this third value. Now Heyting assumes that falsity is the negation of this third value. If we designate truth by '1', falsity by '0', and possibility, as the intermediate value, is '½', we obtain the following table, or 'matrix' of negation, in which the top line gives the logical value of the proposition 'a', and the bottom line the negation 'a' of the corresponding value at the top:

| a | 1 | 0 | ½ |
|---|---|---|---|
| ~a | 0 | 1 | 0 |

Consequently we see that the negation of the truth is falsity, the negation of falsity is truth, the negation of possibility in Brouwer's logic is falsity.

Let us now pass to the implication matrix which is read off using the terms 'if, then'. Two-valued logic lays down the value of this relation only for four possible combinations of the value of the antecedent with that of the consequent, viz. for the combinations 11, 00, 01 and 10. Now we get five additional pairs ½0, ½½, ½1, 1½, 0½. Of course, for the former pairs we preserve the values of traditional logic, while for the new ones we have to adopt such values as would deviate least from intuition. This is what Heyting has done. This is the implication matrix.

| a | b | a ⊃ b |
|---|---|---|
| 1 | 1 | 1 |
| 0 | 0 | 1 |
| 0 | 1 | 1 |
| 1 | 0 | 0 |
| 0 | ½ | 1 |
| ½ | ½ | 1 |
| 1 | ½ | ½ |
| ½ | 0 | 0 |
| ½ | 1 | 1 |

We can see at first glance that the analogy with ordinary logic has been preserved. Here too implication has the value '1' when the values of the antecedent are equal to those of the consequent; next implication in both logics has the value 'one' also for the case when the value of the consequent is 'higher' than that of the antecedent; finally, in both logics the value of implication becomes equal to the lower value when that lower value appears in the consequent.

Heyting constructs a similar matrix also for the logical sum and logical product. For the time being we shall be happy with the two former ones; further remarks will be referred to them. In two-valued logic, there occur between basic ideas such as negation, sum, product, implication, relationships which render it possible to express any one of them by means of the relative product of two other ones, or else to simply define one of them with the aid of another one. Now in Brouwer's logic anything of this kind is impossible. None of the four basic ideas can be expressed by means of another idea, nor be represented by the relative product of two, or even three, remaining ones. This explains partly why Heyting needed as many as eleven axioms for his theory of propositions.

Yet, as shown by Tarski, deductive theory can be constructed in two ways: using the axiomatic or the matrix method. Heyting constructed the theory of propositions by the former method, deducing theorems from the adopted axioms, with the aid of usual directives, or inference rules, such as the rule of detachment and that of substitution. He made use of the matrices only towards the end of his work, in order to use them to prove that it is impossible in his system to deduce the law of the excluded middle. We can however use these matrices as a means of testing which theorems are true in a given system, irrespective of the construction of the theory by an axiomatic method, the method used reminding us 'of the method of zero-one testings of the two-valued logic'. The only difference will consist in that the true formula should attain the value '1' for any arbitrary substitutions not only of the values 0 or 1, but of 0, 1 and ½, for the propositional variables. By testing the laws of two-valued logic in this manner, we shall find that only some of them will be true in Brouwer's logic, while others will be only possible. This will be all those for which the value of ½ is obtained for at least one substitution. In order to carry out such a test we shall also need a table for the logical product and sum, as presented below.

| a | b | a V b |
|---|---|---|
| 1 | 1 | 1 |
| 0 | 1 | 1 |
| 1 | 0 | 1 |
| 0 | 0 | 0 |
| ½ | 0 | ½ |
| 0 | ½ | ½ |
| ½ | ½ | ½ |
| ½ | 1 | 1 |
| 1 | ½ | 1 |

| a | b | a·b |
|---|---|---|
| 1 | 1 | 1 |
| 1 | 0 | 0 |
| 0 | 1 | 0 |
| 0 | 0 | 0 |
| ½ | 0 | 0 |
| 0 | ½ | 0 |
| ½ | ½ | ½ |
| ½ | 1 | ½ |
| 1 | ½ | ½ |

A test of this kind will reveal an important detail which was observed for the first time by Gliwenko, viz. that not a single proposition of two-valued logic will prove to be false, but only that some of them will prove to be possible. Thus, as we already know from Brouwer's work of 1925, the law of double negation is not true in Brouwer; it will only prove to be possible. For Brouwer, only the implication law of double negation is true, with the sign of two negations from the right-hand side; let us call it the right-hand implication law of double negation. On the contrary, as we know from Brouwer's work of 1925, he rejected the left-hand one not as a false one, but as a non-true one, i.e. a possible one, which can be easily tested by way of substitutions with the aid of matrices. This is precisely the law he had given the name of the law 'of the reciprocal of the complementary kind'. Among the roughly seventy-five theorems of the theory of propositions proved by Heyting we find of course those known since 1925:

4.31 $\sim a \supset \sim\sim\sim a$
4.32 $\sim\sim\sim a \supset \sim a$

We will turn our attention to another very important principle of Brouwer's logic:

4.85 a V ~a. ⊃ .~b: ⊃ ~b

The point is that this principle points to the fact that one can, in Brouwer's system of propositional logic, use the law of the excluded middle to prove negative propositions. For we know that the only essential feature for Brouwer's standpoint is that one must not prove the existence of something positive on the basis of the law of the excluded middle.

However, the comparison of Brouwer's logic with two-valued logic does not give us a complete image of its meaning. We shall get a full understanding of this logic only after having compared its matrices with those of Professor Łukasiewicz's three-valued logic. The table for the product and the sum is exactly the same as in Łukasiewicz. However, the matrix of negation and implication looks differently. For a negation of possibility is, in Łukasiewicz, likewise a possibility, and not a falsity as in Heyting. Thus we have

| ~a | 1 | 0 | ½ |
|---|---|---|---|
| a | 0 | 1 | ½ |

Compared with Heyting's matrix, the matrix of implication reveals also a small difference which, however, entails considerable consequences. The point is that Heyting adopted for the implication ½ ⊃ 0 the value '0', while Łukasiewicz adopted for it the value '½'. Hence the table of implications in Łukasiewicz looks as follows:

| a | b | a ⊃ b |
|---|---|---|
| 1 | 1 | 1 |
| 0 | 0 | 1 |
| 0 | 1 | 1 |
| 1 | 0 | 0 |

While according to Heyting it looks like:

| a | b | a ⊃ b |
|---|---|---|
| 0 | ½ | 1 |
| ½ | ½ | 1 |
| 1 | ½ | ½ |
| ½ | 0 | ½ |
| ½ | 1 | 1 |

Both matrices adhere to the right principle (by analogy to intuitionist logic), viz. that implication should not have the value '1' when the antecedent has this value, but not the consequent. Łukasiewicz and Tarski rightly consider this condition to be the condition of matrix normality. One could produce a great number of quite fantastic, abnormal matrices; of course each time we would define a certain set of formulae, but they would hardly deserve the name of logic. Therefore, when constructing multi-valued logic, one should limit oneself to a task for which the matrix should be a normal one, just as we limit ourselves in the construction of multi-valued logics by the postulate that their matrices contained cases established by the matrices of two-valued logic. The point is also that this postulate was always observed in the generalization of the structures of all deductive systems.

However observing this most general postulate, as well as the postulate of normality still does not point univocally to the value of implication in the case ½ ⊃ 0. Let us generalize Łukasiewicz's and Tarski's postulate of normality, so that the implication must not have the value '1' whenever the value of the consequent is lower than that of the antecedent (i.e. without setting the condition that the antecedent is = '1'); even in this case I do not know whether the value '0' or '½' should be adopted for ½ ⊃ 0. Obviously Heyting still adhered to the assumption that, in cases which are not designated univocally by the postulate of normality, implication must be given the same value as the consequent has; hence for 1 ⊃ ½ the value should be ½, and for ½ ⊃ 0 the value should be 0. Now Łukasiewicz adopts the value ½ in both cases.

Let us, however, revert to Heyting's system which revealed some drawbacks within a short time. It turned out quite quickly that Heyting's matrix is inadequate, i.e. that it is satisfied not only by Heyting's axioms and the resulting theses, but also by some formulae that do not result from Heytings axioms, hence do not belong to the system. This can easily be shown in the following example. In Heyting's system, one can prove the law that the law of excluded middle implies the left-hand law of double negation (theorem 45, article H).

$$(a \vee \sim a) \supset (\sim\sim a \supset a).$$

On the other hand the inverse theorem, says Heyting, does not arise. He asserts that, on the basis of his system, there takes place only the inferential equivalence of these two principles, which means that, if one included in the system the law of the excluded middle, then it would become possible to prove likewise the left-hand law of double negation and, conversely, by adding this second law we would prove the former one. But an implication equivalence does not arise, i.e. the formula $(a \vee \sim a) \equiv (\sim\sim a \supset a)$ does not hold, since much as theorem 45 does take place, the inverse implication $(\sim\sim a \supset a) \supset (a \vee \sim a)$ does not. Heyting's matrix does satisfy also this inversion, which can be easily checked by substituting ½ for a:

$$(\sim\sim \tfrac{1}{2} \supset \tfrac{1}{2}) \supset (\tfrac{1}{2} \vee \sim \tfrac{1}{2})$$
$$(\sim 0 \supset \tfrac{1}{2}) \supset (\tfrac{1}{2} \vee 0)$$
$$(1 \supset \tfrac{1}{2}) \supset \tfrac{1}{2}$$
$$\tfrac{1}{2} \supset \tfrac{1}{2}$$
$$1$$

Thus the formula acquires the value '1', much as it cannot be derived from the axioms. And since the axioms did only decide in what manner Brouwer and Heyting wished to reform mathematics, it became clear that the matrix must undergo a correction.

(4) It soon came to light, however, that, in no matrix having a finite number of values are we able to carry out this correction. Kurt Gödel published in 1932 the paper 'Zum intuitionistischen Aussagenkalkül', in which he showed that no matrix having a finite number of values can be adequate for Heyting's system, since a logical formula will always be found which does not result from the axioms of Heyting's logic and which will likewise be verified by the matrices which verify these axioms. Since the proof is extremely short and simple, we shall state it here *in extenso*. Let us form a logical sum of equivalences only, which arise between the propositional variables:

$$(p_1 \equiv p_2) \vee (p_1 \equiv p_3) \vee (p_1 \equiv p_4) \vee \ldots \vee (p_1 \equiv p_n) \vee (p_2 \equiv p_3) \vee (p_2 \equiv p_4) \vee \ldots \vee (p_{n-1} \equiv p_n);$$

Each matrix containing fewer than *n* values, for which Heyting's axioms are satisfied, will satisfy this formula. For, let us try to give a different value to each variable; since, as assumed, there are fewer such values than variables, in one equivalence at least the variables will get the same value, i.e. at least one equivalence will be true, and the truth of at least a single member of a logical sum results in that the entire sum obtains the discriminated value of '1' (the respective law of two-valued logic is also valid in Heyting's logic).

We will now show that the above formula is independent of the axioms; hence it cannot be proved in this system. The proof of independence is usually carried out by means of interpretation. Thus, in order to obtain the independence of our formula which represents the sum of equivalences, of Heyting's axioms, we look for such an interpretation of variables, for which the axioms will be satisfied and our formula will not be satisfied. Now Gödel provides such an interpretative table; it has *n* variables: 1, 2, 3, . . . *n*, the discriminated value being 1. The implication: $a \supset b = 1$ for $a \geqslant b$, and $a \supset b = b$ for $a<b$; $\sim a = n$ for $a \neq n$, and $\sim n = 1$; $a \cup b$ will assume the value of the smaller component, hence min.(a,b), while the logical product – the value of the greater component, hence max.(a,b). For this table of values the axioms are satisfied while our formula, true enough, will also be satisfied but only in the case when the number of propositional variables $p_n$ is greater than n, whereas if this number is equal to, or less than, n, our formula will not be satisfied, and hence it is independent of the axioms.

Consequently the adequate matrix of Heyting's logic must be infinitely multi-valued, the laws of Heyting's logic are satisfied for an infinite series of finite matrices with an ever-increasing number of values. Thereby Gödel has shown not only (1) that no finite matrix is adequate for Heyting's system, but also (2) that, between the system of the common propositional calculus and Heyting's system, there lie an infinite number of systems, hence a series of monotonously decreasing systems in which Heyting's system is included and which are included in the two-valued logic.[13]

(5) More light was cast upon this infinite series of systems, and consequently of matrices, by the studies of Jaśkowski, published in the *Acta* of the 1935 International Congress of Philosophy of Science, viz. in volume VI 'Recherches sur le Systeme de la Logique Intuitioniste'. It has been found that the structure of these matrices is much more complicated than one would assume initially. In addition to the main stem of matrices,

there also exist lateral ramifications each of which supplies an infinite series of new matrices. The main-stem matrices can be embraced into a relatively simple scheme which is obviously different for each logical constant. Let us start with the implications. In all matrices we shall always designate by 1 the discriminated value, 'truth'. The non-discriminated values in the consecutive matrices constantly increase by a single value. Let us introduce a function $\alpha$, such that, when its argument has a discriminated value, one would obtain as a function value a new value which had been absent in the previous matrix, while if the argument of the function $\alpha$ is a non-discriminated value of the previous matrix, the value of function $\alpha$ equalled that previous non-discriminated value. In the uni-valued logic the only value is the discriminated value 1. When passing to the two-valued logic we introduce a new value, the non-discriminated value, called falsity, and we designate it most frequently by the symbol 0. Thus $\alpha$ (1) = 0. We do not make use of the second value of the function because there were no non-discriminated values in the uni-valued logic. If now we pass from the two-valued to the three-valued logic by adding a new value designated by ½ or 2, then $\alpha(1) = 2$ whereas $\alpha(0) = 0$. When passing to four-valued logic we have $\alpha(1) = 3$, if 3 denotes the newly-added fourth value, and we obtain for the previous non-discriminated values: $\alpha(0) = 0$, $\alpha(2) = 2$ etc.

By introducing this function Jaskowski was able to give a general table which indicated how to construct matrix N when one knows the construction of the preceding matrix M.

For the implications this general table looks as follows:

| | $\supset_N$ | second argument 1 | $\alpha(y)$ |
|---|---|---|---|
| First | 1 | $1 \supset_M 1$ | $\alpha(1 \supset_M y)$ |
| argument | $\alpha(x)$ | $x \supset_M 1$ | $x \supset_M y$ |

Since in the uni-valued logic there exists no value besides 1, we must substitute 1 for x and y. We shall obtain

| $\supset$ | 1 | $\alpha(1)$ |
|---|---|---|
| 1 | $1 \supset 1$ | $\alpha(1 \supset 1)$ |
| $\alpha(1)$ | $1 \supset 1$ | $1 \supset 1$ |

However, in the uni-valued logic not only the arguments, but also the function values cannot obtain any value other than 1, hence:

| $\supset$ | 1 | $\alpha(1)$ |
|---|---|---|
| 1 | 1 | $\alpha(1)$ |
| $\alpha(1)$ | 1 | 1 |

Let us remember that, according to the given definition of function $\alpha$, $\alpha(1) = 0$, hence:

| $\supset$ | 1 | 0 |
|---|---|---|
| 1 | 1 | 0 |
| 0 | 1 | 1 |

We have thus obtained the well-known implication matrix of two-valued logic. By taking the first argument from the first column before the vertical line, and the second argument from the first line above the horizontal bar we shall obtain the value of the implication at the crossing of the appropriate vertical column and the horizontal like, thus $1 \supset 1$ will yield 1, and similarly $0 \supset 1 = 1$, $0 \quad 0 = 1$, while $1 \supset 0 = 0$.

If we pass now from this matrix to that of three-valued logic, let us remember that $x$ and $y$ can assume two values: 0 or 1. Hence the passing to three-valued logic looks as follows:

| $\supset$ | 1 | $\alpha(1)$ | $\alpha(0)$ |
|---|---|---|---|
| 1 | 1 | $\alpha(1 \supset 1)$ | $\alpha(1 \supset 0)$ |
| $\alpha(1)$ | $1 \supset 1$ | $1 \supset 1$ | $1 \supset 0$ |
| $\alpha(0)$ | $0 \supset 1$ | $0 \supset 1$ | $0 \supset 0$ |

When remembering that, in accordance with the definition of function $\alpha$, $\alpha(1)$ yields now a new value of 2, while $\alpha(0) = 0$, and taking into account the value of implications from the preceding two-valued table, we obtain:

| ⊃ | 1 | 2 | 0 |
|---|---|---|---|
| 1 | 1 | 2 | 0 |
| 2 | 1 | 1 | 0 |
| 0 | 1 | 1 | 1 |

Proceeding analogously we obtain the matrix of the implication of four-valued logic:

| ⊃ | 1 | α(1) 3 | α(2) 2 | a(0) 0 |
|---|---|---|---|---|
| 1 | 1 | 3 | 2 | 0 |
| α(1) = 3 | 1 | 1 | 2 | 0 |
| α(2) = 2 | 1 | 1 | 1 | 0 |
| α(0) = 0 | 1 | 1 | 1 | 1 |

of five-valued logic:

| ⊃ | 1 | 4 | 3 | 2 | 0 |
|---|---|---|---|---|---|
| 1 | 1 | 4 | 3 | 2 | 0 |
| α(1) = 4 | 1 | 1 | 3 | 2 | 0 |
| α(3) = 3 | 1 | 1 | 1 | 2 | 0 |
| α(2) = 2 | 1 | 1 | 1 | 1 | 0 |
| α(0) = 0 | 1 | 1 | 1 | 1 | 1 |

etc.

However, one ought to remember that the new values do not play the same role as in Łukasiewicz's logic. The value denoted by 2 is not the same as 'possibility' in Łukasiewicz. It denotes the value of propositions which are not false, but whose truth cannot be proved if one adopts the limitations of the new logic. Thus they are propositions which are undecidable on the basis of the decree of intuitionists, but which are considered by ordinary mathematics to be true. Hence it is, so to say, a truth of a worse kind, something only approaching truth. Now Jaskowski took note of some details in the structure of matrices, which entitle us to assert that the new values 2, 3, 4 draw nearer and nearer to the classical truth and merge in infinity.

For, let us equate 2 with 1 in the matrix of the three-valued logic, and the result will be the matrix of the two-valued logic, while this result cannot be obtained from the matrix of Łukasiewicz's three-valued logic.

| ⊃ | 1 | 1 | 0 |
|---|---|---|---|
| 1 | 1 | 1 | 0 |
| 1 | 1 | 1 | 0 |
| 0 | 1 | 1 | 1 |

(in Heyting)

| ⊃ | 1 | 1 | 0 |
|---|---|---|---|
| 1 | 1 | 1 | 0 |
| 1 | 1 | 1 | 1 |
| 0 | 1 | 1 | 1 |

(in Łukasiewicz)

From Heyting's table all values of implications of the two-valued logic will result, i.e. $1 \supset 1 = 1, 0 \supset 1 = 1, 0 \supset 0 = 1$ and $1 \supset 0 = 0$.

On the contrary, from Łukasiewicz's matrix, after the identification of 'possibility' with truth, things arise which are inconsistent with the two-valued logic. The point is that $1 \supset 0$ will yield the value 1, which is nonsense. Similarly, if, in Heyting's four-valued matrix, we identify 4 with 1, we will obtain the three-valued matrix in order, while something analogous will not take place in Łukasiewicz's logic. For it is known that Łukasiewicz had also created logics with optional numbers of the degree of value, and to which he had been inspired, perhaps, by some analogies with the probability calculus, the first idea having appeared as early as 1922 (*Ruch Filozoficzny, 7*, p. 93). If we agree to designate the logical values of propositions by numbers in the interval from 0 to 1, then the value of negation of the proposition *p* in Łukasiewicz's logic of any values will equal one *minus* the value of the proposition *p*:

$$w(\sim p) = 1 - w(p)$$

Thus, e.g., if $w(p) = 1/3$, then $w(\sim p) = 2/3$, and so on. On the other hand, implication is understood in such a way that, if the value of the antecedent equals, or is smaller than, the value of the consequent, then implication has the value of one. Heyting's implication matrix also has the same value, but, if in implication the value of the antecedent is higher than that of the consequent, then the implication is calculated from the formula: 1 *minus* the value of the antecedent *plus* the value of the consequent (Heyting's matrix lacks this property), i.e. briefly:

$$w(p \supset q) = 1 \quad \text{when} \quad w(p) \leqslant w(q)$$
$$w(p \supset q) = 1 - w(p) + w(q) \quad \text{when} \quad w(p) > w(q);$$

Let us then construct the four-valued matrices of Heyting and Łukasiewicz:

| ⊃ | 1 | 3 | 2 | 0 |
|---|---|---|---|---|
| 1 | 1 | 3 | 2 | 0 |
| 3 | 1 | 1 | 2 | 0 |
| 2 | 1 | 1 | 1 | 0 |
| 0 | 1 | 1 | 1 | 1 |

| ⊃ | 1 | 2/3 | 1/3 | 0 |
|---|---|---|---|---|
| 1 | 1 | 2/3 | 1/3 | 0 |
| 2/3 | 1 | 1 | 2/3 | 1/3 |
| 1/3 | 1 | 1 | 1 | 2/3 |
| 0 | 1 | 1 | 1 | 1 |

Let us now equate with 1 in both tables the last values, i.e. 3 or 2/3:

| ⊃ | 1 | 1 | 2 | 0 |
|---|---|---|---|---|
| 1 | 1 | 1 | 2 | 0 |
| 1 | 1 | 1 | 2 | 0 |
| 2 | 1 | 1 | 1 | 0 |
| 0 | 1 | 1 | 1 | 1 |

| ⊃ | 1 | 1 | 1/3 | 0 |
|---|---|---|---|---|
| 1 | 1 | 1 | 1/3 | 0 |
| 1 | 1 | 1 | 1 | 1/3 |
| 1/3 | 1 | 1 | 1 | 1 |
| 1 | 1 | 1 | 1 | 1 |

In Heyting we descend by such an equation to the relations of three-valued logic, though with an unnecessary two repetitions of the same line and the same column, while the equation process taken in Łukasiewicz's table results in nonsense. There the relations of both the three-valued and the two-valued logic are destroyed. $1 \supset 0$ yields 0 on the basis of the first line and, simultaneously it yields 1/3 on the basis of the second line: similarly $1 \supset 1/3 - 1/3$ or 1.

The above-mentioned examples explain that in Heyting's matrix the values which newly enter into the matrix can be assumed to be very close to truth – this does not arise in Łukasiewicz. But why do we assert that, in the same matrix of Heyting's, 3 is nearer to the truth than 2; why the newly entering values are, generally speaking, nearer to the truth than the previous ones? This theorem is justified since, if we had, in Heyting's four-valued matrix preserved 3 and identified the value 2 with 1, we would obtain an absurdity, whereas the identification of 3 with 1, with the simultaneous preservation of the value 2 does not as we had already seen, spoil the previous matrix. The four-valued matrix in which 2 = 1, and 3 remains unchanged, looks as follows

| ⊃ | 1 | 3 | 1 | 0 |
|---|---|---|---|---|
| 1 | 1 | 3 | 1 | 0 |
| 3 | 1 | 1 | 1 | 0 |
| 1 | 1 | 1 | 1 | 0 |
| 0 | 1 | 1 | 1 | 1 |

We see here that $1 \supset 3$ yields 3 on the basis of the first row below the horizontal bar, whereas on the basis of the third line we obtain $1 \supset 3 = 1$. Consequently, three can be identified with truth if we maintain the primary meaning of two, but one cannot deal with 'two' as truth when 'three' remains as a distinct value. This justifies the assertion of Jaskowski, viz. that the further we progress in the succession of new values the nearer we do get to the truth. Had we wished to set it off in symbols, it would be necessary to present truth by means of the symbol of infinity which can be pass for the limit of a natural series of numbers, or else one could retain the symbol of 'one' for the truth and to represent further values by means of proper fractions nearer and nearer to '1'. Thus instead of using for them the values, 2, 3, 4 . . . we would write, e.g., 2/3, 3/4, . . . provided that nothing else would stand in the way.

However, before we proceed further, we shall give still the negation matrix for Heyting's system, and the matrix of logical sum and logical product. The general negation schema for all *n*-valued logics, where *n* is a finite integer, is as follows:

| $\sim_N$ | |
|---|---|
| 1 | $\alpha(\sim_M 1)$ |
| $\alpha(x)$ | $\sim_M x$ |

If M is the matrix of the one-valued logic then $\sim 1$ is to be treated as 1, also $x = 1$ and $\sim x = 1$; hence for two-valued logic we obtain:

| $\sim$ | |
|---|---|
| 1 | $\alpha(1)$ |
| $\alpha(1)$ | 1 |

But, when passing to two-valued logic, $\alpha(1)$, according to our definition of the $\alpha$ function, designates, falsity; hence

| $\sim$ | |
|---|---|
| 1 | 0 |
| 0 | 1 |

Let us now pass to three-valued logic. The argument $x$ can assume now two values, 0 or 1, and the function $\alpha(1)$ is to designate the newly-entering element; hence

| $\sim$ | |
|---|---|
| 1 | 0 |
| $\alpha(0) = 0$ | 1 |
| $\alpha(1) = 2$ | $\sim 1 = 0$ |

For four-valued logic:

| $\sim$ | |
|---|---|
| 1 | 0 |
| $\alpha(0) = 0$ | 1 |
| $\alpha(2) = 2$ | 0 |
| $\alpha(1) = 3$ | $\sim 1 = 0$ |

The negations of all of the newly-added values equal zero, i.e. they yield falsity; we see here why it is possible to treat these new values as the approximations of truth, much as it is not obvious why the higher and higher values are nearer and nearer the truth. We give without comments the matrices of the sum and of the product:

| N | 1 | $\alpha(y)$ |
|---|---|---|
| 1 | $1 ._M 1$ | $\alpha(1 ._M y)$ |
| $\alpha(x)$ | $\alpha(x ._M 1)$ | $\alpha(x ._M y)$ |

| $\vee_N$ | 1 | $\alpha(y)$ |
|---|---|---|
| 1 | $1 \vee M^1$ | $1 \vee_M y$ |
| $\alpha(x)$ | $x \vee M^1$ | $\alpha(x \vee M^y)$ |

On the basis of our explanations for the general tables for implication and negation, the reader will accept these tables as well. As an example we give only the product and sum tables for three-valued logic, which do not differ in the least from the respective tables for Łukasiewicz's three-valued logic.

| . | 1 | 2 | 0 |
|---|---|---|---|
| 1 | 1 | 2 | 0 |
| 2 | 2 | 2 | 0 |
| 0 | 0 | 0 | 0 |

| V | 1 | 2 | 0 |
|---|---|---|---|
| 1 | 1 | 1 | 1 |
| 2 | 1 | 2 | 2 |
| 0 | 1 | 2 | 0 |

This agreement is also retained for matrices which are higher than the three-valued ones. The question arises whether it would not be better to replace these tables with general verbal rules, just as in Łukasiewicz, viz. by saying that the logical product assumes always the value of the lower component, and the sum the value of the higher component, that the negations of all values are zeros, except zero, for which the negation is 'one', and finally that for implication the rule is valid that

$$w(p \supset q) = 1 \qquad \text{when} \qquad w(p) \leqslant w(q)$$

just as in Łukasiewicz, while

$$w(p \supset q) = q \qquad \text{when} \qquad w(p) > w(q);$$

(here already in a different manner than in Łukasiewicz where $w(p \supset q)$ in the last case equalled $1 - w(p) + w(q)$. Why yes, we could proceed like that, but the matrices thus far discussed are not all of the matrices which are satisfied in Heyting's system. There exist other branchings whose discovery is the exclusive merit of Jaskowski; it was Gödel who conjectured the matrices discussed, as can be seen from his interpretation table for demonstrating the independence of the formula which is the sum of equivalences. Conversely, one ought to note that, independent of Gödel, Jaskowski reached the result that no matrices with a finite number of values can be adequate for Heyting's system.

The process of the transition from matrix M to the following matrix N is called by the author the operation Γ carried out on matrix M:

N is Γ (M)

Since the system is designated by the matrix as well as by a set of axioms, we can say that an operation Γ carried out on one system yields the following one. The operation Γ carried out on the system $L_1$ of one-valued logic yields the two-valued logic $L_2$. Thus $L_2$ is Γ ($L_1$), etc. The set of logical expressions satisfying the matrix M is denoted by the author by using the symbol L(M). Since one-valued logic includes the laws of two-valued logic, and the latter includes the laws of three-valued logic, etc., we have $E(\Gamma M) \subset E(M)$. An inverse relation generally does not arise.

In order to pass to matrices of a new type which are satisfied by Heyting's system the author introduces the idea of the multiplication of matrices. If we have two matrices, one of which has *m* elements and the other *n* elements, we form at first pairs of elements by combining each element of the first matrix with each element of the other matrix. Let us assume that, e.g., we wish to multiply one matrix of the two-valued logic by a second matrix of the two-valued logic. Thus by combining 0 and 1 of one matrix with 0 and 1 of the other one, we obtain four pairs: 00, 01, 10, 11. The pair of discriminated elements 11 will be considered to be the discriminated element of the new four-valued matrix. Now we have to determine implication, multiplication, addition and negation for it. It will be understood that the implication of two pairs is the pair of two implications, the sign of implication combining the first element of the first pair with the first element of the second pair, next the second element of the first pair with the second element of the second pair. Thus, e.g.,

$$(0,1) \supset (1,1) = (0 \supset 1, 1 \supset 1) = (1,1)$$

We shall proceed similarly with the addition and multiplication. The negation of the pair is the pair of negations:

$$\sim(0,1) = (\sim 0, \sim 1) = (1,0)$$

Let us now form a table of implication for a new four-valued matrix:

| ⊃ | 11 | 10 | 01 | 00 |
|---|---|---|---|---|
| 11 | 11 | 10 | 01 | 00 |
| 10 | 11 | 11 | 01 | 01 |
| 01 | 11 | 10 | 11 | 10 |
| 00 | 11 | 11 | 11 | 11 |

This four-valued matrix differs from that which can be obtained by using the operation Γ to the lower matrices. We can thus form the products of any two matrices, e.g. by multiplying the two-valued matrix by a three-valued, four-valued one etc. In the former case we obtain a six-valued matrix, in the second an eight-valued one, etc. Since it is possible to form the products not only of two, but also of three and more matrices, there are so many possible products as there are possible subsets among the infinite number of finite matrices. What is more, each of them may undergo the operation Γ and give rise to a new infinite sequence. The discovery of the matrices of this type which satisfy the system of Heyting (with small exceptions which are dealt with below) is the most important achievement of Jaskowski. Let us however pay close attention to this matrix which was obtained by multiplying two matrices of the two-valued logic. This matrix is interesting because, though a four-valued one, it satisfies all laws, and only those laws, of two-valued logic. This is due to the fact (as we shall see below) that it is an integral positive power of the ordinary two-valued matrix. Let us verify it on the example of the law of the two-valued logic:

$$p \supset .q \supset p$$

Let

$$p = 01, q = 10 \qquad 01 \supset .10 \supset 01$$

While remembering that the implication of the pairs of values yields a pair of implications of the single values, we obtain:

$$0 \supset .1 \supset 0 \qquad 1 \supset .0 \supset 1$$
$$0 \supset 0 \qquad 1 \supset 1$$
$$1 \qquad 1$$

If one finds it cumbersome to calculate the values of logical values for the pairs of values, we can substitute new single numbers for these pairs. Thus, when adopting: 11 = 1, 10 = 2, 01 = 3, 00 = 0, one will obtain instead of:

| ⊃ | 11 | 10 | 01 | 00 |
|---|---|---|---|---|
| 11 | 11 | 10 | 01 | 00 |
| 10 | 11 | 11 | 01 | 01 |
| 01 | 11 | 10 | 11 | 10 |
| 00 | 11 | 11 | 11 | 11 |

the matrix

| $\supset$ | 1 | 2 | 3 | 0 |
|---|---|---|---|---|
| 1 | 1 | 2 | 3 | 0 |
| 2 | 1 | 1 | 3 | 3 |
| 3 | 1 | 2 | 1 | 2 |
| 0 | 1 | 1 | 1 | 1 |

which satisfies the same functions as the preceding one does. Thus the formula $p \supset .q \supset p$ proves true at present for the new value symbols as it did previously.

Let e.g. $p = 3$, $q = 2$; we shall then get:

$3 \supset .2 \supset 3$
$3 \supset 3$
$1$

For $p = 2$, $q = 1$ we get:

$2 \supset .1 \supset 2$
$2 \supset 2$
$1$ and so on.

In connection with the new matrix two things deserve attention. It was necessary to substitute 1 for 11 because we have so defined the product of two matrices that the pair of discriminated values is the discriminated value of the matrix and, when the matrix value is represented by one digit, the discriminated value was always designated by 1. It is quite natural and understandable that 0 was substituted for 00; but it was not necessary to designate 10 by 2 and 01 by 3. We could have done so conversely, in which case such a matrix would be equivalent to the previous one, and it would satisfy the same functions as the previous one.

The second remark is even more important. Let us compare the four-valued matrix just obtained with the four-valued matrix obtained by the successive application of the operation Γ to the matrix of one-valued logic, or of the two-valued one (if one does not wish to start from one-valued logic).

| ⊃ | 1 | 2 | 3 | 0 |
|---|---|---|---|---|
| 1 | 1 | 2 | 3 | 0 |
| 2 | 1 | 1 | 3 | 3 |
| 3 | 1 | 2 | 1 | 2 |
| 0 | 1 | 1 | 1 | 1 |

| ⊃ | 1 | 2 | 3 | 0 |
|---|---|---|---|---|
| 1 | 1 | 2 | 3 | 0 |
| 2 | 1 | 1 | 1 | 0 |
| 3 | 1 | 2 | 1 | 0 |
| 0 | 1 | 1 | 1 | 1 |

We see at once that the matrices which were formed by multiplication do not satisfy the conditions set for the matrices obtained by operations Γ only. For the first it is not true that, when the antecedent has a value higher than the consequent has, the implication acquires the value of the consequent, since we have $2 \supset 0 = 3$, $3 \supset 0 = 2$; likewise it is not always true that, when the value of the antecedent is lower than that of the consequent, the implication has the value of 1, because $2 \supset 3 = 3$. Someone can observe that this matrix was obtained by multiplying two matrices of the two-valued logic, and satisfies, as has been said, only the laws of two-valued logic; hence it would be senseless to lock in it for satisfying the rules formulated earlier for the three- and more-valued systems. However this exceptional nature of the matrices, which is the square of the matrix of the two-valued logic, is here immaterial. We could make the same remarks when comparing, e.g., two six-valued logics, one of which was formed by the multiplication – let us call it operation ∏ – of a matrix of two-valued logic by the matrices of the three-valued logic, while the other was obtained by the successive application of operations Γ only. It would be burdensome to write out these matrices. But let us do something else; we shall apply operation Γ to a matrix which is the square of the matrix of two-valued logic, and the five-valued matrix thus obtained will be compared with the five-valued matrix obtained by using only the operations mentioned above.

Here is the Γ of the matrix of the logic $(L_2)^2$:

| ⊃ | 1 | α(1) 4 | α(2) 2 | α(3) 3 | α(0) 0 |
|---|---|---|---|---|---|
| 1 | 1 | 4 | 2 | 3 | 0 |
| α(1) = 4 | 1 | 1 | 2 | 3 | 0 |
| α(2) = 2 | 1 | 1 | 1 | 3 | 3 |
| α(3) = 3 | 1 | 1 | 2 | 1 | 2 |
| α(0) = 0 | 1 | 1 | 1 | 1 | 1 |

We see here again the same deviations from the rule which had been formulated for matrices obtained exclusively by applying the operations Γ. Here also: $2 \supset 3 = 3$, $3 \supset 2 = 2$, $3 \supset 0 = 2$, $2 \supset 0 = 3$. That is the reason why it is impossible to give as simple a rule for the construction of matrices of Heyting's multi-valued systems as for Łukasiewicz' systems.

We shall now prove the theorem: $E(M \times N) = E(M) \cdot E(N)$.

This means that the logical formulae which satisfy the matrix obtained by the multiplication of two matrices form a class which is identical with the class which is the logical product of the system based on the matrix M and of the system based on the matrix N, i.e. which form a class identical with the common area of the classes E(M) and E(N).

Jaśkowski's paper is written in a highly condensed style, since it was reckoned to be read in the course of roughly one hour; that is why the proofs were either adduced in a highly abbreviated form or else omitted altogether. For this reason the paper became, for general readers, a book closed under seven seals. With the author's permission, and basing myself on his oral explanations. I am taking the liberty of setting forth somewhat more detailed proofs of some theorems.

The proof of the mentioned theorem is to consist of two parts, viz. that the left-hand side is comprised in the right-hand one and *vice versa*, the right-hand in the left-hand one. We shall prove firstly that $E(M \times N) \subset E(M) \cdot E(N)$. It suffices to demonstrate that $E(M \times N) \subset E(M)$, since the same can be repeated for E(N), from which the result will be obtained that the class $E(M \times N)$ is contained in the logical product of both classes E(M) and E(N). We shall prove indirectly that $E(M \times N) \subset E(M)$ by proving that, if any logical formula is not an element of the class E(M), it cannot be an element of the class $E(M \times N)$. Let us then assume that one can insert, for the propositional variables $p_1, p_2, \ldots p_k$, contained in the logical formula *a*, such values: $m_1, m_2, \ldots m_k$ which are the values of the matrix M and at the same time of such a nature that the formula *a* obtains for them an undiscriminated value. Let us now substitute for $p_1, p_2, \ldots p_k$, as values, the pairs $(m_1, 1), (m_2, 1) \ldots (m_k, 1)$. It is obvious that, for such pairs of values, formula *a* cannot obtain the discriminated value of 11, for the first '1' depends only on the values $m_1, m_2, \ldots m_k$, and for these values the formula did assume the undiscriminated value.

Let us now pass to the second part of the proof, that

$$E(M) \cdot E(N) \subset E(M \times N)$$

Let us assume that the logical formula $\alpha \varepsilon E(M) \cdot E(N)$. Let this formula include again $p_2, p_1, \ldots p_k$ of propositional variables. Let the arbitrary substitution of the value of the matrix M×N have the form $(m_1, n_1)$, $(m_2, n_2) \ldots (m_k, n_k)$. If we substitute, for the variables $p_1, p_2, \ldots p_k$ the values of the matrix M only, i.e. $m_1, m_2, \ldots m_k$, the formula will obtain for them the discriminated value 1. This follows from the definition E(M) as a set of formulae which satisfy any substitutions of the matrix M values.

The same takes place when the values $n_1, n_2, \ldots n_k$ are substituted for the propositional variables $p_1, p_2, \ldots p_k$ – the formula *a* acquires the discriminated value '1'. And in the case of the product substitution $(m_1, n_1)$ $(m_2, n_2) \ldots (m_k, n_k)$ it will acquire the value 11, i.e. a discriminated one. Hence the assumption that $\alpha \varepsilon E(M) \cdot E(N)$ implies that $\alpha \varepsilon E(M \times N)$.

From the proved theorem that $E(M \times N) = E(M) \cdot E(N)$ there follows the specific consequence that, when M = N, we get $E(M^2) = E(M)$. Therefore the logical laws which satisfy the square of the two-valued matrix are, as mentioned before, identical with the laws which satisfy the two-valued matrix itself.

The question arises, how does one ascertain that each of Heyting's formulas does satisfy this infinite number of matrices. Of assistance in this case is the theorem which predicates that, if Heyting's axioms, hence also the entire system – let us call it briefly the H system – does satisfy a certain matrix M, then it does satisfy also the consequent of that matrix, i.e. the matrix Γ(M). The axioms of the system H have been already written out by us. There were eleven of them. However, in addition to them Heyting admits also the directive that, if it is true that *p*, then if it is true that *q*, then the conjunction of these two propositions is likewise true. This directive can be replaced with the axiom: p ⊃ q ⊃ pq, and so we would get in all 12 axioms; Jaśkowski reduces this number to ten.

(1) p ⊃ .q ⊃ r: ⊃ :p ⊃ q. ⊃ .p ⊃ r
(2) p ⊃ .q ⊃ p
(3) p ⊃ q ⊃ pq
(4) pq ⊃ p
(5) pq ⊃ q
(6) p ⊃ r.q ⊃ r: ⊃ p V q. ⊃ r
(7) p ⊃ .p V q
(8) q ⊃ .p V q
(9) ~p ⊃ .p ⊃ q
(10) p ⊃ ~p. ⊃ ~p

We will not carry out the proof in a general way. We shall limit ourselves to showing it for the example of the first axiom

$$p \supset .q \supset r : \supset : p \quad q . \supset . p \quad r.$$

We assume that this axiom, just as the others, does already satisfy a certain matrix. And they undoubtedly satisfy the matrices of one-valued logic, because this logic is satisfied by all formulae. They satisfy the matrix of two-valued logic. That they must also satisfy the following matrix, we shall show in such a manner that we will first prove that our axiom will not assume the value $\alpha(1)$, i.e. the last value which we must append to the matrix M in order to obtain $\Gamma$ (M). When passing from $L_1$ to $L_2$, $\alpha(1) = 0$, when passing from $L_2$ to $L_3$, $\alpha(1) = 2$ etc. as has been already mentioned. Now the implication assumes the value $\alpha(1)$ only when the antecedent has the value 1, and the consequent has the value $\alpha(1)$. Hence, in our axiom, $p \supset .q \supset r$ ought to be equal to 1, and the entire consequent $p \supset q \quad .p \supset r$ ought to be equal to $\alpha(1)$. Hence in the consequent itself $p \supset q$ ought to be equal to 1, and $p \supset r = \alpha(1)$. The next inference is that in this final term there must be $p = 1$, and $r = \alpha(1)$. However, since $p = 1$, and the preceding term should also $= 1$, i.e. $p \supset q = 1$, it follows that also $q = 1$. Let us now remember that the entire side of the formula to the left of the main functor of the implication $p \supset .q \supset r$ ought also to be $=1$. And since we had found that $p = 1$, then $q \supset r$ should be also $= 1$. But we have also found that $q = 1$, and this yields the final result $r = 1$. Now we have obtained before that $r = \alpha(1)$.

Consequently the assumption that our axiom could assume the value $\alpha(1)$ did lead us to a contradiction. We have therefore to assume that our formula can never assume the value $\alpha(1)$. One deals similarly in the case of other axioms.

However, in order to prove that our axiom must be true for $\Gamma$(M), it does not suffice to demonstrate that it cannot assume the last value $\alpha(1)$; it is also necessary to demonstrate that it cannot assume any other undiscriminated value. We shall indicate here only the direction of further reasoning. Were it the case, i.e. that our axiom assumed this undiscriminated value for certain values of the matrix $\Gamma$ (M), e.g. for the value $x$ substituted for $p$, $y$ for $q$, $z$ for $r$ – which we will shortly denote as p/x, q/y, r/z – then our axiom would assume this undiscriminated value for the matrix N, since we can carry out the following substitutions: if $x$ differs from $\alpha(1)$ we shall substitute $x$ for p and, if it is equal to $\alpha(1)$ we shall substitute 1 for p.

Thus symbolically:

p/x . . . if $x \neq \alpha(1)$
p/1 . . . if $x = \alpha(1)$

and similarly for q:

$$\begin{array}{ll} q/y \ldots & \text{if } y \neq \alpha(1) \\ q/1 \ldots & \text{if } y = \alpha(1) \end{array}$$

and for r:

$$\begin{array}{ll} r/z \ldots & \text{if } z \neq \alpha(1) \\ r/1 \ldots & \text{if } z = \alpha(1) \end{array}$$

In this manner we have identified in this substitution α(1) and 1 and have found ourselves in the matrix M. From the definition of the operation Γ we infer that if our axiom (and it is possible to extend this reasoning to any other sensible expression) with the variables *p*, *q*, *r*, had obtained after the first substitution an undiscriminated value other than α(1) for the matrix Γ(M), it has also the same value in the second substitution; in order to bring it out in better relief that it must be so, let us take as an example the expression which obtains in the three-valued logic the value 0, i.e. an undiscriminated value other than α(1) for, in this case, α(1) = 2.

Let the formula be represented as follows (p = 2, q =0, r = 1):

$$\begin{array}{l} p \supset q. \supset r: \supset :p \supset q \\ 2 \supset 0. \supset 1: \supset :2 \supset 0 \\ 0 \supset 1: \supset :0 \\ 1 \supset 0 \\ 0 \end{array}$$

Let us now carry out the second substitution in which q and r preserve their value, and the value p, as equal to α(1), will pass into 1. Thus p = 1, q = 0, r = 1. We are here on the solid ground of the two valued logic

$$\begin{array}{l} p \supset q. \supset r: \supset :p \supset q \\ 1 \supset 0. \supset 1: \supset :1 \supset 0 \\ 0 \supset 1: \supset :0 \\ 1 \supset 0 \\ 0 \end{array}$$

The reasoning which would reassure us that the passage from matrix Γ (M) to matrix M by way of the substitutions indicated would retain the value of the formula also for this matrix M, would require certain complements in the case of complex expressions. However, it is always possible to be assured that, had the logical formula for the matrix Γ(M) assumed a non-discriminated value other than α(1), then this formula would acquire the same value also for M; it would not be true also in M. We have proved before that a formula which is true for the matrix M cannot acquire the value $\alpha(1)$ in $\Gamma$ (M). (True, in that case the reasoning did have gaps, because it referred to one axiom only, so it would be necessary to repeat it for all of Heyting's axioms, and then to point out that the property is hereditary, i.e. that it is transferable onto all theses that can be derived from these axioms). This results in the theorem that the logical laws of the Heyting's system are characterized by the property that, whenever they satisfy the matrix M, they also satisfy the matrix $\Gamma$(M). Let us give the name $L_1(\prod\Gamma)$ to the set of matrices which can be obtained both by applying operations $\prod$ and $\Gamma$ to $L_1$, and the name of $E[(L_1)\prod\Gamma]$ to the set of logical formulae which satisfy this matrix. We have obtained in such a case the result that $H \supset E[(L_1)\prod\Gamma]$. However in order to show that Heyting's system is identical with the set of formulae $E[(L_1)\prod\Gamma]$, one has to prove that $E[(L_1)\prod\Gamma] \supset H$. We will put an end here to our historical review and refer the reader to the next work of Jaśkowski, in which the author promises to give this proof in a more complete form.

In the last part of this paper the author points also to the possibility of replacing the infinite sequence of matrices satisfying the system of Heyting by one infinite matrix $I_\infty$ so that finally the set of logical laws which satisfy this matrix, $E(I_\alpha)$ will prove identical with the set H and with the set $E[(L_1)\prod\Gamma]$.

Lastly, we shall mention that, one year before the outbreak of World War II in 1938, a highly interesting work by Alfred Tarski appeared, 'Der Aussagekalkül und die Topologie' (*Fund.Math.*), in which the author provides the topological interpretation not only for ordinary two-valued propositional calculus, but also for the intuitionist system of Brouwer-Heyting.

CHAPTER 3

# METHODOLOGY

## 3.1. RELATIONSHIP BETWEEN THE PRINCIPLE OF CAUSALITY AND THE PRINCIPLE OF RELATIVITY (A HABILITATION LECTURE)

*Sine doctrina, vita est quasi mortis imago.*

In his book on *The Foundation of the General Theory of Relativity* Einstein gives two motives which called for the extension of the primary relativity postulate upon which the special theory was based. The first was the well-known fact of the equality of inertial mass and gravitational mass, i.e. a fact of a purely physical nature, whereas the other was a consideration of an epistemological nature, a consideration about the principle of causality. It is precisely that philosophical argument drawn from epistemology, an *'erkenntnistheoretisches Argument'* that is placed in the foreground by Einstein. In order to see what it looks like, let us present the relevant passage of Einstein's book, in which he develops his idea using a very simple example:[14]

In classical mechanics, and no less in the special theory of relativity, there is an inherent epistemological defect which was, perhaps for the first time, clearly pointed out by Ernst Mach. We will elucidate it by the following example:– Two fluid bodies of the same size and nature hover freely in space at so great a distance from each other and from all other masses that only those gravitational forces need be taken into account which arise from the interaction of different parts of the same body. Let the distance between the two bodies be invariable, and in neither of the bodies let there be any relative movements of the parts with respect to one another. But let either mass, as judged by an observer at rest relatively to the other mass, rotate with constant angular velocity about the line joining the masses. This is a verificable relative motion of the two bodies. Now let us imagine that each of the bodies has been surveyed by means of measuring instruments at rest relatively to itself, and let the surface of $S_1$ prove to be a sphere, and that of $S_2$ an ellipsoid of revolution. Thereupon we put the question – What is the reason for this difference in the two bodies? No answer can be admitted as epistemologically satisfactory, unless the reason given is an *observable fact of experience*. The law of causality has not the significance of a statement as to the world of experience, except when *observable facts* ultimately appear as causes and effects.

Newtonian mechanics does not give a satisfactory answer to this question. It pronounces as follows:– The laws of mechanics apply to the space $R_1$, in respect to which the body $S_1$ is at rest, but not to the space $R_2$, in respect to which the body $S_2$ is at rest. But the privileged space $R_1$ of Galileo, thus introduced, is a merely *factitious* cause, and not a thing that can be observed. It is therefore clear that Newton's mechanics does not really satisfy the requirement of causality in the case under consideration, but only apparently does so, since it makes the factitious cause $R_1$ responsible for the observable difference in the bodies $S_1$ and $S_2$. (Einstein, 'The foundation of the general theory of relativity', in Einstein, *et al. The Principle of Relativity*, Dover Publications, 1923, pp. 112–113.)

While attempting to formulate the standpoint of classical mechanics as generally as possible, Einstein gave it such an abstract form that it was not comprehensible to everyone. However, we can present it more simply and intelligibly in the following manner. To be sure, relative motion can be attributed equally well to both bodies, but, since only one of them shows a flattening at the poles, this means that in fact only that body moves performing rotations with respect to absolute space, whereas the other one executes only an apparent movement. In this way only one of the spaces, with respect to which each of the two bodies can be accepted by its observers as stationary, plays the role of a privileged space, say, absolute space. However, a movement with respect to absolute space is not itself something that could be observed in any way; hence it is not a phenomenon, an experimental fact, and what is not a phenomenon likewise cannot be the cause of any other phenomenon, according to the formulation of the principle of causality. This principle had previously been enunciated by Einstein and we shall of course unanimously assent to it today, viz. if this principle is to be useful at all in natural sciences, some phenomena must always figure as cause and effect.

Thus classical mechanics, while giving a fictitious cause for the difference in the shape of both bodies, cannot as a matter of fact, according to Einstein, provide a cause that would be in agreement with our currently acceptable formulation of the causality principle. Prior to giving much thought to this statement of Einstein, let us first see what according to him, should be the correct answer.

The only satisfactory answer must be that the physical system consisting of $S_1$ and $S_2$ reveals within itself no imaginable cause to which the differing behaviour of $S_1$ and $S_2$ can be referred. The cause must therefore lie *outside* this system. We have to take it that the general laws of motion, which in particular determine the shapes of $S_1$ and $S_2$, must be such that the mechanical behaviour of $S_1$ and $S_2$ is partly conditioned, in quite essential respects, by distant masses which we have not included in the system under consideration. These distant masses and their motions relative to $S_1$ and $S_2$ must then be regarded as the seat of the causes (which must be susceptible to observation) of the different behaviour of our two bodies $S_1$ and $S_2$. They take over the rôle of the factitious cause $R_1$. Of all imaginable spaces $R_1$, $R_2$, etc., in any kind of motion relatively to one another, there is none which we may look upon as privileged *a priori* without reviving the above-mentioned epistemological objection. *The laws of physics must be of such a nature that they apply to systems of reference in any kind of motion. (Ibid.)*

A well understood causality principle requires this amplification.

Now we see what an extraordinarily important role in the construction of the theory is played by a motive of a quite general nature, viz. the consideration of a certain principle of epistemology. Up to now this detail had attracted little attention from the philosophers. As far as we know, it was only Schlick who had considered this matter closely, maintaining rightly that the fulfillment of the causality principle is the *conditio sine qua non* of cognition in natural science. Thus if Einstein was right, then classical mechanics ought to be eliminated as violating this law, and scientific truth should be attributed exclusively to our contemporary relativistic mechanics. However Schlick does not wholly bear Einstein out, mentioning this question once in a brief note in his book *Raum und Zeit in der gegenwärtigen Physik* (1917), and then discussing it in more detail in the paper 'Naturphilosophische Betrachtungen über das Kausalprinzip', published in *Die Naturwissenschaften* (1920). In this paper, Schlick concludes to be sure that we cannot say that classical mechanics contradicts the causality principle, but we have to admit that it limits the range of its applicability. However, in principle we do not see any essential difference between the standpoints of Einstein and Schlick, if we stick to his latest enunciations (in *Raum und Zeit* he spoke in a different way). To infringe upon the causality principle means to us to adopt a certain change, which has no cause, in another change, and, as we shall see, it is to this that, according to Schlick, the standpoint of classical mechanics is reduced.

Besides we ought to admit that Schlick reproaches Einstein correctly for expressing himself inappropriately here and there. Einstein presents the problem as if the causality principle was meant to give an account of the difference between the forms of the bodies as such, whereas the causality principle refers to processes only. The form of the body itself, or even the difference of the forms, is not a process, unless we want to consider that the form of a body is only a sign of some processes taking place within the body. However, we do not doubt that, in this case, what Einstein did mean was the difference of processes. Of course instead of the process of flattening one could take into account some other symptom of the centrifugal force, which would have to take place simultaneously and which, in a more obvious way, would have the nature of a process. Let us then assume that our masses $S_1$ and $S_2$ are rigid masses and not liquid ones as assumed originally by Einstein, and let us imagine, as proposed by Schlick, small blocks attached to them by means of clamps. If these clamps are suddenly removed, then, on one of these bodies the block will remain at rest, but on the other it will fly off at a tangent, this direction being judged from the position of the former body. If we now ask what is the cause of this phenomenon, it is obvious that we must not look for it in the relative velocity of these bodies, since we can attribute it both to the former or to the latter, and even with the same angular velocity. However, says Schlick, Newton could give the following answer:

Motion relative to body 1 is not in fact the same as motion relative to body 2! For experience teaches that the one body is *singled out* in nature, and hence it becomes practical to relate the motions to a system of coordinates at rest therein: the motions so related I *call absolute*. (Moritz Shlick, *Philosophical Papers*, vol. I, eds. H.L. Mulder and F.B. Velde-Schlick, Feidel Publishing Co., 1979, p. 311.)

Of course we could suggest such an answer to Newton although he himself had never used this expression. The point is that Newton did not have to stick to some *a priori* presuppositions that there must not exist any singled out, privileged systems in nature; so he could accept a principle that experiment ought to play the decisive role. Now if two bodies, under identical conditions, still reveal different properties, we are bound to admit that some bodies are singled out in nature, thus lending themselves better as reference systems than others. They can precisely be called rest systems, while those in which we shall observe the signs of centrifugal force will be considered by us to be in a state of motion. Thus, according to Schlick, Newton's dynamics will be quite in order with respect to the principle of causality since it will not be obliged to treat the symptoms of centrifugal force as some results, but merely as a certain property of the systems, by means of which the dynamic rotation can be defined. "Absolute rotation must not be denoted as the cause of flattening, one can only say: that the latter is defined by the former".[15] Similarly in the other work he says: "The appearance of centrifugal acceleration should be conceived as a definition of dynamic rotation but not as an action of kinematic rotation".

However, Schlick cannot assume that, likewise, a situation would be inadmissible in principle, in which, on the contrary, the bodies for which centrifugal accelerations do appear, would be assumed to be resting, if the question was indeed merely that of a definition; at most we could note that such procedure would be very impractical. However another reason also exists why such a stand, which we could suggest to Newton, should be considered to be unsatisfactory in the highest degree. The point is that the property by means of which we describe the dynamic rotation is not a dead, motionless property; it has the character of a process! Wherever we have to deal with a process, we should speak about causes and effects. That is why Einstein, when reverting to this question for the second time in his popular discourse, stresses always: "with things as they now stand no logically thinking man can be satisfied". The question, what is the cause of the singling out of some systems, does not allow us to leave the matter alone. The state of our mind when faced with this question is illustrated by Einstein in this discourse by means of the following apt comparison:

In front of me, on a gas cooker, there are two completely identical pans, half-filled with water. I see that steam escapes continuously from one pan, but not from the other one. This amazes me, even had I not seen any gas cooker before. However, when I perceive under the first pan something emitting blue light, which I do not see under the second pan, my astonishment disappears even when I had never seen previously a gas flame. The point is that I can say that this bluish thing does certainly, or probably, bring about the emission of steam. However, had I seen nothing bluish under any pan, and had steam been escaping continuously from one pan, but not from the other, I would have remained astonished and dissatisfied until I should observe a circumstance which I would make responsible for the different behavior of both pans.

It was already Mach who had noted that the 'singled-out' systems of Newtonian mechanics are precisely those which are at rest in relation to the system of fixed stars, which is distinguished among other systems by its enormous magnitude and extent. He also was the first to postulate a connection between the one and the other, when discussing in his *Mechanics* the example of a pail of water in the case of Newton,[16] which, to Newton, was supposed to be the proof of the existence of absolute movements. Einstein's example is nothing but a certain modification of Newton's experiment. The role of the body $S_1$ is taken by water in the vessel so long as we can attribute to water a movement with respect to the bucket walls, while the same water takes the role of the body $S_2$ as soon as it starts rotating, along with the pail, with respect to the surrounding masses. It is precisely then that the form of its surface reveals the appearance of centrifugal forces.

However, while, according to Newton these centrifugal forces prove that, in the latter case, the movement is absolute, Mach wants us to look for the source of these not in the absolute nature of the movement, but in the fact that water executes a relative motion with respect to the enormous masses of the stellar system, while, in the former case, it moved with respect to the mass of the thin wall of the vessel. Right from the moment when this postulate was advanced Mach's achievement begins, but it also ends there since Mach was unable to go any further.

However, Einstein showed how to formulate this connection in accordance with the causality principle. It became evident that it does not suffice to refer to the presence of masses only, but that one has to alter the idea of the gravitational field and to know how to present it by means of differential equations in such a way as to enable the states of the gravitational field to be understood as processes which spread with a certain finite velocity in accordance with the postulate advanced already by the special theory of relativity, which excludes *actio in distans*. It is only with such a treatment of the gravitational field that the laws of nature could be given a form that is wholly independent of the systems of reference. The generally invariant form of nature's laws – such is this extremely important positive result achieved by the rejection of absolute movements, which puts an end to the fruitless search for the absolute system of reference. As Cassirer stressed rightly, one achieved here the solution of a certain scientific problem, viz. in a completely analogous manner as the squaring of the circle problem or the search for *perpetuum mobile*. The negative result of some investigations was rewarded by a valuable positive gain which explains at the same time why some efforts had to remain fruitless. The squaring of the circle was found to be impossible, but, at the same time, the irrationality of $\pi$ was discovered, just as the fruitless efforts involved in the search for *perpetuum mobile* to the founding of the energy conservation principle. Similarly in the present case, we failed to find the absolute systems of reference but instead found the invariant form of nature's laws, which explains at the same time why such a search for privileged systems must appear to be absurd. Should such systems be found, the laws of nature would lose their generally invariant character. It is fair to assume that it is only from the creation of the general theory of relativity that we know what is the really general and commonly binding form of the law of nature!

We see that a consistent observance of the causality principle in its empirical formulation has contributed to obtaining these extremely valuable results. Now this formulation is nothing absolutely new from the philosophical point of view. It was already Hume and Kant who spoke of the limiting of the causality principle to phenomena only, and even Leibniz, in his polemics with Clarke, Newton's disciple, precisely on the subject of absolute space, had formulated a principle which is similar to that to which Einstein does refer. In his case this principle is called the 'principle of observability', and it states that there is no change where there is no perceivable change "quand il n'y a point de changement observable, il n'y a point de changement du tout". Hence one cannot give the name of change to something that would be only a change introduced in the mind, but not a change that could be detected by measurement, if this word has to denote something physically real. The movement of the universe with respect to absolute spac,e says Leibniz, if not observable in principle by our measurements, cannot therefore denote any real change.[18]

However, as mentioned above, this principle did not suffice for Einstein. He had to supplement the causality principle by the statement that all changes are propagated continuously with a certain finite velocity. This presupposition whose value is also determined only by experiment, throws new light upon various philosophical difficulties which have been connected up to now with the idea of causality, starting with the period when Sextus Empiricus, was first to criticize this notion and ending with Mach's objections. It was frequently maintained that cause and effect were simultaneous. Now at present, on the basis of the achievements of physics, i.e. at least as far as physical phenomena are concerned, one can say with respect to the causal connection of spatially distant changes that these changes cannot absolutely be simultaneous and that a time must elapse between them, this time being however not greater than the time during which the distance in question is covered by light, whose velocity cannot be infinitely great. To be sure, in the case of regions which are directly accessible to us, this lapse of time becomes practically so small that it is difficult to be distinguished from a momentary simultaneity.

Moreover, the old presupposition of the *actio in distans* made one assume that it is the whole universe that influences every phenomenon. We now know that this is not so. For every phenomenon one can carve out in the world those phenomena with which it is causally related and separate it from those with which it surely does not remain in any causal relationship. This is also possible because of the magnitude which we call the velocity of light. Besides, the presupposition that all changes are propagated from one place to another continuously and with a finite velocity also provides an easy practical means for investigating the causes of any phenomenon which takes place in a certain defined place. It suffices to enclose this phenomenon within a boundary surface and only to investigate the changes which pass during a certain time across this surface. The point is that, irrespective of whether such changes come from a short or long distance, they must always pass through all intermediate places, which means that, to reach a certain range, they must pass through a boundary surface which is, besides, arbitrarily outlined. Hence it suffices to observe what takes place on this surface in order to determine, after having got to know the state of the inner space at a certain time, all further processes taking place in it. Physicists have known for a long time that, if one has the given laws of nature in the form of differential equations (in order to be able to determine the course of phenomena in a certain concrete case), one has also to know the so-called initial and boundary conditions. Just as, in the temporal flow, the immediately preceding state is decisive for every phenomenon, so also for the spatial disposition for every phenomenon the processes taking place in its immediate neighborhood are decisive.[19] Now, according to the new theory, the above point of view ought to be applied to all processes without exception, hence also to the processes taking place in the so-called gravitational fields. And for these fields one should construct differential equations resembling the equation of wave motion. The ultimate aim of natural-science cognition is to reduce all processes to a minimum quantity of such differential equations. The laws of nature presented in this form can be called, following Schlick, the micro-laws, as opposed to the integral laws as macro-laws which can be obtained from the former ones by way of integration processes. Only macro-laws are directly within the ambit of the experiment, while micro-laws are already the product of a certain mental construction adapted to the empirically confirmed macro-laws.

The objection comes across whether there is no contradiction between this postulate of continuity in the transfer of changes and the postulate formulated right at the beginning with respect to the causality principle, viz. to limit oneself to the observable changes only. The point is that changes of states in a purely gravitational field, or any other field, are not things equally perceptible by senses as the movements of visible bodies. Consequently, is there here no violation of the accepted and so strongly emphasized principle, viz. that only observable changes are always to play the role of cause and effect? A similar remark is made by Oskar

Krauss in his work 'Fiktion and Hypothese in der Einsteinschen Relativitätstheorie'.[20] When adopting the continuity principle we rather go beyond the sense data, yet this transgression is of another nature than in the case when one attributes to something absolute reality, independently of any sensory data. When assuming that changes propagate continuously we make a certain interpolation, and often an extrapolation, of empirical data, and we subject the value of the mental construction thus obtained to experimental test, while in the case of a purely metaphysical hypothesis there can be no question of such an experimental test. When operating with the notion of the field, the physicist has always in mind the totality of certain symptoms which might have been unconfirmed in practice, but which could have been verified if we had inserted certain test bodies (*Probekörper*) into the places involved, just as this is presented in a very instructive manner by some physicists (in many discourses within the grasp of the non-physicist) such as, e.g., Thirring or Mie.

However, one cannot take exception to classical physics because it did not take a suitable account of the continuity principle, since at that time the facts were still unknown which compel the physicists to adopt that principle, by forcing them to admit the existence of a certain finite limiting velocity with which all changes can be propagated. On the other hand, the objection that classical physics did not duly observe the empirical formulation of the causality principle is valid since the experiment with the bucket of water ought to have been, according to that principle, explained in fact by Newton in another way. The defence of Newton, as suggested by Schlick, does not exonerate him, we think, from this objection, because, in accordance with this defence, we would have, in the case of centrifugal accelerations, to do with changes that lack any cause. We also think that another statement of Schlick is just as inaccurate, viz. that the general theory of relativity does widen the applicability range of the causality principle only insofar as it forces us to view the states of the gravitational field as processes. The only truth consists in that the general theory of relativity could not be deduced from the causality principle had we not combined it at the same time with the continuity principle – and also if we had not taken into account other empirical arguments such as the equality of the inertial mass and gravitational mass, which Schlick fails to mention at the relevant place. On the other hand the non-violation of the causality principle, or else its 'widening', as Schlick calls it, would take place also if, according to Mach's remarks, we saw the cause of centrifugal accelerations in the movement of bodies with respect to the masses of the universe, even if we were unable to carry out the connection of these two phenomena according to the continuity principle.

In his work, adduced by us previously ('Naturphilosophische Betrachtungen über das Causalprinzip'), Schlick discusses the relationship between the principle of causality and the principle of relativity from yet another point of view, which however, had Schlick studied it in a consistent way right to the end, would reveal a number of new difficulties in the behavior of this relationship to its usual formulation. We usually say that the causal relations between phenomena can be discovered owing only to the fact that some processes in nature repeat themselves. However, the same processes are never repeated, even processes which are exactly identical are not repeated, at most such processes can repeat themselves, between which there exists a far-reaching similarity. It is known that Bergson did negate the possibility of using the causality principle with reference to mental life, since in mental life phenomena are never repeated. But it may well be that the same is true also in the physical world, much as we identify in it a certain monotony which enables us to discover the so-called laws of nature and permits us to discuss the problem of the arbitrary number of cases of their applicability. Let us therefore consider more closely what this monotony consist in. It has been first established that place and time themselves do not influence the course of the phenomenon; for let us consider, what would follow in the opposite case. This would mean that, e.g., a certain chemical reaction would take place quite differently with changes in time and space, in one room in a different way than in another room, also in a different way now and a few hours later. Now we do not find in our world such a far-reaching chaotic disorder. In a world where something of this kind would occur, the place itself *in abstracto*, i.e. the thing we call space in abstraction from the phenomena, as well as time, would play the role of real factors, and would have the significance of absolute realities. One would not be able then to talk of any monotony and of a generally binding form of laws, although a *sui generis* regularity could still exist provided one could devise certain mathematical formulae with which one would be able to co-ordinate the course of such phenomena. However, in such laws the co-ordinates of time and space would be unable to appear only as certain differential ratios, but they would have to figure *expliciter*, e.g. as the arguments of certain aperiodic functions. A change in the system of reference would result in a change of the form of these laws. In short in such a world the time-space definitions would not be relative; they would be something characteristic for every phenomenon, as if touching their inner content. Nothing of this kind does take place in our world. Time and space display a certain homogeneity – at least as long as we contemplate everything from the standpoint of classical mechanics – no place differs in any respect from another, and, as regards space we have, besides, to speak of a certain isotropism, since here, in addition, no direction distinguishes itself from another one. It is this homogeneity that, in accordance with the above remarks, was identified with the time and space relativity. Thus, e.g., Poincaré says: "la relativité de l'espace et son homogénéité sont une seule et même chose" – the relativity of space and its homogeneity are identical. This relationship between time and space and the regularity of nature and the principle of causality, if it finds its expression in this regularity, likewise draws the attention of the English physicist Maxwell at the end of the first chapter of his popular scientific work *Matter and Movement*, where he suggests replacing the proposition 'the same causes always bring about the same results' with the following one: "The difference between two events depends not on the pure difference between times or places where they take place, but only on differences in the essence, configuration or motion of the bodies considered. It follows that, if the event has taken place at a definite time and in a definite place, a completely identical event can take place at a different time and in a different place."

Now the general theory of relativity destroys precisely this homogeneity and isotropism of space, by pointing to the fact that the metrics of space and its curvature vary with the accumulation of masses. Such a consequence of the principle of relativity would be simply incomprehensible if one would admit that the essential meaning of the principle of relativity becomes exhausted by the mere ascertaining of the homogeneity of space and time. Consequently one has to admit that Maxwell's and Poincaré's remarks are correct only with regard to the principle of relativity of classical mechanics. The true sense of the principle of relativity is different, viz. it consists in the generally covariant form of the laws of nature. From the former connection between the principle of relativity and the homogeneity of space and time, as considered by Poincaré, there remains true only the fact that there are no distinguished systems of reference; all of them are equally entitled to the mathematical formulation of the laws of nature. In spite of everything we have no right to assert that there exists no place which would differ in anything from another one, nor that no direction is different in anything from other directions. This results from the fact that the idea itself of place or time had lost its original meaning, and that therefore one cannot, from the standpoint of the new theory of gravitation, connect unconditionally the relativity of the space-time definitions with the idea of any homogeneity of space and time. For the same reason likewise the inhomogeneity of time and space cannot be identified with their absolute character viz. that of the reverse of relativity. Indeed, in Newton the empty infinite space, or similarly time, were understood as something absolutely homogeneous, although he did understand them as something absolute, something endowed with reality irrespective of any bodies and phenomena occurring in them.

In Aristotle, on the contrary, space undoubtedly did not have the character of something homogeneous; the contrast between above and below there was absolute and affected the behavior of bodies. Indeed, for some of them, the heavy bodies, the natural place was below, for others, the light ones, above; if the body did not occupy its natural place, each body would behave differently, as each would tend to return to its natural place. In the case of this inhomogeneity, space and time had real meaning, if we interpret Aristotle's stand realistically, as this is usually done.

For Kant, just as for Newton, space and time preserve the character of a homogeneous medium, much as they do not play the role of absolute realities. Temporal/spatial descriptions have a relative character, although this relativity does not go as far as in the new theory and enables Kant to stress a certain difference between the reality of linear and rotary movements.

On the contrary, in the theory of relativity, at least if we adopt right from the beginning the standpoint of the general theory, we are obliged again to attribute inhomogeneity to space and time, but at the same time they cannot be understood as realities independent of the sense data in connection with which they appear. It would have been impossible to attribute, without them, a certain structure to the space-time continuum. Thus the question arises in what sense can one speak here of the repetition of certain processes in different places and times, when this repetition appears to be indispensable for discovering the laws of nature and consequently certain causal relations. As we have mentioned before, Schlick, in accordance with the quotations of Poincaré and Maxwell, considers that the homogeneity of space and time is indispensable for the detection of causal relations; he identified this homogeneity with their relativity and sees therein precisely the proof of a more profound relation between the principle of relativity and causality. True, he realizes that the general theory of relativity introduces certain space-time inhomogeneities, but he dispatches this matter briefly, adding a parenthetical remark that they do not destroy this fundamental homogeneity which is required for ascertaining causal laws. However, this remark is insufficient. It is known that Whitehead, one of the eminent English mathematicians and a collaborator with Bertrand Russell, gives such an interpretation to Einstein's theory of relativity for the purpose of maintaining in this theory the Euclidean character of space, since the space-time inhomogeneity Einstein wants to introduce would render impossible any induction, and hence also the detection of causal laws by way of experiment. Thus the same principle which, for one philosopher becomes a basis of a theory, serves another one as a tool and means, if not for the theory, then at least for a different interpretation. We have therefore to consider more closely how far does this space-time homogeneity go which is indispensable for the formulation of causal laws, and to what extent it is accepted by the theory of relativity.

In his work 'On the Notion of Cause'[21] Russell correctly draws the reader's attention to the fact that, when speaking of the repetition of the same facts, we use this word 'fact' in a fairly general and abstract way, and, the more accurate and concrete the definition, the less repeatable it becomes. A fact taken in its full specificity and individuality could be repeated only if it were something completely independent of the remainder of the universe or if such a dependence yielded a certain resultant stable property of the environment so that the fact could be treated as a relatively isolated system; now we can often treat as virtually isolated such systems which theoretically are not isolated, when the probability of the influences coming from the remainder of the world exceeds a certain boundary value which is not too high. The occurrence of such isolated systems is a condition of the possibility of detecting any regularities and relationships, both functional and causal.[22] In addition one has to remember that the same system can be treated in one respect as isolated, and in another as non-isolated, since a law of nature concerns always only the consistency of certain relations, and, after all, as we know already, what is at stake is always the question of relations which can always be expressed as differential equations. Any absolute homogeneity of space and time as a common natural environment for all phenomena of the world is not at all required for the possibility of detecting the laws of nature; it suffices only for this homogeneity to occur in a sufficiently limited area and only for a certain category of relations, and the latter is guaranteed also in the general theory of relativity.

However, Schlick, when speaking about a certain fundamental homogeneity which is necessary for the detection of the laws of nature had something different in mind. The question in the case of Schlick was that there are in nature no privileged systems of reference, and that a change in the position does not alter the mathematical form of nature's laws. There exist in nature no points of Archimedes which would uncover to us the one true face of nature, as opposed to another, the false ones, and, wherever we stand, we are always at the center of nature. Yet one must introduce some reservations here in order to avoid misunderstandings. The point is that, when the reference system is changed, when the time-space parameters are arbitrarily replaced, the detailed form of the laws of nature undergoes a continual change,[23] there change in them also changing the forms of the motion of bodies, and even the forces appearing in the systems (at least as far as the name is concerned), but the construction of all these detailed forms of nature's laws is such that all of them must be able to be subordinated under common schemas of differential forms, and are the singular cases of one common invariant form. Now among these different singular, covariant forms, some are made more conspicuous than others by their simplicity, and this is why, in practice, we always give priority to some systems rather than to others.

It is, however, difficult to consider, as Schlick does, that the fact that theoretically no reference system is distinguished from the other ones, is a condition enabling the constant causal relations to be detected empirically. At best the generally invariant form of the nature's laws can make a contribution as a certain heuristic principle to the detection of the mathematically most general form of these laws. However, if the objective is to formulate conditions enabling constant relations to be discovered empirically, it suffices to assume the existence of virtually isolated systems in accordance with Russell's remarks; their existence is recognized also by the general theory of relativity, by admitting that, in sufficiently small areas not only are the laws of the special theory valid, but so also are the laws of classical mechanics.

Consequently Whitehead's objections against Einstein's 'non-homologous space-time', which allegedly makes induction impossible, also falls through. The point is that there always exist sufficiently large areas in which the space-time medium does preserve homogeneity and, besides, in the entire universe and in all areas a certain squared differential form of space-time parameters does preserve a constant value. Therefore the reviewer of Whitehead's work in *Mind*[24] rightly stresses: "I would be greatly astonished if one could obtain such a concrete result (as a certain modification of the theory of relativity) from such abstract premises (as the possibility of using the principle of induction) . . . It seems to me that the most this argument could prove would be that there exist some uniform relations." If we assume that these uniform relations must be time-space relations "the question could arise how much uniformity they must possess". Here the author draws the attention to the fact that Einstein's theory preserves a certain uniformity of these relations in the invariant squared differential form of the time-space intervals. "Would it not be perfectly compatible with the epistemological argument to say that only this amount of uniformity was necessary, and that the particular coefficients of these differentials are contingent and variable?"

We will add ourselves that, in order to apply induction to the usual meaning of this word, it is necessary to distinguish the temporal coordinates from the spatial ones, and this is not expressed by the squared differential form itself, when written in the most general form. In areas where our inductive investigations are really carried out, there occur indeed always, at least in approximation, the conditions of classical mechanics where the invariant referred to before breaks down into two independent invariants, the space invariant separately and the temporal invariant separately i.e. everything remains unchanged.

In connection with the relation of the principle of causality and the theory of relativity, contemporary scientific literature broaches one more group of problems, which we have avoided thus far, viz. the problem of the complete elimination of the idea of causal relation to the advantage of the functional relation. It is known that recently, quite independently of Mach, the idea of causality was criticized by Bertrand Russell, who arrived at results partly consistent with those of Mach. Russell's criticism is undoubtedly more profound than that of Mach, though one cannot admit that Russell is right in every case. Firstly, Russell is wrong in asserting that only philosophers discuss causality. Indeed, the authors referred to before, who are concerned with the principle of causality, such as Maxwell, Einstein, as well as Laue and Minkowski are physicists of high quality, and one can add to this group mathematicians dealing with the foundations of physics, such as Hilbert and Weyl, who are also in a position to discuss the principle of causality. Yet, in the case of the above-mentioned authors, one cannot find in their works any trace of mystical ideas of action, or of their ideas being governed by analogy with a human arbitrary action, or by any remains of anthropomorphism.

Undoubtedly, one can so widen the idea of functionalism that it will include both the reversible relations and the irreversible, unidirectional ones.[25] But in such a case it will be found that the entire issue in the dispute consists of words, because within the so-called functionalism one has to distinguish between causal relations which take place unidirectionally from functional relations in which the temporal order is absolutely irrelevant. This is the meaning which was combined with these words by the authors who rose up to speak about the relation between the principle of causality and the theory of relativity. The construction of the light cone, introduced by Minkowski, aims at separating precisely such world lines in which the temporal order cannot be reversed from those where this order can be reversed. This is what Hilbert had in mind, when stating the condition which must be satisfied by the components of the gravitational potential if a certain constant temporal order of phenomena is to be preserved under the presuppositions of the general theory of relativity.

The shrinking of the segments in the direction of the motion and the lengthening of time, which take place in the systems for which the Lorentz transformation formulae are valid, belong to functional changes which are mutually attributed to themselves by the observers of both systems. On the other hand, the defenders of causalism reject Einstein's interpretation and adopt the original explanation of Lorentz himself, according to which these changes are not mutual, but take place only in one system, viz. only in that one which moves with respect to the stationary ether. Even today one cannot consider this stand to be completely surmounted; the Danish scientist Helge Holst recently defended it. Lorentz's contractions are, for him, not the result of relative motions of the systems themselves, regardless of their magnitude, but the result of motion with respect to the neutral homogeneous gravitational field related to the system of fixed stars. The gravitational fields of the system of fixed stars go to make in our region of the world a certain 'homogeneous neutral field' upon which, just as for Mach, the smaller neighboring masses exert no perceptible influence. It is precisely the movement relative to this neutral gravitational field that is the cause of the Lorentz contractions, just as, in the general theory of relativity, the space-time deformations have their causes in real gravitational fields. This is undoubtedly some correction of the Lorentz theory according to the spirit of Mach, since one speaks here not of a movement with respect to the absolutely stationary ether, but of one with respect to the system of fixed stars. In this way Helge Holst tries to occupy an intermediate position between the purely phenomenological descriptive physics and the previous explanatory mechanistic physics which manipulated the idea of a rigid elastic ether and adopts as the device the 'causal description' (*kausale Beschreibung*) instead of the 'causal explanation' or a pure phenomenological description. Similarly the German physicist Laue, who continually voiced the superiority of Einstein's theoretical interpretation with respect to Lorentz's interpretation, does not however exclude the possibility of a causal explanation of the contraction.[26] After all one ought not to identify the contrasts of causalism and functionalism with the contrast between realistic physics and purely phenomenological physics. This is demonstrated also by the example of Russell whose anti-causalism – as it appears – does not arise from any phenomenological tendencies but solely from the effort of reducing all scientific ideas of natural science to mathematical-logical ideas. This commendable tendency, we think, can undoubtedly be satisfied without eliminating the idea of causal relation completely.

In the case of relative movements we are, as a rule, not inclined to treat them as determining factors along with other ones – we think, quite wrongly. Indeed, there takes place in them a change in the relation of bodies, which must bring about other changes. Now if, in accordance with Einstein, we treat Lorentz contractions as only functional ones, this is only because the other observer does not see them and mutually attributes identical changes to the former observer. Now this is not the case with rotary movements. Here both observers agree in observing the occurrence of centrifugal symptoms on one of the bodies only, and can at most give them different names. Hence there is no reason for hesitation here when looking for the real cause of centrifugal accelerations in the relative movement.

There arises here another question. If I am at liberty to state that the centrifugal symptoms occur on earth because the masses of the entire world rotate round the earth, why do not they appear on the rotating masses of the universe? The only possible answer is that the former understanding of centrifugal forces must be rejected from the standpoint of the new theory. True, not all of the adherents of the theory of relativity are able to give a concordant reply to this question. It seems that the best answer was provided by Thirring and Petzoldt. Only great masses can, owing to their movement, induce by their movement, in a visible manner, the gravitational field of small masses, but not the other way round. Therefore, the earth mass, vanishingly small compared with the gigantic masses of the universe, cannot induce in them by its movement such centrifugal symptoms as those brought about by them in it. If somebody would like to interpret these processes purely functionally but not causally, then, we think, nothing would change in principle in Einstein's reasonings. Instead of the postulate advanced by the principle of causality, viz. 'there exists no change which does not have a cause in other preceding changes', one ought to advance only an analogous more general principle for the functional relation, viz. 'there are no changes that could not be connected functionally' with some other perceptible changes.

### 3.2. OBSERVATIONS ON THE METHOD OF NATURAL SCIENCES

The method of testing hypotheses in natural sciences seems to be a matter of greatest importance for a philosopher interested in the method of such sciences. A hypothesis is either a general proposition or a singular statement dealing with some elements of reality, which are inaccessible to direct observation. When provisionally assuming a hypothesis we draw concrete conclusions from it and try to check whether these conclusions do agree with the relevant facts. If one finds an agreement between the predictions of the hypothesis and the facts observed directly, the hypothesis becomes verified and obtains the freedom of science. On the other hand, if disagreement is found, the hypothesis is overthrown and rejected.

It has been stressed recently quite often that the negative result of the testing of a hypothesis is more important than the positive one. The verification of a hypothesis does not prove its truth; it can be altered at any time when the progress of science requires this. On the other hand, a negative result is of a decisive nature – the hypothesis then becomes completely overthrown. The rejection of the hypothesis corresponds with the law of logic (from deduction theory) known as *modus tollendo tollens:* If the hypothesis $H$ implies the conclusion $p$ and if conclusion $p$ is false, then hypothesis $H$ is likewise false.

$$[(H \supset p).\sim p] \supset \sim H$$

Thus asymmetry exists between the positive and the negative results of the testing of the hypothesis. The positive result merely increases the probability of the hypothesis whereas a negative result completely refutes it.

Nevertheless some reservations are indispensable here. The total refutation of the hypothesis takes place only when no change had been introduced into the ideas entering into the composition of the hypothesis and playing a role in the description of the experiment. The point is that, if one changes, even to a very small degree, the meaning of the terms applied, the refuted hypothesis can be taken up again. Thus, for example, the corpuscular theory of light, which was rejected after Foucault's experiments, was again reintroduced into physics owing to the quantum theory when, along with various changes in the theory of light, the meaning of the word 'corpuscle' was changed. The same applies to the case of *experimentum crucis*, and *instantia crucis*, i.e. to the case when, out of the two competing hypotheses or theories equally suitable for being adopted, the one selected is that which predicts a newly discovered fact. Out of the two hypotheses equally suitable for being adopted on the basis of the same phenomenon discovered, one hypothesis will be verified, while the other one will be falsified. There exists an analogy between the testing of a singular hypothesis and the solution of a conflict between two competing hypotheses. In either case we have to choose one of the several possibilities which form an alternative. In the latter case the alternative is only more complex. In the former case one expects to obtain an affirmative or negative answer to one question only. In the latter case a positive answer with regard to one hypothesis is, at the same time, a negative answer with regard to the other hypothesis. Nevertheless, one can easily note an analogy between the two cases and notice the difference between them, if one becomes aware of the content and the complexity of the hypothesis. Let us assume that the hypothesis is a logical conjunction of three propositions 'p', 'q', 'r' and that it is rejected in view of the falsity of its consequences. One obtains a negation of this conjunction. However, this negation '$\sim(p,q,r)$' can be transformed in accordance with the laws of logic in such a way as to yield a seven-member alternative:

$$\sim(p,q,r)=[(\sim p.q.r) \vee (p.\sim q.r) \vee (p.q.\sim r) \vee (\sim p.\sim q.r) \vee (\sim p.q.\sim r) \vee (p.\sim q.\sim r) \vee (\sim p.\sim q.\sim r)]$$

Having rejected the conjunction 'p.q.r' we face the necessity of choosing one of the seven remaining possibilities. At the same time, the more complex the hypothesis the more possibilities we would have at our disposal. It is only in the cases of very simple hypotheses that the rejection of one hypothesis is equivalent to the verification of only one other hypothesis. However, such cases are rare. Even the problem of the world being finite or infinite is not such a case, since the meaning of the term 'finite' is often confused with the meaning of the term 'limited' and, therefore, owing to this reason only, one gets, instead of the simple two-member alternative, a four-member alternative: $(p.q) \vee ({\sim}p.q) \vee (p.{\sim}q) \vee ({\sim}p.{\sim}q)$ (let *p* denote the statement that the world is finite, and *q* that it is limited). The physical sense of the terms used, it is true, reduces the alternative to a certain extent, since the case of the world being infinite and limited is physically impossible, though logically admissible. Thus the member '$({\sim}p.q)$' is no longer taken into consideration, and yet, while negating the possibility of the world being 'finite and bounded', we obtain the alternative that it can be finite and unbounded (a case omitted in classical cosmology) or infinite and unbounded. In the case of the so-called *experimentum crucis* it occurs infrequently that the rival hypotheses are exactly contradictory statements. That is why one should pay attention rather to the complexity of hypotheses than to their number, which often depends on accidental historical conditions.

Duhem had reservations with respect to the possibility of the existence, in physics, of a case of *experimentum crucis*. He maintained that "a physical experiment cannot at all refute one singular, isolated hypothesis, but only a complete set of hypotheses, which forms the entire theory of the phenomenon" and therefore "the experimentum crucis is impossible in physics". This reservation is absolutely right. The formula

$$[(H \supset p){\sim}p] \supset {\sim}H$$

can be used, for all intents and purposes, only in very rare cases when the theory almost does not exist at all. The conclusion *p* is more often deduced not exclusively from one hypothesis 'H', but also from the entire set of statements '$T_1$', '$T_2$', '$T_3$' . . . , '$T_n$', which form the theory to which we add also the hypothesis 'H'. Hence the schema of reasoning should be the statement:

$$[((T_1.T_2.T_3 \ldots T_n.H) \supset p).\quad p] \supset (T_1.T_2.T_3 \ldots T_n.H)$$

whence we get the statement

$$[((T_1.T_2.T_3 \ldots T_n.H) \supset p).{\sim}p] \quad T_1 \vee {\sim}T_2 \vee {\sim}T_3 \vee \ldots \vee .{\sim}T_n \vee {\sim}H$$

The negative result of the experiment proves only that the conjunction '$T_1.T_2.T_3 \ldots T_n \ldots H$' is false. Hence either something must be changed in this conjunction, or the hypothesis H or one of the laws thus far accepted, should be rejected. Had we taken into consideration the degree of complexity of the hypothesis H itself, the number of possibilities would increase even more.

Nevertheless, much as Duhem is right in theory, the situation is not as hopeless in practice. We always choose the simplest road. Theory is not only a set of experimental laws; it is a set containing also many conventions and definitions. Sometimes it suffices to alter the definitions only. In other cases the situations are more complex and can bring about a revolution in science. Such cases occur when it is difficult to decide which road is really the simplest one. Such a situation existed during the period when the theory of relativity came into being and also when the quantum theory originated. The classical laws were found to be at variance with experiment. Therefore one could either doubt classical physics in its entirety or supplement it with some auxiliary hypotheses. At first, it was the latter choice that seemed to be the simpler one. However, as the number of negative experiments increased continuously and as, in this connection, more and more mutually independent auxiliary hypotheses appeared, the thought of reconstructing the edifice of classical physics was found to be the simplest way out from the troublesome situation.

On the other hand, the history of quantum physics represents something not met with so far, viz. that two originally mutually inconsistent theories may be complementary and capable of co-existence. A few years ago we suggested in this connection the use of three-valued logic in quantum physics. However, prior to doing this it was necessary to examine thoroughly whether the ideas of wave and corpuscle are really contradictory. It was found that they are not. The wave and the corpuscular theories are, to a certain extent, mutually inconsistent, but one of them is not a negation of the other. That is why one has rather to adopt a wait-and-see attitude than to find a more specific answer, viz. that the $h$ constant and, consequently, the law of Heisenberg create an ultimate limit for the testing of the nature of the world microstructure.

CHAPTER 4

# TIME: SELECTIONS FROM *THE EVOLUTION OF THE NOTION OF TIME**

## Part One

## INTRODUCTION: A BRIEF SURVEY OF THE TIME THEORIES

1. When investigating the development of theories of time in chronological order, we are exposed to the danger of being completely lost and disoriented, in a mass of the most varied opinions, which are sometimes highly capricious. For this reason we will preface our historical review with a general overview of the literature, which allows us to classify these theories from the logical point of view. But what shall we adopt as a principle for this classification? Evidently, we should use, as the distinctive features of the systems, those characteristics of the concept of time which have played the most important role in the history of this problem. There is no doubt that all of our systems are grouped around two fundamental problems: the reality of time and the origin of the idea of time.

It is very probable that these problems have been badly posed, and it is certain their solutions do not exhaust all specific features of time theories. Nonetheless, our classification will be of use to us, and we do not believe that it is entirely artificial.

From the point of view of the problem of time's reality, we can divide all theories of time into those idealistic and those realistic. According to the idealistic theories, time is only an appearance, as in the case of the Eleatics, or an *a priori* form of our sensibility, of our inner sense, as in the case of Kant and the German idealists, and the being of the thing-in-itself has nothing in common with this temporal form. In the realistic theories, time has a reality that is independent of our senses and our intellect. It is something not less real than our own lives. However, we should distinguish between two groups of realistic theories – extreme or exaggerated realism, and moderate realism.

2. The extreme realism, represented in history by certain neo-Platonists and by Newton, conceives of time as a reality which is independent of other objects, it is a reality which flows 'in itself and according to



**L'Évolution de la Notion du Temps*, Cracovie 1936, Librairie Gebethner et Wolff.

its own nature', without any relation to a real object of some kind, and which would indeed continue to flow even if the world did not exist. In opposition to this extreme concept of time as substance, moderate realism postulates another conception of time as a derived reality, dependent upon and connected with the reality of other objects, *viz.* the events which come to pass in time. Time is merely the measure of movement or 'the number of movement' as Aristotle put it; hence this time would itself cease to exist if there occurred no movement and no change.

This moderate realism can assume different forms. Here we might introduce the idea of 'accidental time', named thus in opposition to 'substantial time'. However, this term does not appear to us to have a very precise meaning, since time is not a property of objects, the way a colour is: we can therefore say, as Leibniz did, that time is a relation, an order of succession. This definition is no less imperfect than the first, being almost a tautology, but it is also an expression of moderate realism, since relation cannot exist without the elements to which it refers.

Here belongs the nominalist concept of time, according to which the word 'time' is merely an apparent name,[27] since it does not designate any real thing. We might likewise say that it is an abstract word which serves to name 'a class of concrete durations' (the view of Jean Nicod and the nominalist mathematicians). We find this opinion very congenial: 'it is the only means of risking nothing'.

The moderate realist may indeed allow that time does not exist. But this thesis does not have the same significance for the moderate realist that it had for Kant or the Eleatics. Time does not exist because there is no concrete, individual object which bears that name; it is not a separate reality as it was for Newton. Yet events exist, objects last over time, and these things would last and events would follow one another even if our consciousness did not exist. On the other hand, according to Kant, the temporal aspect of the world results from the application of our human form of intuition, and the world in itself 'is neither variable nor invariable'.[28]

The difference between these three groups of systems is very clear. Kant affirmed that, were our inner sense not endowed with its *a priori* form, the world would not have a temporal aspect. On the other hand, the moderate realist affirms that our consciousness does not add anything to the reality of time taken in its objective sense. It is only in the case where all motion and all change disappeared that the world could no longer possess its temporal aspect. Now an extreme realist, such as Newton, ventures still further when he affirms that movement adds nothing to the reality of time and that time would exist even if the world did not. God's existence alone suffices for the existence of time.

To our mind, Newton's opinion is that of common sense. However, most philosophers have opted for moderate realism concerning time. This was the opinion of Aristotle, of St. Thomas Aquinas, and it is that of relativistic physics. Bergson's theory is also that of moderate realism. But in order to better comprehend this moderate realism, we must now consider our second principle of classification: the origin of the idea of time.

3. This second principle can be considered as wholly independent of the first one. We should apply it to each of the three aforementioned theories. But the object here is to present a quick overview of all of the historical systems and, as the idealistic and ultra-realistic systems appear very rarely in the history of time theories, we shall content ourselves with the application of our second point of view to the group of theories which come close to that of moderate realism.

Whence comes our idea of time then? Doubtless, the faculties of our mind must take part in its formation, as is the case with any other idea. But where do we find the original, temporal data? Is the life of the mind their unique source? Is the physical world their source, or is it the combination of both of these together? In accordance with our answer to this question, theories of moderate realism can take on either the form of a mainly psychological doctrine, or that of a mainly naturalistic one, or else the form of a universal doctrine in which the two sources complement each other. The relativistic theory of time is, to be sure, a naturalistic one, but so too is the theory of Aristotle. However, it must be admitted that, in the hands of neo-scholastics, the theory became a universal one. The only objection which can be raised here is that this theory unduly emphasized the quantitative characteristic in its definition of time. On the other hand, the psychological theories, such as that of Bergson, stress too exclusively the qualitative properties of time, unmindful of its quantitative or metric properties.

The controversy between these two orientations has continued throughout the entire history of philosophy. The Platonists, St. Augustine, and Locke mark the stages of an evolutionary line which ends with Bergson. Aristotle and St. Thomas lead us toward the orientation of modern physics. Need we add that these two trends must be reconciled since time reveals qualitative properties as well as quantitative or metrical properties?

Each of these two tendencies in moderate realism can degenerate into an absolute realism of 'substantialized time'. Newton's theory is the product of the degeneration of the psychologistic line, Proclus and his theory of divinized time are tied to the naturalistic path. The idealism of the Eleatics and of Kant also appears to us to be an extravagance of human thought.

In addition to these three groups of systems we encounter several mixed or poorly specified theories in the history of philosophy. For example, it is very difficult to classify Descartes' point of view; his definition of time appears to be idealistic, but his attitude is entirely realistic. Despite these difficulties, we are convinced that our classification affords an adequate orientation within the multitude of the systems. The classification approximates that of Désiré Nys as expressed in his work, *La Notion du Temps*, with one small but significant difference. Nys's classification likewise leads toward the three groups of systems, but the two principles of division – the principle of the origin of the idea of time and that of its objectivity – are intertwined and confused in his schema. As a result, the group of subjectivist and idealist theories is opposed to the groups of extreme and moderate realism in such wise that next to Kant, we find Bergson, and even Locke! In our classification, the psychological theories of time belong to moderate realism and have nothing in common with Kant's idealism.

A classification which is altogether identical with ours can be found in the book of a Polish philosopher M. Wartenberg[29] *Le problème du temps*, in which our three groups figure as: the idealistic point of view, that of extreme or 'substantial' realism, and that of moderate, accidental realism.

4.1

# ANTIQUITY

## DEFINITION OF TIME ACCORDING TO ARCHYTAS OF TARENTUM, THE PYTHAGOREAN

4. One of Antiquity's best elaborated theories of time, is, to be sure, the theory of Aristotle, which we find in Book 4 of the *Physics* (chapters 10–14). Before Aristotle, we must mention a definition of time attributed to the Pythagorean, Archytas of Tarentum, and also everything Plato writes concerning time, especially in his dialogue on nature, the *Timaeus*. After Aristotle, we find a highly interesting chapter 'On eternity and time' in Book 3 of Plotinus's *Enneads*. Finally, the *Commentaries* of Simplicius on Aristotle's *Physics* and his categories represent a very precious source of information concerning the opinions about time of the Greek philosophers, starting with the Pythagorean Archytas of Tarentum and ending with neo-Platonists of the sixth century A.D. These *Commentaries* were very well analysed by Duhem in his paper, 'Le temps selon les philosophes hellènes'[30], of which we shall avail ourselves in turn.

Archytas of Tarentum defined time by the following formula: 'Time is the number of a certain motion, or else it is, generally speaking, the interval peculiar to the nature of the universe'.[31]

The beginning of this definition reminds us of Aristotle's definition, but it is less clear than the latter. When taken separately, this definition might appear to be quite obscure and ambiguous; fortunately however, Simplicius, who had before him the treatise of Archytas, added to it commentaries capable of casting light on it. "All of the movements that are produced in the world possess a primary cause, a prime-mover; in Aristotle's philosophy this prime-mover insofar as it moves itself: this is the soul of the world. The *Commentaries* of Simplicius assume implicitly the accord, on this point, of Plato's metaphysics with that of Archytas; they assume, in the doctrine of the Pythagorean of Tarentum, the existence of a mobile soul, principle of all the successive movements in the universe".

In these words, Duhem expresses well the interpretation of Simplicius, but this interpretation is none other than the commentator's personal supposition. Here is how Simplicius tries to establish it:

Aristotle declared that time is the number of motion; on the other hand, among the Stoics, Zeno said that time was simply the interval of each movement, while Chrysippos affirmed that it was the interval of the motion of the universe. Archytas . . . does not say that time is the number of each movement, according to the definition of Aristotle, but that it is the number of a particular movement . . . Our author seems to designate by these words the substantial motion of the world soul . . . this unique motion is that particular motion about which he affirms a relation to time.

Such a justification seems to us very weak and unreliable.

5. Duhem says that the theory of Archytas of Tarentum is a theory of absolute time. However this is not demonstrated by the fragments of Archytas, as Simplicius reproduced them. Here is what Archytas, himself, says,

Time is continuous, and not at all formed of distinct elements, like number, discourse and melody . . . it is not a subsistent nature (*physis*), since it is time; neither is it movement because the present does not come; the present has always been, and always will be; the present will never be absent as, forever new, it is ceaselessly engendered and though numerically different, it is specifically the same . . .; in time, the parts that have already been generated are annihilated, and those that will be generated will be annihilated. This is why time either has absolutely no existence at all, or else its existence is extenuated and difficult to understand.

We can see (from his remarks) that Archytas, himself, is hardly resolute in his opinions. The last sentence quoted above can be found again in Aristotle; viz., when he points out the unintelligibility of the problem about to be tackled. This proves that the doctrine of Archytas was known to Aristotle and that it probably influenced his own doctrine of time.

'The interval peculiar to the nature of the universe', which we find in the definition of Archytas, is interpreted by Duhem as that period of nature, at the end of which all of nature would again find itself identical to that which it was at the beginning. Since the Pythagoreans believed in the 'eternal return' of the universe, it is probable that the interval in this definition is linked to the great year adopted as the natural unit of time.

## NEGATION OF BECOMING AND OF TIME ACCORDING TO THE ELEATICS

6. The Eleatic school taught that there exists but one substance, eternal and immutable, whose illusory appearances are finite, perishable and movable things. The senses can be duped by these appearances; however, reason discerns their nothingness. The negation of the reality of time emerge from all these assertions as a necessary consequence. Now, it

seems that Xenophanes did not go as far as denying all change and becoming in the universe. But Parmenides was more audacious. He denied every becoming and every multiplicity, and left but one unique being. Zeno tried to confirm the thesis of Parmenides by reducing *ad absurdum* the assertions of common sense. He became famous as the author of the four paradoxes, arguments against the existence of motion. Of these the best known is the paradox concerning Achilles and the tortoise. Achilles, who runs twenty times faster than the tortoise, will never overtake the latter.[32]

Imagine a tortoise which is twenty paces in front of Achilles, that fleet-footed hero, and assume that Achilles progresses twenty paces during the time interval required by the tortoise to advance one pace. Now when Achilles has completed his twenty paces, the tortoise is ahead of him by one pace. This one pace can be divided into twenty parts. In the course of time, when Achilles progresses by these twenty parts, the tortoise advances by one-twentieth of the next pace. This one-twentieth part which, for Achilles, is the first part of the twenty-second pace, can be divided into twenty new, equal parts. During the time Achilles needs to pass them, the tortoise goes over the twentieth part of the second part of the twenty-second pace, and so forth without end. Consequently, if we assume the reality of movement, with the infinite divisibility which it implies, we find that Achilles, who runs twenty times as fast as the tortoise, will never overtake it: which is absurd.

The paradox relating to the arrow can be summarized as follows: at any moment of time, a flying arrow occupies a space equal to itself. Thus, at each moment of its flight, it is at rest. Now, if we suppose that time is composed of indivisible moments, motion is impossible.

The third paradox is that of the runner (or that of dichotomy). To pass from one point to another, a body must first cover one-half of the space between the points; but first it would have to traverse one-half of that one-half, and so forth. Thus, in order to pass from one point to another, a body would have to traverse an infinite number of spaces, but infinity cannot be travelled in a given time.

The last paradox is that of the stadium.

If two runners move at the same speed in opposite directions and meet while passing in front of a stationary object, they will move, each with respect to the other, twice as fast as they do with respect to the stationary object. Now, if we assume that bodies are composed of points and that the interval between one point and another will be traversed in an indivisible moment, then it follows that, for one runner, the moment required to pass from one point at the stationary body to the next point will amount to one-half of the moment required for passing from a point of the other runner to the next point.[33]

The alleged consequence – according to Aristotle – is that 'half the time is as long as the whole time'.[34]

7. We shall be satisfied with a brief remark. If we suppose that all of the above arguments are correct (the last one is not) still it does not follow at all that motion and becoming are impossible or unreal; it only follows that our ways of accounting for them are wrong. This consequence was indicated by modern criticism, while Aristotle's criticism is not always satisfactory. In any case, the school of Elea exerted a tremendous effect not only on the development of philosophy but also, as it is supposed nowadays, on the development of mathematical sciences. If the Greek mathematicians avoided the idea of infinity, this was probably because they were intimidated by the Eleatics' criticism. This influence was bad; it hindered the development of infinitesimal calculus. Greek mathematicians were not very far from the discovery of the calculus; what they did lack was the general method into which it was necessary to introduce the idea of infinity. Instead of doing this, as in the method of limits, they hid the idea, making use – in particular cases – of an indirect proof.

## PLATO. TIME: MOVING IMAGE OF ETERNITY

8. One cannot say that Plato created a theory of time, but everything he says in this connection is of the greatest interest. It is the description of the creation of the universe, in the *Timaeus*, that gave him occasion to speak on this subject.

When the Creator of things had formed the soul of the universe to his liking, he placed in it the body of the universe and joined the two by attaching their centres to one another . . .

And when the construction of the Soul had all been completed to the satisfaction of its Constructor, then He fabricated within it the Corporeal, and uniting them centre to centre He made them fit together.[35]

And when the Father that engendered it perceived it in motion and living this image of the immortal gods and alive, He too rejoiced; and being well-pleased He designed to make it resemble its Model still more closely. Accordingly, seeing that the Model is an eternal Living Creature, He set about making this Universe, so far as He could, of a like kind. But inasmuch as the nature of the Living Creature was eternal, this quality it was impossible to attach in its entirety to what is generated; wherefore he planned to make a movable image of Eternity, and, as He set in order the Heaven, of that Eternity which abides in unity. He made an eternal image, moving according to number, which we have named Time. For simultaneously with the construction of the Heaven He contrived the production of days and nights and months and years, which existed not before the Heaven came into being.

And these are all parts of Time; even as 'Was' and 'Shall be' are generated forms of Time, although we apply them wrongly, without noticing, to Eternal Being. For we say that it 'is' or 'was' or 'will be', whereas, in truth of speech, 'is' alone is the appropriate term; 'was' and 'will be', on the other hand, are terms properly applicable to the generation Becoming which flows in time, since both of these are motions; but it belongs not to that which is ever changeless in its uniformity to become either older or younger through time, nor ever to have become so, nor to be so now, nor to be about to be hereafter, nor in general to be subject to any of the conditions which Becoming has attached to the things which move in the world of Sense, these being generated forms of Time, which imitates Eternity and circles round according to number.[36]

Time, then, came into existence along with the Heavens, to the end that, having been generated together, they might also be dissolved together, if ever a dissolution of them should take place; and it was made after the pattern of the Eternal Nature, to the end that it might be as like thereto as possible.[37]

9. This passage invites us to reflection. If eternity is the model of time, and time is only its image, a shadow of the perfect reality, what is the cause of its mobility? The answer can be found in the same text. It comes from the motion of the soul of the universe. God saw the soul of the world moving and, in order to give more perfection to this universe, He created time. It follows from this answer that there exists, or has existed, a mobility outside of time, a motion which is non-temporal, and which occasioned the creation of time. This answer prompts us, however, to pose another question. What perfection has the universe gained by the creation of time? In this case, the answer is more difficult, but it seems to us that it can likewise be found in the text. Thanks to time, motion became numerable and measurable; this image of eternity 'advances according to number'.

With time the Heavens and the stars had to be created, and it is because of them that time became a number. Plato repeats several times, that stars are 'necessary for the existence of time'. Their motion gives to numeric expression its regularity and power, and these are the characteristic features of time.

Is this an absolute time that enjoys the privilege of a separate existence? The answer is sooner 'no' than 'yes'. Although our reply is *no*, we do not wish to pass over in silence the fact that this metaphysics invites us to absolute realism vis-à-vis time. Every metaphysics is inclined to grant reality to all objects. Where there are two words, we can very easily come to believe in the reality of two entities. It is for this reason that Platonism led ultimately to the Neoplatonists' absolutism about time. In Plato, however, we find a moderate realism in this regard. It sufficed to remove from it all its poetry to reach the time theory of Aristotle.

Plato speaks also of that perfect unity of time which Duhem saw in the definition of Archytas of Tarentum.

Nevertheless, it is still quite possible to perceive that the complete number of Time fulfilled in the Complete Year when all eight circuits,[38] with their speeds, the revolution of the Same and Similarly-moving. In this way and for these reasons were generated all those stars which turn themselves about as they travel through Heaven, to the end that this Universe might be as similar as possible to the perfect and intelligible Living Creature in respect of its imitation of the Eternal Nature thereof.[39]

To be sure, Plato's theory was conceived under the influence of the Pythagoreans; the fact that the title of the dialogue dedicated to the philosophy of nature bears the name of a Pythagorean proves that Plato was aware of this dependence, or even of his own indebtedness to these philosophers. The passage quoted above leads us to believe that Plato perhaps saw the imitation of eternity in nature's capacity to revert to its primordial state. The returning of all the planets to their starting point was, for the Pythagoreans, the sign of the return of the entire universe to a new life.

## ARISTOTLE. TIME IS THE NUMBER OF MOTION ACCORDING TO THE PRIOR AND THE POSTERIOR

10. Aristotle's theory of time likewise carries the profound trace of the influence exerted by the doctrine of the Pythagorean of Tarentum, but we see herein a greater perfection. First, Aristotle raises the difficulties concerning the existence of time, as occasioned by its division into the present, the past and the future, and by the relation between moment and continuity. Indeed, its parts: past and future, do not exist. Even the instant which, moreover, is not a part of time, appears incapable of existence. But the most important thing is the relation we establish between motion and time. Time is not motion since motion is in the changing, the mobile, and also in a place, while

time is current alike and is in relation with everything. And further, all changes may be faster or slower, but not so is time; for fast and slow are defined by time, 'faster' being more change in less time, and 'slower' less in more. But time cannot measure time thus, as though it were a distance or a qualitative modification.[40]

However, if time is not motion, it does not exist without motion. It is inseparable from motion since, to perceive time, we have to perceive change. Therefore, time is 'something of motion'. That is why the analysis of motion becomes for Aristotle the starting point of his research into the essence of time. Motion is continuous since it obeys the category of quantity, and all quantity is continuous. Now, 'it is because magnitude is continuous that motion is so also, and because movement is continuous so is time'.[41] Aristotle affirms what is doubtless an astonishing thing that both beforeness and afterness are also originally in space and, consequently, in magnitude.

But, since there is before-and-after in magnitude, there must also be a before-and-after in movement in analogy with them. But there is also a before-and-after in time, in virtue of the dependence of time upon motion. Movement, then, is the objective seat of before-and-afterness both in movement and in time; but conceptually the before-and-afterness is distinguishable from movement. Now, when we determine a movement by defining the first and last limit, we also recognize a lapse of time: for it is when we are aware of the measuring of movement by a prior and posterior limit that we may say time has passed.

And our determination consists in distinguishing between the initial limit and the final one, and seeing that what lies between them is distinct from both; for when we distinguish between the extremes and what is between them, and the mind pronounces the 'nows' to be two – an initial and a final one – it is then that we say that a certain time has passed.

This is just what time is, the calculable measure or dimension of motion with respect to before-and-afterness.[42]

To this definition Aristotle adds an explanation. One should distinguish between 'the numbers that are counted in the thing in question, and the numbers by which we count them and in which we calculate. Now, time is the countable thing that we are counting, not the numbers we count in.'[43]

11. Further, Aristotle examines the nature of the instant and its relation to time. 'Time taken as an entire piece is the same everywhere; it retains its identity in one sense, but is continually changing in another.'[44] In this respect Aristotle compares the instant with a moving object. This object is also in one sense the same, and in another it is not the same. Coriscus in the Lyceum may be considered (the Sophists did this often) to be different from Coriscus in the Agora, since he is, at one time, here and at another time, there.

'Now' measures time insofar as beforeness and afterness; that is why one can say that 'neither would time be if there were no 'now', nor would 'now' be if there were no time'.[45] But 'time owes its continuity to the 'now', and yet is divided by reference to it'.[46] This property corresponds also in a certain way to that of the point. And there is an analogy between a body that is being moved and a point; 'for it is a point that both constitutes [by its movement] the continuity of the line it traces and also marks the end of the line that is behind, and the beginning of the line in front.' One ought not to say that 'now' is a portion of time. Insomuch as it is a limit, 'now' that is, the instant, is not time.

In another chapter, Aristotle adds further remarks concerning the role of 'now'.

The present instant is the instant which produces the continuity of time; it joins indeed past time with future time, being the beginning of the latter and the end of the former. Obviously this does not take place in the way that as it does in the case of a permanent point; present produces a division *in potentia*. Insofar as such a present does exist, the present is unceasingly different; insofar as it establishes the continuity of time it is always the same.[47]

Let us briefly summarize once again everything Aristotle said concerning the instant. The instant is in a sense the same, in a sense always different, it 'measures time', it divides time and makes it continuous, it is not a portion of time, it is a division of time *in potentia* and not *in actu*, it limits and unifies the two parts of time.

12. After having established the definition of time and explained the role of the instant, Aristotle considers several consequences of this definition. Here are the most important consequences. Movement and time are so linked to one another that they determine one another reciprocally. Consequently, 'not only do we measure the length of uniform movement by time, but also the length of time by uniform movement.'[48] This statement of Aristotle, as we will see, already provoked many objections in Antiquity.

'Time measures both the movement and its duration by the same act, and its duration being so measured constitutes it as existing in time.'[49] But 'to be in existence while time is in existence does not constitute being 'in time', just as neither is a thing constituted as in movement or in a place because movement and place exist while it does.[50] It seems to us that this sentence best expresses Aristotle's idea, viz. that time does not possess any separate existence. Aristotle states that we attribute to time a certain influence upon objects; it is our custom to say that time consumes, that everything grows old under the effect of time. However, it is more movement than time which brings about destruction.

'. . . things which exist eternally as such are not in time, nor is their duration measured by time'.[25] Neither has time any hold on them, nor is their duration measured by time. On the contrary, it is rest that can be measured by time, hence rest is within time. The immobility of the eternal beings and rest are two things different from one another. At rest is only that 'which, by its nature, is capable of moving but now lacks its actual movement.'[51]

Aristotle likewise denies any existence in time to ideal objects which form the field of interest of mathematicians, such as, for example, the commensurability of the diagonal and the side (l.c., p. 407). However, whatever is destructible or generable, or (more broadly) 'sometimes existing and sometimes not, must be embraced by time; for there must be some time which measures their being . . .' 'Those things which are not in any way embraced by time neither were nor are nor will be.'[52]

Time is infinite. There can be no exhaustion of time since motion always exists. '. . . since the 'now' is the end and the beginning of time, but not of the same time, but the end of the time past and the beginning of the time to come . . . so time will be always engaged in beginning and in ending.'[53]

13. The last chapter is devoted to the solution of some difficulties. The most important are the three following ones: the relation of time to the human soul, the unity of time, and its relation to the movement of the celestial sphere. As to the first difficulty, Aristotle's response is a bit astonishing: 'there can be no time without soul since, by nature, nothing but the soul can count.' But, at the same time, Aristotle adds that motion can exist without a soul, and that beforeness and afterness do exist in motion.

The unity of time appears to be threatened by the fact that more than one motion occurs at a time. Now, each one of these ought to have its own number. Are there consequently, several times? No, time is everywhere the same, because number is equal and simultaneous.

However, not all movement is uniform. 'Neither change, nor increase, nor generation is uniform, but only transportation, local motion.' And amongst these motions, it is the rotary motion of the celestial sphere that is foremost and 'what is first is a measure for what is of its kind.' It is also the rotary motion of the heavens that is best known. 'That is why time appears to be the motion of the sphere, because it is this motion that measures other motions and also measures time.'

14. The theory of Aristotle has been criticized for various reasons. Most, though not all of these objections are justified. We ask why Aristotle linked time to physical motions, but not to the motions of the soul. In principle, however, Aristotle does not exclude from any motion its relation to time. The term 'motion' signifies for him a change of any kind. He distinguishes between three kinds of motion (*kinesis*): qualitative change (*alloiōsis*; alteration), increase and diminution (*auxesis*, *phthisis*), and transportation, or local motion (*phora*). True, the latter lends itself best to the measuring of time, but this priority of local motion does not appear in his definition of time. The measurement of time

requires a continuous, uniform, uninterrupted motion, comprising all the others and being the best known of them all – one can concede him this. But why do the metric properties appear in the foreground of the definition? Does not time possess more important properties, of which the metric is but one of the consequences? The order of succession, the before and the after, cannot be deduced from the metric, and one need not analyse local motion to discover it.

Now, Aristotle's greatest error consists in his statement that it is space which conveys to time these two characteristics: the beforeness-and-afterness relation and continuity. As to the former, this is an unpardonable error. The beforeness and the afterness, applied to the points on a line, have no literal sense. When applied to magnitude, they simply signify: before and behind. As regards the second feature, continuity, it is at least an arguable thing. The continuity of our mental life appears to us to be closer to that of time, but it is quite possible that the image we give ourselves of the continuity of time is rather of a spatial nature.

What appeals to us in Aristotle's theory is what we have called its moderate realism. However, we find that his definition is unsuccessful; we have to come to a clear understanding of this notion. To a certain extent, every definition of time is unsuccessful if it attempts to reduce the meaning of temporal concepts to the meaning of the non-temporal concepts (viz. terms which, originally, have nothing in common with time). This is why, already in Antiquity, Galen observed that, if one takes the purely temporal meaning of before and after, the formula of Aristotle can be expressed as follows: time is the number of the motion according to time.

True, logical analysis can discover a more general idea which might have some common features with time, but the specific difference will always be something temporal. For instance, we can find an order in time; but what distinguishes this order from an infinity of other orders is that it is a temporal order, an order of terms called 'before', 'after'.

Having stated that this is a continuous order, Aristotle is fully entitled to treat it as measurable. However, his definition expresses this new property in a coarse and awkward way, calling time 'a numbered number'. We will leave aside all these already over-criticized imperfections, however. It is likewise of little interest to us what Aristotle says about the role of the instant which limits the time intervals. All this, as demonstrated by Duhem, has already been expressed by Archytas the Pythagorean. Modern analysis, founded upon set theory, has worked out another concept of the continuum, which can also be applied to time, but we shall come back to this subject later on. Here, we shall limit ourselves to the remark that neither Plotinus nor Bergson are right when stating that the metric properties are completely accidental or even foreign to time. On the contrary, being an unavoidable consequence of the continuity of time, they constitute its essential properties. Aristotle's only error in this respect consists in having placed this property before the others in his definition, to the detriment of the qualitative features of time.

15. It is a great pity that Aristotle omitted, in the *Physics,* everything he had said with respect to time in his article, 'On memory and reminiscence'. The consciousness of time is therein shown to be closely linked to memory. We do not learn that, to perceive time, it is necessary to count or to measure it. 'Each memory is linked to time. Only those animals that perceive time are endowed with memory. The organ of memory is also that of the perception of time.'

However, we read in the *Physics: adunaton einai chronon psuches me ouses* – it is impossible for time to exist without the soul. This proposition obviously has nothing in common with Kant's idealism, as Aristotle adds immediately that motion does exist without the soul and that the before and the after exist within motion. Since this before and after is the temporal order with its continuity, the participation of the soul is only expressed in its numerical estimation. This is what Aristotle intends when, justifying the aforementioned comment, he remarks that 'nothing counts but the soul, and in the soul, the intellect.' Now, how to reconcile all this with the passage from his 'On memory and reminiscence', wherein we learned that it is memory, or the organ of memory, that perceives time? Obviously, this time which memory perceives is not number.

How can we extricate ourselves from these difficulties? To be sure, according to Aristotle, time is partly real and objective, and partly subjective. Continuity and the order of succession belong to the reality which is independent of our consciousness. However, any appreciation of duration – be it acquired through memory, or expressed in number by intellect – is nonetheless our own creation, and thus subjective. But one restriction is necessary here. The metric *per se* of time, its ability of being expressed in numbers, of being measured, this belongs to real time, because the metric is a consequence of time's continuity which indicates an order; on the other hand, the measurement attributed effectively to time is only subjective and conventional. This agrees very well with the point of view of modern physics and even with that of the mathematical sciences. In the theory of groups, where measurement is treated in an abstract way, one has already to distinguish between two kinds of measure, that of Jordan and that of Lebesgue. An ordered group can be measurable in the former sense and non-measurable in the latter. The evaluation depends upon our definition. Physical sciences widen this point of view. The number attributed to a measurable quantity depends upon the choice of the unit of measure. In relativistic physics, still other particularities come into play. This is why the number attributed to a duration cannot be identified with time as such, but the metric expresses something essential to time.

The greatest value of Aristotle's theory consists, however, in that he does not attribute to time a separate reality. This is expressed by him in a decisive manner, whereas Plato and Archytas of Tarentum had not been clear in this respect. 'Time does not exist without changes'. This affirmation of Aristotle finds corroboration in other of his treatises. In his treatise, *On the heavens*, he states, '*oude chronos estin exō tou ouranou*', beyond the sphere of fixed stars time does not exist, because nothing changes there.

We shall return again to some details of Aristotle's theory, after having become acquainted with Plotinus's criticism thereof.

## TIME AND THE ATOMISTS

16. Certain historians have claimed that the theory of absolute time, such as we rediscover it, e.g., in Newton, had its source in atomism. But, for all of the ancient atomists, time does not seem to have played the same role as did space. At any rate, this claim is not true as concerns Lucretius, who had said explicitly:

Time does not exist by itself; it is the state of objects, which makes us feel that which occurred in the past, that which is present, and that which will follow; and we must admit that no one experiences the sensation of time in itself, independently of the movement of bodies and of their rest.

Tempus item per se non est . . . . . . . . . . . . . . . . . . . . . . . . . . . . . . . . . . . . . . . .
. . . . . . . . . . . . . . . . . . . . . . . . . . . . . . . . . . . . . . . . . . . . . . . . . . . . . . . .
Nec per se quemquam Tempus sentire fatendum est
Semotum ab rerum motu placidoque quiete.

A few verses later, he adds,

So you see that, without exception, completed events are not like the bodies of objects which exist by themselves, nor do they have an existence like that of the void, rather it is more justified to call them accidents of matter, or of the space in which the events occur.

However, the opinion of Epicurus was not so decisive. In all likelihood, he attributed to time the existence of an incorporeal being; such that incorporeal time, (*asomaton ton chronon einai*), was the companion of motion (*parakolouthema kineseon*), and a phantom resembling the night (*nuktoeides phantasma*). Now one may interpret this expression as the belief in an existence. Gassendi, who studied Epicurus's philosophy, probably borrowed from the latter his doctrine according to which 'time was an incorporeal being occupying the middle place between spirit and matter, while belonging neither to the category of substances nor to that of accidents.[54]

However, Sextus Empiricus further informs us that Epicurus called time an attribute of attributes or, rather, a property of properties (*sumptōma sumptomatōn*), which does not well agree with the preceding interpretation. Epicurus probably hesitated between these different conceptions of time, just as did Archytas, the Pythagorean, who said that time either had no existence, or else that it had only an attenuated existence, difficult to understand.

## THE SCEPTICS: ALL OPINIONS REGARDING THE NATURE OF TIME HAVE THE SAME FORCE

17. Greek skepticism, which called into question the possibility of knowledge, also applied its methods to the problem of the existence and the nature of time. Sextus Empiricus devoted an entire chapter to this, our question. First he enumerates the different definitions of time. Some philosophers state that time is an expanse of movement; others that it is movement, or the number of movement. According to some it exists, or seems to be something; according to others, it does not exist. For certain philosophers time is corporeal, for others still, it is incorporeal. Either all of these opinions are true or all are false, or finally some are true while others are false. However, it is impossible to decide which of them is true because all these opinions have the same force; moreover their proofs and doubts are equally compelling on each side.

18. We discover the same diversity of opinions with respect to the properties of time. It is impossible to decide whether time is limited and finite, or unlimited and infinite; whether it is divisible or indivisible,

engendered or not engendered, perishable or imperishable. If time is finite, then it began and will end one day, whence follows the absurd consequence that there was a time when time did not exist, and there will come a time when it will no longer exist. So we must admit that time is infinite. If it is infinite, then it is the past or the future which is infinite since the present is always very brief. But if the past were infinite, it would be present, and we would have to say the same with respect to the future. Now it is absurd to admit that the past and the future are present. 'Hence, if time is neither finite nor infinite, it does not exist at all.'

19. Time is neither divisible nor indivisible. It is not indivisible because we divide it into present, past and future. But it cannot be divisible because each divisible thing can be measured by one of its parts. If it were the present that should measure the past, then the present would be found in the past and, instead of being the present, it would have become the past. We reach the same, absurd consequence if we allow that the present measures the future. In an analogous way, it can be shown that neither the future nor the past can measure the other parts of time. Thus, time is not divisible. Whence the conclusion: if it is neither divisible nor indivisible, then it does not exist at all.

Sextus Empiricus gives us one more proof that the present is neither divisible nor indivisible. It cannot be indivisible because there is a moment at which a variable being changes its state, such as hard iron which becomes soft. Nothing, however, can undergo change in a time without divisions. Thus the present is not indivisible: but it is not divisible, either. Were the present divisible, it would itself contain past or future moments. So, indeed, the present is neither divisible nor indivisible. Being neither the one nor the other, it does not exist, and it can be neither the end of the past nor the beginning of the future. Nothing can be composed of what does not exist.

20. Time is neither engendered and perishable, nor imperishable and unengendered. Now, the latter is inadmissible since we cannot adduce some thing which is imperishable and unengendered, of which one part has gone and no longer exists, while the other is to come and does not yet exist. The former alternative must also be rejected since, if time arose and perished, it would arise from a non-being and would become a non-being. Yet, no thing can be engendered from nothing nor turn into nothing.

Finally, Sextus Empiricus offers us a special proof of the fact that time cannot be engendered. If it were engendered, this would be within a time. Now, this time would be either the same time or another one. If it were the same time, the contradiction would be obvious, since the same thing would, at the same time, both exist and not exist. Time cannot arise in another time, either. The point is, that if the present arises in the past, it becomes past; if this happens in the future, then present becomes future, and is it not another time? However, it was demonstrated above that time is not unengendered. If it is neither engendered nor unengendered, it does not exist at all since of these two contradictory attributes, every real object must necessarily possess one or the other.

These conclusions are astonishing. We would be inclined to believe that the skeptic ought to refrain from any dogmatic affirmation relative to the existence or the nature of time. However, what we see is the opposite: all of these considerations aim at proving to us the non-existence of time, just as the reasoning of the Eleatics or of Kant. They remind us of Kant's antinomies; especially, of the first of the antinomies. In both instances the sophistic character of the reasoning is evident. The arguments make use of the equivocity of the terms used; for example, when it is claimed that a thing cannot be called unengendered or imperishable if it possesses parts which pass and no longer exist. The same word play is found in the last passage wherein Sextus speaks of an 'other' time.

But not everything is sophistry. The difficulties relative to the divisibility or indivisibility of the present are discussed throughout the entire history of our problem; as are the difficulties tied to the measurement of time. We shall come back to these in the critical section of our work.

## PLOTINUS. CRITICISM OF ARISTOTLE'S THEORY. DEFENCE OF PLATO'S THEORY

21. The Stoics mostly recover the Pythagoreans' definition of time, presenting it as the interval of the motion of the universe. They call time incorporeal, as do the Epicureans, but they understand that, when separated from motion, time is only an abstract idea (*nooumenon pragma*), produced by our thought. The neo-Platonists, on the other hand, give us a searching defence of Plato's theory, accompanied by a rigorous criticism of all of preceding definitions of time. The Seventh Book of the Third Ennead, 'of Eternity and Time' therefore plays a remarkable role in the history of time systems. At the beginning of Book Seven, Plotinus informs us that the authority of the ancient philosophers will, in no manner, hinder him in his personal efforts.

Eternity and Time; two entirely separate things, we explain, 'the one having its being in the everlasting Kind, the other in the realm of Process, in our own Universe'; and, by continually using the words and assigning every phenomenon to the one or the other category, we come to think that, both by instinct and by the more detailed attack of thought, we hold an adequate experience of them in our minds without more ado.

When, perhaps, we make the effort to clarify our ideas and close into the heart of the matter we are at once unsettled: our doubts throw us back upon ancient explanations; we choose among the various theories, or among the various interpretations of some one theory, and so we come to rest, satisfied, if only we can counter a question with an approved answer, and glad to be absolved from further enquiry.

Now, we must believe that some of the venerable philosophers of old discovered the truth; but it is important to examine which of them really hit the mark and by what guiding principle we can ourselves attain to certitude.[55]

Since for both Plotinus and Plato time is only the image of eternity, there is no wonder that the prototype, or the model, ought to be studied first. Only after the study of eternity can that of time be made; we can here distinguish two parts. The former is devoted to the criticism of the predecessors (chaps. 7–10), the latter to the defence of Plato's theory (chaps. 11–13). The doctrines of time are divided into three groups.

Time is others that it is the motion of the universe. Those who believe it to be a moving body say that it is the sphere of the universe. Finally those for whom it is something belonging to the motion maintain, some of them, that this is an interval of the motion, others that it is its measure, and others still that it is something that accompanies it; moreover it is said sometimes that it belongs to all the motions, sometimes that it belongs to the regular motion only.

The first doctrine, which identifies time with motion, is countered by the remark that motion differs from the duration of this motion. Motion would be able to cease or only not take place by intervals, as opposed to time. The falsehood of the latter doctrine is obvious because, if the name of time would be attributed to the sphere itself, it would be because the sphere is in motion. There only remain the different forms of the third group. Is time the interval of the motion? But the interval as such does not imply any idea of time, being simply a certain length traversed during a time. If one lays stress upon this point, that it is the interval of the motion and not simply an interval, time will be in the motion as in a subject, and one would have to revert thus to the first thesis which asserts that time is a motion 'since the interval of the motion is by no means outside the motion'. Lastly, where would one place the duration of the rest?

The second form of the third kind of theories is the theory of Aristotle according to which time is the number, or the measure of the motion.

Now, either the number of the motion is not more connected with the motion than the number 'ten' is connected with the horses and dogs it counts, in which case it is a simple number without any right to the name of time, or else it is inseparable from the motion it measures because it increases and advanced together with it; but in this case how would one distinguish time from the measured motion itself? If this distinction is possible, this means that the measure is outside the motion and the number it expresses is then, as in the former case, an arithmetical number which keeps nothing from time: if one speaks of three days, the number 'three' does not tell us anything about the time interval designated by the word 'day'. The measure retains only the number or, at most, the space.[56]

On the other hand, why should one need a number for time to exist . . . This would be tantamount to saying that a magnitude remains undetermined unless there were somebody who would see what measure it had. Moreover, since it is said (and it is so in reality) that the time is infinite, what would be the relation between time and number? For this end it is necessary to extract from it a certain part which, besides, does possess the being itself before it had been measured . . . It is not at all necessary to have it measured in order that it does exist; a thing has its duration, even if it is not measured.

Finally the last form, which is restricted to the affirmation that time is something that accompanies the motion, cannot be satisfactory since 'it is not possible to understand or even to say what it is before one explains the meaning of the word 'to accompany'.

22. Thus Plotinus demonstrates the inadequacy of all the theories of time that had been founded solely upon the consideration of the physical world and presented time independently of its relation to the soul and to eternity. Let us now see in what way does Plotinus pass from eternity to time.

'We have to refer now to that state which, let us say, was that of eternity: immutable life, given wholly at one and the same time, infinite, absolutely fixed, at rest in One and directed towards One. There existed still no time, at least no time for intelligible beings; this does not mean that time would be engendered after them, but that it is subsequent to them logically and by nature. Since these beings remain in themselves in absolute tranquility, from what fall time had been born? To answer this one cannot invoke the Muses since they themselves did not exist then. Perhaps one could ask time itself in what way it did appear and how it was born. Now this is what time itself would tell us:

"Prior to having engendered beforeness and attached to it the afterness which beforeness does claim, it rested on the being, it was not time, it retained its complete immobility in the being. But the nature interested in action, which wanted to be the master of itself and a being for itself, chose a way of finding something better than its present state. Then it moved, and time likewise started moving; they directed themselves towards the future always novel, a state not identical with their previous state, but different from it and ceaselessly changing. And after having passed a certain distance, they made time which is an image of eternity. There was indeed in the soul a restless power which wanted always to change the place of the objects it saw in the intelligible world: but the soul acts like the spermatic reason which leaves a motionless germ, develops by evolving little by little, it appears, towards plurality by way of division: instead of preserving its internal unity, it squanders it at the outside and loses its force during this very progress. Similarly, the soul made the perceivable world according to the image of the intelligible world, and it made it movable not as a result of the intelligible motion, but due to a motion resembling it and yearning to become its images . . ."

". . . *Just as the world moves in the soul (because the perceivable world has no other place but the soul), it also moves in the time which belongs to this soul* . . . To say that time is the life of the soul, consisting in motion by which the soul passes from one state of life to another state of life, is it not to say something?"

The relation of time to eternity corresponds to that of the perceivable universe to the intelligible world. Instead of an intelligible life we have a life of the soul, which is called life only by homonymy. To the permanence and identity of the intelligible world there corresponds the change and the always different activity of the perceivable world; to the indivisibility and the unity corresponds 'an image of unity, one which is in the continuum'; in the infinity, an incessant progress.

"However, let us not take time beyond the soul, and no eternity outside being; it does not accompany the soul, it does not succeed it; but it manifests itself in the soul, it is in the soul, and is unified with it, just as eternity is with the intelligible being".

To this action of the soul, which manifests itself as time, Plotinus attributes uniformity and continuity. The stopping of this life would bring about the stoppage of the time. 'Hence, since time is annihilated when the soul goes to unit with the intelligible, it is clear that it is produced through the initiative of the motion of the soul towards perceivable things and through the life of the soul, which then begins.'

23. However, the most important thing for us is, how does Plotinus understand the metric nature of time. 'Time is not essentially a measure of motion, it is in the first place something else, and by accident it makes known the quantity of motion' . . . 'Time is measured, i.e. made manifest by the revolution of the sun; it is not engendered by this revolution, it is only known by it.' 'Thus the revolution of the sun makes time known to us, because it takes place in it. But the time itself must not still have something in which to exist. It is in the first place by itself what it is.'

According to Plotinus, Plato's conclusion is, in any case, erroneous. Time is not the measure of motion, but motion is the measure of time, in so far as time is measured by motion, 'however without saying that it is there for him a purely accidental feature'.

On the contrary, he defends Plato and protests against a false interpretation of a passage in the *Timaeus*, viz. which the doxographers had utilized in order to attribute to Plato a direct link between time and heaven. If Plato says that time is born at the same time as heaven, this is because the same life of the soul produces heaven and time. Beforeness and afterness also exist originally in the soul of the world; it is highly absurd to attribute 'to an inanimate motion the beforeness and afterness and, consequently, time, and to deny it to the motion, in the likeness of which the inanimate motion does nevertheless exist, to the motion from which beforeness and afterness originally proceed.' This motion of the soul of the world is spontaneous (*autourgos*).

The universe is in time since it is enveloped by the soul; but it is senseless to say that the soul is in time, because 'before the soul there is only eternity which does not accompany the course of life of the soul . . . Thus the soul is the first to go up to the time which it engenders and possesses with its proper acts.' 'Why is time everywhere? The point is that the soul is not absent from any part of the world, just as our soul is not absent from any part of our body.'

Finally the last question: 'And in ourselves? Is time also in us? It is in the universal soul and, by conformity, in all souls, since all make one; consequently time will not be dispersed among them, just as eternity does not scatter in the beings of the same species, which, all of them, do contain it.'

24. Plotinus showed a lot of spirit in his criticism of theories of time, but, strangely, solely for the purposes of defending a myth! It is time itself that reveals the mystery of its origin. However, the way in which he wants to obtain, from the nature of the being, the time and the soul of the world,

demonstrates best the absolute impossibility of any logical passage from eternity to time. 'Nature stirred, it sought again.' But how is one to find this new thing if it possesses everything in eternity? Yes, but it did not want all of being to be presented to it at one go, it did want it little by little, and besides, not in the intelligible world. Now this is a logic!

Of course, Plato's myth is rather simplified here. In the *Timaeus* God creates the soul of the world, this soul begins to move, God is delighted and so creates time in order to give more perfection to the motions of the soul. One sees that all of these acts of creation are superfluous. Time is the life of the soul, and the exit from eternity is not a temporal act but an act subsequent to the intelligible beings 'logically and by nature'. To be sure, this presents progress, but the exit from eternity, though not a temporal one, remains still mysterious.

In his criticism of the systems, Plotinus does not name any authors, and for good reason. The same authors admitted several definitions at the same time. The Pythagoreans called time the interval of the motion, but they called it likewise the number of the motion, and the doxographers attributed these definitions to Plato himself. For us the principal point of this criticism is the question of whether the metricity of time is its essential property or an accidental one, as Plotinus affirms. According to Plotinus there is no need for time to be measured in order to exist. Of course not, but this is not the question. It matters little whether anybody does or does not measure time; what we want to know is, whether this feature of being measurable is peculiar to it. For Aristotle, the measurability of time was a consequence of its continuity. Now, Plotinus attributes continuity to the life of the soul; he had no alternative but to admit its logical consequence. It is quite possible that psychologists use the word 'continuity' in a sense differing slightly from that of the physicists, but Plotinus did not even put this question.

Moreover, he attributes to the soul of the world not only continuity but also uniformity, and speaks of the progressive 'extension' of its life. Now all these properties of its nature, viz. that it is continuous (*syneches*), uniform (*homales*) and that it possesses a statement that it is an accidental property and not an essential one? However, Plotinus goes further in these negations! According to him it is absurd to concede beforeness and afterness to an inanimate (*apsuchos*) motion while denying it to the life of the soul where this order is given originally. But who has denied it? Certainly Aristotle is not clear in this respect. However, to us it is immaterial whether the order of succession is that of the physical or psychical phenomena. To be true, this order is given differently in various places, and an analysis should begin with the internal life. Yet does it follow that it is absurd to attribute this order to the movement of an inanimate body? One finds here a presentiment of Bergson's theory. We shall revert to it later.

Plotinus raises one more objection to Aristotle. He asks whether it is movement that measures time, or the time that measures movement; in other words, does time measure movement or is it measured by it. However, to Aristotle this is the same thing. Why? Strictly speaking, time can only be measured by time, and movement by another movement. However, since the flow of time is nothing else but the flow of changes, the difference is non-existent. It would exist if we attributed to time a separate existence, as according to Newton. If this question did exist in the case of Plotinus, this would demonstrate that he did not see clearly the problem of the reality of time. This is borne out by various passages in his work. In one place he says that time ought not to be taken outside the soul; 'it does not accompany the soul, but it is manifested in it, it is in it and is connected with it.' In another passage, where he wants to prove that it is wrong to identify time with the revolution of the sun, he says on the contrary: 'time itself ought not to have anything in which it would exist. It is from the outset and by itself that which it is.'

To be sure, this contradiction gave impetus to Plotinus's successors, the neo-Platonists of the Syrian and Athenian schools, to consider time as a substantial being.

## TIME AS A SUBSTANTIAL BEING IN IAMBLICHUS AND PROCLUS

25. Iamblichus was first to make of time a substance possessing an autonomous existence. In his 'Commentary on Categories' we find, as regards this subject, the following consideration: 'It is not, as some thinkers believe, that time is generated according to the natural order of our actions; on the contrary, time is the principle according to which our actions are ordered; indeed, it would not be possible to compare, in our actions, the preceding state with the following state, if time would not exist by itself; the order of actions relates precisely to it.'

We make use here of the study of Duhem who had extracted from the Commentaries of neo-Platonists the passages relating to our topic. According to these Commentaries it is obvious that time was created by the Demiurge who, in turn, engendered the soul of the world and the heavens.

Consequently Proclus, just as Iamblichus, places time above the soul and makes out of it a kind of God.

26. More interesting is what the last 'scolarch' of the Academy, Damascius, introduced into the doctrine of his masters. He places an intermediate term between eternity and time. This intermediary is sometimes called '*chronos*' and sometimes '*aion*', from whence the Latin name of '*aevum*' derives. The subject is important since the Christian scholastics introduced also in the course of Middle Ages this intermediary between time (*tempus*) and eternity (*aeternitas*). Damascius himself gives sometimes to this intermediary the name of primordial time (*chronos protos*), and sometimes that of substantial time (*chronos en hupostasei*).

The existence of time consists in a continued becoming, while the '*aevum*' is withdrawn from the outflow; it exists simultaneously in totality, and possesses thereby a permanent existence analogous to that of the eternity. Simplicius, pupil of Damascius, explains the reasons which his master did assert in order to bring him over to his doctrine. There are beings which exist in a permanent manner, and beings whose existence consists in perpetual becoming. Now, '*aevum*' ought to be attributed to the engendered beings, whose existence, however, is not in the becoming. To these being there belong: the universal soul, the celestial substance. We find it again in the case of scholastics who attribute eternity to God (*eternitas*), the 'aevum' (*aeternitas*) to human souls and to angels.

Our intelligence, just as Simplicius does explain it, is able to catch in an adequate manner neither the '*aevum*' which exists in its entirety at one and the same time, nor the time which flows continously; it catches the time in a form which partakes at the same time from both, and which bears the mark of its mixed nature.

4.2

# MEDIEVAL CHRISTIAN PHILOSOPHY

## SAINT AUGUSTINE EXAMINES TIME FROM THE PSYCHOLOGICAL POINT OF VIEW

27. In the study of medieval philosophy we shall deal in the first place with the opinions of St. Augustine and of St. Thomas Aquinas. The theory of the latter is only a development of theory of Aristotle, while the opinion of St. Augustine resembles rather that of Plotinus. It is probable that St. Augustine was familiar with the chapter of the Enneads dealing with time. We find in his writings the same criticism of the definition of time as a measure of movement, and even the introduction reminds us of that in the paper by Plotinus. 'Now what is the time? If nobody asks me this question, I know it very well; but if I am asked about it and try to explain it, I find that I do not know.' Plotinus said something similar when he indicated that we seem to have a lucid impression of time and of eternity but, when we try to subject them to an attentive examination, we become embarrassed in our reflections.

The reflections of St. Augustine, contained in the 11th book of his *Confessions* (chapters 13–28) are rather of the nature of a prayer, but nevertheless one finds there very interesting and novel observations. We shall not give here their systematic analysis, but we shall try to bring out the most important points. St. Augustine agrees with the point of view that, if nothing would pass, there would be no past time, and similarly, if nothing would come, there would be no future time. He rejects the opinion of those who affirmed that it is the movement of the sun or of the moon or of the fixed stars that is time. The movement of a body is not time. If the movement lasts long, 'I can only signify it to be a long time, but not how long: because when we pronounce how long, we must do it by comparison: as for example: This is as long as that; or This twice so long as that, or the like . . . Seeing therefore the motion of a body is one thing, and that by which we measure how long it is, another thing; who cannot now judge which of the two is rather to be called time?'

Most important is the problem of how time is to be measured. St. Augustine makes a very correct observation that we measure a longer time by a shorter time, e.g. the duration of a long syllable by that of a short syllable. 'Thus measure we the spaces of the staves of a poem, by the spaces of the verses; and the spaces of the verses, by the spaces of the feet; and the spaces of the feet, by the spaces by the syllables; and the spaces of long syllables by the spaces of short syllables. I do not mean measuring by the pages; for that way we would measure places, not times: but when in our pronouncing words pass away, we say, It is a long stanza.'

A new difficulty arises since 'it may so fall out that a shorter verse, if it be pronounced leisurely, may take up more time than a longer verse pronounced roundly' . . . 'Upon which ground it seems to me, that time is nothing else but a stretching out in length: but of what, I know not, and I marvel if it be not of the very mind.'

Here, St. Augustine ought to have taken into consideration the conditions which guarantee the absolute stability and uniformity of the flow of time. However, his interest carries him towards another problem viz. how can one measure something that does not exist? The past disappears for ever, having passed it does not exist any more, the future does not exist yet, and the present does not possess any extension; 'if it had any, one could divide it into the past and the future'. Thus, how can we measure time?

'We measure times, for all this; and yet neither those which are not yet come; nor those which are now no longer; nor yet those which are not lengthened out by some pause; nor yet those which have no bounds. So that we neither measure the times to come, nor the past, nor the present, nor the passing times; and yet we do measure times.'

Where is then this extension which we measure? This is the answer given by St. Augustine: solely in our memory; 'I measure something which is in my memory and which remains there strongly engraved. In thee, o my mind, it is that I measure the times. Do not interrupt me now, that is, do not interrupt thine own self with the tumults of thine own impressions. The impression, which things passing by cause in thee, and which remains even when the things are gone, that is it which being still present, I do measure; not the things which have passed by that this impression might be made. This do I measure when as I measure times. Either therefore I do not measure times, time is nothing but these impressions which are formed in my memory.'

It is also by the mind that the relation and linkage between the past, the present and the future is explained. 'How comes that future, which as yet is not, to be diminished or wasted away? Or how comes that past, which now is no longer, to be increased? Unless in the mind which acteth all this, there be three things done. For it expects, it marks attentively, it remembers; that so the thing which it expecteth, through that attentively it marketh, passes into that which it remembereth. Who therefore can deny, that things to come are not as yet? Yet already there is in the mind an expectation of things to come. And who can deny past things to be now no longer? But yet is there still the memory of things past. And who can deny that the present time hath no space, because is passeth away in a moment? But yet our attentive marking of it continues so that that which shall be present proceedeth to become absent. The future therefore is not a long time, for it is not; but the long future time is merely a long expectation of the future. Nor is the time past a long time, for it is not; but a long past time is merely a long memory of the past time.'

'I am about to repeat a psalm that I know. Before I begin, my expectation alone reaches itself over the whole: but so soon as I shall have once begun, how much so ever of it I shall take off into the past, over so much my memory also reaches: thus the life of this action of mine is extended both ways: into my memory, so far as concerns that part which I have repeated already, and into my expectation too, in respect of what I am about to repeat now; but all this while is my marking faculty present at hand, through which, that which was future, is conveyed over, that it may become past: which how much the more diligently it is done over and over again, so much more the expectation be at length vanished quite away, when namely, that whole action being ended, shall be absolutely passed into the memory. What is now done in this whole psalm, the same is done also in every part of it, yea and in every syllable of it; the same order holds in a larger action too, whereof perchance this psalm is but a part; this holds too throughout the whole course of man's life, the parts whereof be all the actions of the man; it holds also throughout the whole age of the sons of men, the parts whereof be the whole lives of men.'

28. St. Augustine's merit is that it is he himself who first considered time from the psychological point of view. Aristotle's observations on this subject, found in the treatise 'On memory and reminiscence' were completely insufficient and remained in disagreement with what he had said in '*Physics*'. The considerations of Plotinus were of a rather metaphysical nature. The idea of a soul of the universe was merely a metaphysical hypothesis, which did not explain at all the properties of time, such as uniformity and continuity. Since the metricity of time was, in his eyes, a merely accidental property, he did not take pains to examine its

conditions. Similarly, Aristotle also made a blunder here. In order for one to express time numerically, one has first to establish in what way we can recognize the equality of two time intervals. Taking as an example the local movements, he hardly saw a difficulty here.

But let us consider auditory data, just as St. Augustine has done. In this case we have no spatial medium to help us. One perceives immediately that only our consciousness decides the equality of two time intervals. It is only our intuition that guarantees the equality to be true – of course just to a certain extent – and not conventional. Besides, the same difficulty exists when we compare two motions, since the equality of the traversed spaces does not prove the equality of two spaces of time. One has, first, to assume uniformity, as Aristotle did; but can it be verified?

It must be said to the credit of St. Augustine that he did draw our attention to this point, though only in an incomplete manner. He mainly stresses the role of memory in order to explain how to measure that which does not exist. Some of his expressions are in this respect quite erroneous, for example the phrase 'a long past is nothing but a long memory of the past' (*longum praeteritum longa memoria praeteriti est*). The past exists in our memory only in an incomplete manner, and, strictly speaking, it does not exist at all, except as a form of recollection which comes and goes. Just at the moment of apperception, the consecutive images and the effort of attention prolong the present state and produce a continuous link between the past and the present, and this has been well stressed by St. Augustine.

Nevertheless, St. Augustine's reflections do merit the good opinion they had acquired, thanks to their lively, clear and accessible representation of various difficulties relating to reality and to the measure of time.[57]

## SAINT THOMAS DEVELOPS ARISTOTLE'S THEORY OF TIME

29. St. Thomas's theory of time is the theory of Aristotle, but with a supplement which was introduced doubtless under the influence of neo-Platonism and concerned the existence of an intermediary between eternity and time. This innovation entailes also another consequence: the relation of time to substances is considered from another point of view than that of Aristotle. According to the latter, it is only God and the eternal beings that are exempt from the flow of time. According to St. Thomas, each substance, though created in time, exists to a certain extent

*beyond* the time. Much as this statement is probably not contradictory to Aristotle's doctrine, it is nowhere expressed explicitly by Aristotle.

The problem of time has been examined by St. Thomas in many respects in his main works: *Contra Gentiles* and *Summa Theologiae*, but he had devoted to it also several special short treatises, including *De Tempore* and *De instantibus*. Since his doctrine is only the repetition of Aristotle's theory, we do not need to analyse it systematically; we shall direct our attention in the first place to the points where St. Thomas appears to say something novel.

The pamphlet *De tempore* stresses very strongly that substantial being is not subjected to time. 'A moving object can be considered in so far as a substantial being or in so far as it moves. If it is considered from the former point of view, it is not measured in time, but if it is viewed in accordance with its motion, it is measured in time. We find the same opinion again in *Summa Theologiae*. 'The objects that move, for example a man or a stone, are not measured by time in so far as they are, because their essence does not possess any succession, but is found in no matter what present moment of time; now, it is not time that corresponds to them, but rather the present (*nunc temporis*).' 'A being is subjected to changes in so far as it withdraws from eternity'. This is why the neo-Scholastic philosophers, for example D. Nys, see in time a measure of the relative imperfection of the beings. The imperfection of a creature is the greater, the greater the strictness and the extent to which the idea of time is applied to it.

And the life of a human, a being substantial and, at the same time, spiritual, is it not determined in time by the fact of birth and of death? Yes, but this is a determination in an exterior time, and not in the interior time, inherent in this being. One should distinguish – just as one does in modern physics – between time which is intrinsic or internal in a being, a body in motion for instance, and external time. The motion of a sphere of fixed stars measures in the first place its own time, but it also provides us with an external measure for all other motions, and consequently even for human life.

But why does not a spiritual being possess its internal time? Because our psychical acts are not continuous, and continuity is indispensable for the measure of time. However, one has to distinguish between two series of our psychical activities, the series of activities appropriate to animal life, and that of activities characteristic of rational life. In the former, we are able to find again a continuity of images, and this continuum is admissable to temporal measurement. But the activities of rational life cannot, by their union, give rise to a continuum. If they acquire a temporal determination, it is only indirectly, due to their junction with the images, since 'there is no thought without an image'.

In the life of the angels the former series does not exist any more, and there only remains a series of discontinuous acts which do not form a time similar to ours.

30. We see that the theory of time in the case of the Scholastics is closely connected with their entire metaphysics of substances. One introduces a radical difference between the persistent duration of the substance and the continuous and successive duration, the example of which we find in motion. It is only this successive and continuous duration that is called time. The temporal determination of the duration of substances is only indirect, and if the sequence of changes in its accidents is not continuous, this determination becomes as well incomplete.

We can now understand the role of the '*aevum*' as an intermediary between eternity and time. This is the duration of the angels and of spiritual beings. It partakes of the nature of time and eternity. The tenth question of the first part of *Summa Theologiae* is devoted to an elucidation of these three concepts and of their relation. Eternity is all at the same time, time is successive; the aevum is also all at the same time (*totum simul*) but the before and the after can be joined to it (*prius et posterius ei coniungi possunt*) while eternity excludes any succession.

The criticism of this theory of time would involve the criticism of the entire system of scholastic metaphysics. We shall refrain from this until we become familiar with the opinions of modern philosophers. In any event, we think it too dangerous to dig up such a radical difference between the duration of a substance and the successive temporal duration. Besides, everything depends on the meaning we want to attribute to the words. It is worthwhile to note some details. For example, the distinction between the external measure of time and the internal measure can be made use of in modern theories, much as it had served the Scholastics for a completely different purpose.

4.3

# TIME IN MODERN PHILOSOPHY AND SCIENCE

31. We can distinguish four periods in the evolution of the idea of time over the course of modern philosophy. The first period comprises the criticisms and corrections of the theory of Aristotle and of the Scholastics, as well from the side of rationalists as from that of the empiricists. This period brought forth nothing essentially new. The new definitions which were substituted for the ones criticized were not better than the former. It may even be said that little pain was taken to understand well the object of criticism itself. The theory of Newton inaugurates the second period. It is the product of modern science, but is influenced by metaphysics which, in its turn, gave rise to further criticism. The third period begins with the advent of the theory of Kant and its influence. Finally, the theories of modern physics and the philosophy of Bergson mark the beginning of the fourth period.

## FIRST PERIOD: THE SEVENTEENTH CENTURY THROUGH NEWTON

### *Suarez Distinguishes Material Time from Spiritual Time*

32. The book of Julius Baumann, *Die Lehren von Raum, Zeit und Mathematik in die neueren Philosophia (The Doctrines of Space, Time and Mathematics in Modern Philosophy)* will be as useful to us, as we make our way through the first two periods of modern times, as was the work of Duhem on Antiquity. Baumann had the good idea to summarize the system of the Scholastic philosopher, Suarez, which is prior to the systems of Descartes and Spinoza. Indeed, as we shall see, it would be impossible to grasp Descartes' theories of time without knowing the Scholastic philosophy of time.

The time theory of Suarez (1548–1617) is scarcely different from that of St. Thomas, but certain details are worth noting. We find therein the same three concepts of eternity, of the *aevum* and of time, but all three are considered as three species of a same genus. This proximate genus (*genus proximus*) is duration in general, understood as a perseverance of being. Eternity is a noncreated duration, all the others are created; that is, the *aevum* and time. time is a successive, continuous duration which takes its dimension from movement and which is measurable. Up to this point, we have nothing new, except perhaps a new terminology. In Suarez's work on time, however, we do find a distinction between material time and spiritual time which does not exist in St. Thomas. Spiritual time exists not only for the angels, but also within the soul united to the human body. Some have affirmed, states Suarez, that the activities of reason and of the will are measured by material time due to the concomitance of images and reason and will's dependence upon these. This is the opinion of St. Thomas who says, '*In actionibus autem animae, licet non sit transmutatio ex parte phantasmatorum, per quorum naturam in cogitationibus animae est continuum et tempus.*' The actions of intellection and of will indeed form, according to St. Thomas, a successive series, but continuity, that fundamental element of time, can neither be found in actions taken in isolation, nor in their concatenation.

Suarez imagines, on the contrary, that these actions are not momentary and indivisible, as are the acts of the angels, but that they too form a continuous succession, that their intensity increases with time in such a way that they create an immaterial time, independent of the continuity of images. The passions of the soul likewise participate, thanks to their successive and continuous expanse.

Baumann perceives herein a great resemblance to the theory of time in Leibniz.

### *Descartes: Opposition to Scholasticism*

33. It has been objected that the Descartes' theory of time is not clear and, indeed, that it is full of contradictions. This is what Descartes states in his *Principia philosophiae*, chapter 55. 'We represent duration clearly to ourselves, provided that we do not identify it with the concept of substance, but rather if we consider duration as a manner by which we represent a being which conserves its existence.' In chapter 57, he continues,

If we distinguish time from duration and say that time is the number of movement, this is only a manner of thinking. In effect, we grasp no other duration, in movement, than that of immobile things. Consider, for example, two bodies animated by different movements which, for one hour, pass through space. For everyone, these movements will have the same temporal duration, although their quantity be of unequal values. In order to measure the duration of all things, we compare a duration with that of the greatest and the most uniform of them, which produces the days and years, and we call this duration time. In so doing, we add nothing but a manner of thinking to duration, considered in its largest sense.

In one of his letters we find the same opinion: '*Tempus non est affectio rerum, sed modus cogitandi.*' Spinoza repeats this phrase in his *Cogitata Metaphysica*: 'Tempus non est affectio rerum sed tantum merus modus cogitandi.' In his *Ethics*, however, Spinoza studies the problem more deeply.

34. Two things are worth remarking in the definition of Descartes: 1) his somewhat curious subjectivism and 2) his opposition to Scholasticism. We will attempt to show that his subjectivism is only an apparent one and that it is merely the result of his opposition to the definition of time given by the Schoolmen. Scholasticism, as we have seen, traced a radical difference between the duration of a substance and time as a measure of movement. Now, if Descartes claims that time is only a mode of thinking, he means thereby that the difference between time, as something opposed to the duration of a substance, is only subjective. The duration of the substance is just as temporal as the time of movement. Further, as the Schoolmen wished to reserve the word 'time' for movement or changes alone, Descartes protested this usage of the word, deeming it subjective.

Though we may be mistaken, it is our personal opinion – and indeed from this point of view all the above difficulties are eliminated – that the system of Descartes is fully realist; yet how to reconcile it with an subjectivist, idealist interpretation of time? If it seems that Descartes expressed himself unclearly on this question, it is because we have forgotten the meaning ascribed to the word 'time' by the Scholastics with whom he was arguing. This is what Descartes said, in one of his letters, on this matter, 'All that has been said on duration and on time rests upon the opinion of the Schools; but, personally, I cannot agree that the duration of movement be of a nature other than that of immobile bodies.'

One might object the following. 'Descartes says not only that time, conceived as something different from duration, is none other than a mode of thinking, but he says the same for duration. Duration itself is none other than a mode by which we represent to ourselves a being which conserves its existence.' This is true, but Descartes does not say that these two 'modes of thinking', taken together, are only subjective. The conservation of the existence of a being is incessant continuation; the world must be created each moment by God in order to continue its existence. So, evidently duration has nonetheless retained its intuitive sense, even for Descartes. It is a mode of thinking, to which there corresponds an objective meaning, the same one which corresponds to the meaning of the word 'time'. It is only the difference between the meaning of these two words, 'time' and 'duration', as created by the Scholastics, which is subjective and artificial.

35. The criticism of D. Nys, appearing in his work, *La Notion du Temps*, is erroneous, whatever else we might say of it. The latter sees only a lack of logic in Descartes, because he assumes that Descartes had accepted the fundamental principle of the Thomist theory of time: the identification of duration and substantial existence. Chapter 55 of the *Principia* shows, however, that Descartes placed important restrictions upon the Thomist theory. The criticism of Baumann, in his work *Die Lehren von Raum, Zeit und Mathematik in der neueren Philosophie* is less severe. Baumann sees the proof of the subjectivism of time in Descartes in so far as the latter affirms in his letters that the origin of the idea of temporal succession is found in our consciousness. But this is a problem of psychology, Descartes may be right or not about it; the problem of the reality of time is not thereby prejudiced.

Baumann also objects to Descartes in that he destroys the continuity of duration and of time in admitting the incessant intervention of God toward the conservation of the world. We do not see, however, how the dependence of the world upon the first cause – which is likewise accepted in Medieval philosophy – could destroy this continuity. This question of the dependence of the world has nothing to do with the problem of the reality of time. Finally Baumann finds unconvincing Descartes' argument in Chapter 57, wherein he wishes to demonstrate that the duration of immobile things can be equal to the duration of movements. It is difficult to accept this. At the most we may agree that Descartes is being a bit sly in saying that two bodies, animated by different movements and crossing through space for one hour, possess different quantities of movement though these movements have the same temporal duration.

Descartes evidently takes the words, 'quantity of movement' in a sense entirely different from that of Aristotle. According to modern physics, movement is the product of the mass and the speed of a body, which implies the intensity of movement, whereas Aristotle and the Schools spoke of the quantity of movement according to beforeness and afterness.

Which of the two are correct, Descartes or Thomist philosophy? Must we make a radical distinction between the permanent duration of a substance and temporal duration? First of all, we have to know whether substances, in the classical sense of the term, exist. But suppose that they do exist; Descartes speaks, moreover, of the duration of immobile things in general; evidently, it is indifferent to him whether it be a question of the substance or of a body in rest. The partisans of the Thomist thesis invoke this argument in their favor, that substantial existence is not subject to the vicissitudes of change; 'being conserves it (substantial existence) unchanged from the moment of first appearance to the moment of its disappearance. Duration always remains the same, whatever be its length; it contains no real succession nor is any gradually acquired.[58]

But it is not a question of this: no one can deny that a substance is, by definition, always identical to itself. If real succession must signify the succession of change, then succession does not exist in substance. However D. Nys, who energetically defends the Thomist thesis, allows nonetheless that this duration be 'virtually successive.' It is further impossible to deny that it is measurable if it is found concomitantly with a series of changes. If the foregoing is true for D. Nys, then the rest is no more than a play of words. The Thomist thesis may only create misunderstandings and there are numerous Catholic theologians who dispute it (Isencrahe, Zigon). If succession and continuity exist here and there in a sense which entails measurability, then this difference is purely verbal.

There exists no succession in substance: we can accept this provided we wish to open an abyss between substance and its attributes. But is it good to do this? Do we not thereby create misunderstandings?

### *Spinoza. The Foundations of his Theory borrowed from Descartes and the Scholastics*

36. Spinoza's conception of time is in agreement with that of Descartes in so far as Spinoza likewise conceives the meanings of the words 'duration' and 'time' as being closer than those of 'duration' and 'eternity'. Spinoza connects 'duration' and 'time' in a much more explicit fashion than did Descartes. The following are his definitions of eternity and of duration. 'By eternity, I understand existence itself, in as much as it is grasped as only a necessary consequence of the definition of the eternal thing.' He adds to this definition the following remarks, that eternity cannot be explained by duration or by time, even if one chose to consider this duration without beginning and without end.

'By duration, I mean an indefinite continuation of existence.' The demonstration of the Eighth Proposition of the Third Part of Spinoza's *Ethics* explains why it is that the effort deployed by a thing in order to persist in its being implies an indefinite time. If time were finite, the destruction of being would follow from the same power by which the thing existed; which is absurd. Thus this effort does not imply a finite time but rather an indefinite time, because the thing – in so far as it is not destroyed by an external cause – tends to exist always by the same effort, thanks to which, indeed, it exists today.

'We cannot possess an adequate science of the duration of our own bodies', because this duration depends upon the universal order of nature and the constitution of things. The same is applicable to the duration of an external object of any kind.

'To the human mind we attribute no duration which might be defined by time, save in so far as it expresses the existence of the body, which itself may be explained by duration and defined by time.' Here the system of Spinoza differs from that of Descartes. Humans have consciousness of their eternity, states Spinoza, but this is confused with duration, which one attributes to imagination or to memory, and one then supposes that this imagination and this memory remain after the death of the body.

From all of these passages one consequence is manifestly clear. Duration is not opposed to time as it was in the Scholastics, but duration – with the time which measures it – is opposed to eternity. Thus, the doctrine of this Cartesian confirms our own interpretation of Descartes since there can be no doubt that Spinoza borrowed from Descartes the foundations of his time doctrine.

As to the measure of duration, Spinoza is aware that we frequently measure duration in our imagination, and that we can thereby represent to ourselves temporal distance, as well as spatial distance in a distinct manner; but this only up to a certain limit. As all objects, whose distance surpasses a certain limit, are represented by us in the same way, so also those objects, whose time of existence is too distant from present time, are imagined by us in a fashion which approximates them all as being at one and the same temporal distance. This law of temporal perspective is perhaps not so rigorous that the intellect might not free itself from it. Baumann has rightly observed that the 'contraction of the past' depends upon the contents which fill this time. The more time is empty, the more it is contracted.

The third representative of continental rationalism, Leibniz, already belongs to our second period by virtue of his polemic with Newton's disciple, Clarke, Therefore we will now turn our attention to British empiricism.

### *Hobbes. Time:* phantasma motus numerati

37. Hobbes declares, in his work *De corpore*, that in his concept of time, he wishes neither to depart from the common conception of all people, nor from Aristotle's own definition of time. But what Hobbes says is not so common as he claims. Time is, for him, but an image of the movement produced in our minds by a moved body. No one will admit, he states, that a year be a mode, or the attribute, of a body; therefore, we should suppose that time does not exist outside of us in objects, but only in the thoughts of the mind. Days, months and years are but the names of representations which exist solely in our mind. Hence time is a '*phantasma motus*'.

But this definition is unsatisfying, continues Hobbes, because by the word, 'time', we designate the before and the after, or a succession in movement. The more complete definition of time will therefore be: '*Tempus est phantasma motus, quatenus in motu imaginatur prius et posterius.*' Hobbes is convinced that his definition agrees with that of Aristotle for, 'to say that time is the number of movement according to the before and the after, and to say that it is '*phantasma motus numerati*', is the same thing.' Another definition, viz. that time is the measure of movement, is not as exact, 'because we measure time by movement and not movement by time.'

This last remark of Hobbes and the tone of the article itself, lead us to believe that he was under the influence, direct or indirect, of St. Augustine. Otto Liebmann has pointed out that we find the same Hobbesian definition of time already in the Middle Ages, in the Scholastic nominalist, William of Ockham. There is no doubt that all movement is represented in our minds by images. But it is impossible to deny that these images aim toward an objective state. If time does not exist within bodies, if it is not an attribute of these bodies, neither is it an attribute of the soul. It is sooner, as one said in Antiquity, an attribute of the attributes of corporeal or spiritual beings. The past and the memory of the past in our consciousness are two, absolutely different things. But the fact that Hobbes sees no difference between Aristotle's definition and his own tells us that the word '*phantasma*' holds a particular sense for him. Perhaps we do not understand it so well; it is quite likely that Hobbes sought to reconcile the psychological and the physical perspectives on time.

*Locke. Duration: Distance which separates two ideas in our mind*

38. Locke's study of duration, time, eternity, and the relation of time to space belongs among the most detailed and deep studies thereof. While we do not share his point of view, we must admit that Locke made great efforts to clarify all of these notions. Whereas Descartes, Spinoza and Hobbes devoted only a few lines to our theme and thereby contributed to the creation of misunderstandings, Locke leaves us in no doubt as to what he thinks. From Locke, the problem of time receives the careful and scrupulous analysis that it deserves.

The Scholastic opposition between duration and time disappears completely in Locke, for there is no doubt that duration contains a real succession, and again, if he distinguishes duration from time, it is only in order to reserve the word 'time' for a duration determined by measure. Duration, then, is the distance of those elements of a series of ideas which follow upon one another, or the distance which separates two ideas in our mind. Further we find in Locke two additional definitions of duration. One of them recalls the definition of Descartes, although only in its first part. Duration designates 'our existence or its continuation, or the existence of an object of some sort, which is commensurate with the succession of our ideas'. Finally, duration is also termed a 'fleeting extension'.

'To me there is no doubt that all men take their idea of duration from their reflection upon the succession of the ideas that they observe in their minds.' When this succession disappears, our perception of duration likewise ceases. 'Everyone may persuade himself of this by the experience of a deep slumber.' But we may likewise rightfully apply the notion of duration to objects during our slumber and when we have not thought about them, in the same way that we attributed extension to objects which are neither seen nor touched.

The idea of succession, Locke continues, does not come to us from external movement. We acquire it in reflecting upon the succession of ideas which present themselves to us, one after another. If external movement can – in us – give rise to the idea of succession, it is only in so far as external movement itself produces in our mind a continuous succession of ideas, each distinct from the others. There are times when movements are imperceptible and provoke no succession of ideas in us, for example in a vessel which moves without shock or vibration; movement is, likewise, often very slow, or too rapid to be clearly perceived as a continuous series.

Neither may the flow of ideas in our minds surpass a certain limit of speed. In the contrary case, we do not perceive a succession of distinct ideas. 'The part of the duration, during which we observe no succession, is that which we call moment; it corresponds to the time of the course of a single idea through our mind.' 'It seems to me that a constant and regular succession of ideas, in a wakeful consciousness, forms the measure of all the other successions.' If some ideas anticipate this natural course, or if they insert themselves between two moments in the natural course, we lose the impression of stable and continuous succession, or we perceive it only with gaps. Neither can the mind stop too long at a single idea without provoking changes in the perception of the course of the ideas.

It is not out of the question that all these ideas be caused by only a few movements. Locke does not wish to discuss this problem, however; in any case, he is certain that these ideas imply no impression of movement.

39. Up until now Locke spoke only of duration. As to time, it is only a duration determined by measure. We cannot measure time by applying a standard of measure to it, since we cannot keep a standard for a time that is immobile and immutable. There only remains the division of duration into equal periods, through repeating phenomena. The movements of the sun and the moon, therefore, give us the most proper measure of time. But this measure is not indispensible. We have even confused time with movement because of the fact that the distinction between day and night is caused by the apparent movement of the sun. Every other phenomenon which is repeated in a periodic manner could serve the same function of measurement; the increase of the heat and brilliance of the sun by equal intervals for example. Still, it is not movement which gives us the measure of the time, but the periodic appearance of the celestial bodies.

The equality of two lapses of time cannot be recognized in an absolute fashion. We have no certainty whether the revolutions of heavenly bodies occur with exact uniformity or not. The same must be said of the movement of pendulums. Although we cannot prove that our measurements of duration be exact, the notion of duration remains nonetheless clear. 'Duration itself must be regarded as flowing in a stable course that is equal and uniform.' Locke is dissatisfied with the definition of time as a measure of movement, for to measure movement we must have knowledge not only of time, but also of the space traversed and the mass of the body which is moving.

Locke adds to these considerations on time another remark on the notion of eternity. Having acquired the notion of succession and of duration, and then that of a certain length of duration, we may add these lengths, the one to the next, as many times as we choose to do so. We can prolong this operation without limit, and thereby arrive at a notion of infinite duration, or of eternity.

Locke presents a summary of his doctrine in the following six points: (1) we receive the idea of duration and of measure from the two sources from which all human knowledge is born: sensation and reflection. By observing the flow of our ideas, we arrive at the idea of succession; (2) by observing the distance between two parts of this succession of ideas, we come to the idea of duration; (3) through the observation of certain sensations which appear at certain regular periods, which seem to us equal, we acquire the ideas of certain lengths or measures of durations such as minutes, hours, days and years; (4) by repeating these measures of time as often as we choose, we come to the representation of duration there where, in reality, nothing lasts, nor exists; (5) having the power to add these lengths of time without ever coming to the end of this operation, precisely as in the addition of numbers, we are brought to the idea of eternity, as of the eternity, for example, of our souls, or the eternity of that infinite Being which, of necessity, always existed; (6) by considering some part or another of the infinite duration as being determined by periodic measures, we arrive at the idea of what we call 'time'.

40. A separate chapter is devoted to the comparison of duration and extension. These two notions contain the common idea of length, which like all quantities may be diminished or augmented. Extension is not limited to bodies, nor duration to movement, for both the one and the other can be mentally prolonged outside of all reality. Yet we accept more easily a duration which is infinite than we do an infinite extension. The cause of this is that we think without any difficulty of the infinite duration of God and that we have doubts as to the infinite extension of matter, regarding extension as an attribute of matter. It is for this reason that we call all that which is outside the limits of the world a 'fictive space', but we never call 'fictive' that duration which precedes the existence of bodies or movements. Time is for duration, then, what place is for extension. Time and place are only parts of the unlimited oceans of duration and space; we use them to determine the position of real objects. In this sense, time

begins and ends with the structure of the visible world, for time and place are the names of that part of duration and of space which can be determined by the existence and the movement of bodies. Sometimes we use the word 'time' in a larger sense, and we apply it also to a part of the infinite duration which is different from that part which is filled by the periodic movements of objects in the world.

Time and place are proper to all finite beings; we measure them by the distances between certain stable points in the sensible world, and by certain time periods marked by visible movements. The questions: 'where', and 'when' are applied to all finite beings. All the parts of an extension are extended and all the parts of a duration form the duration. Each part of the duration lets itself be multiplied or divided to infinity. The smallest part of duration, of which we have a clear and distinct idea, may be named, 'moment'. The moment corresponds to an idea within our minds, when the flow of the mind's ideas is normal.

Extension and duration likewise resemble each other in that their parts may not be separated out even by thought, although the succession of ideas itself may be broken up and separated by rest, or by sleep which can also be called rest. Duration resembles a line in the way in which extension resembles bodies. The present moment is common to all things which exist in the present and we can say that they exist all at the same moment. The two parts of duration can never exist together, the one exists when the other disappears. It is only God who can encompass in His sight all things past and future.

This consideration of the differences between and the resemblance of space and time ends with an interesting remark: each part of space exists in each part of duration, and each part of a duration exists in each part of space. Nowhere but in Locke is found such a combination of these two ideas, and this can become the object of more extensive speculation.

41. We now take the liberty of adding some critical remarks on the theory of Locke taken as a whole. His theory that we may arrive at the idea of succession and of duration just as easily through the observation of external sensation as through reflection upon the internal states of our consciousness, this appears to us wholly accurate. It is only when it is a question of the measurement of time that we give preference to the sensations of movement and to the movements themselves, understood as independent of our consciousness. Consciousness, alone, cannot give itself to this end because of its alternating states of sleep and wakefulness, and because of variations in our attention. The defenders of Aristotle and

St. Thomas emphasize the fact that direct perception precedes reflection; now, 'cosmological time must, it seems, precede psychological time'. It is far from our intention to deny the capacity of the sense data of consciousness to give us the idea of time. We believe only that the knowledge of psychological time is not primordial but follows that of cosmological time, although psychological time may, thereafter, play an important role in our practical life.[59]

We shall be obliged to return to this point, in the critical part of this work, in order to decide in what sense we may say that one time is primordial and the other, consequent. We here limit ourselves to the remark that only one part of the truth lies with the Thomist doctrine of time, and one part with the ideas of Locke. The latter's considerations are much deeper and more detailed than those of Hobbes or of St. Augustine. Doubtless, he too committed errors. For instance, in his criticism of Aristotle's definition, such that time is the measure of movement, Locke forgot, (as had Hobbes and Descartes before him), that this is not the measure of movement *tout court*, but a measure according to the before and the after. At worst, we could object to Aristotle that in speaking thus he has established a vicious circle, since time would be present both before and after itself, and we come back to time as the measure of movement according to time; or, if we want to deliver Aristotle from his vicious circle, we come back to the idea that time is something which can be measured as it measures some aspect of movement – a definition which shows us the specific, but undefinable, character of time. As we have already point out, time contains an order, and as such is an assymetrical and transitive relation; but what distinguishes this particular order from an infinity of other orders, is its temporal character.

There exists no mysterious connection between time and local movement. Any series of changes could be useful to us, it is true, but this series should be continuous, accessible to all human consciousness independent of the fluctuation of our attention, or of our fatigue and states of wakefulness or sleep; and inasmuch as possible, the series would have to be periodic. Now it so happens that such a series is realized in local movement. This is the only reason to give it preference here; Locke himself was obliged to recognize this and we too must acknowledge the accuracy of all his remarks, notably that local movement could be replaced by any other series having the same character, if such a series existed. Unfortunately however, there exist no other such series.

We must also agree with Locke in that the stable and regular succession of the ideas in our minds is the foundation of measure in general. Each perception, according to him, is the acquisition of an idea. Now it is our consciousness alone which decides, within certain limits, that the equality of two lapses of time can really exist and that the role of error and of convention bc not too great. Locke made the remark, indeed, that in spite of the imperfection of our estimates and our measures, time must be conceived as flowing in a uniform manner. But what does this mean? It is a pity that he did not look deeper into the meaning of this question. It means nothing other than this: our judgments of the equality of two lapses of time are true. If time perhaps did not flow uniformly, this would mean that periods which we take to be equal are perhaps not equal, and we have no means of discovering our error, nor of even convincing ourselves whether the two periods are equal or not. In the latter case, we may well wonder whether this question has not lost all meaning entirely.

With the advent of Newton's theory, the problem of the uniformity of time becomes the fundamental point of scientific discussion. Similarly, the problem of the realism of time begins to worry philosophers. The opinion of Locke, that we can prolong time beyond all change and movement, and that this time is not a fiction, indicates to us that Locke tended toward an absolute realism of time, notwithstanding his psychological point of departure.

## SECOND PERIOD: FROM NEWTON TO KANT

### *Newton's theory of absolute time*

42. One could ask why we incorporated Locke's theory into the first period, when his *Essay on Human Understanding* had been published two or three years earlier than the *Philosophiae naturalis principia mathematica* of Newton. It is obvious that, with the advent of Newton's philosophy, the entire tone of the discussion on time changes. In his theory, Locke attacked the doctrine of Aristotle and of the Scholastics, as well as of Descartes and Hobbes. He borrowed from the Scholastics the distinction between duration and time (although this distinction was no longer as important in his time as it was in the Middle Ages) in just the same way as Descartes and Spinoza did. Now, with the advent of Newton's philosophy, this distinction disappears even as a logical distinction. To modern science it becomes superfluous. One of Locke's definitions of the duration was merely the repetition of that of Descartes, the

other (time as distance or moment) was borrowed from the Ancients. From Newton on, it is no longer definitions which are important, but rather the understanding of certain essential properties, such as the uniformity of time and its relation to reality. Now it is the ideas of Newton that become the subject of criticism, instead of those of Aristotle and of Thomist doctrine.

It is highly probable that the development of modern science contributed to the formulation of an absolute realism of time. It existed in the philosophy of Epicurus, though not in a completely decisive manner, and it was renewed by Gassendi with the atomic theory. One finds it likewise, as we have seen, in neo-Platonists such as Iamblichus and Proclus. If it reappears with Newton, it is not merely under the influence of his philosophical pursuits but also under that of modern physics. In any case, the opinion of Mach that Newton had borrowed his theory of time from Medieval philosophy is completely erroneous, because the Scholastic doctrine represented a moderate realism completely opposed to the absolute realism of the atomists, the neo-Platonists and of Newton.

The entire theory of Newton is summarized in the following passage of his *Principia*:

Absolute, true, and mathematical time, of itself, and from its own nature, flows equably without relation to anything external, and by another name is called duration: relative, apparent, and common time, is some sensible and external (whether accurate or unequable) measure of duration by the means of motion, which is commonly used instead of true time; such as an hour, a day, a month, a year.

Absolute time, in astronomy, is distinguished from relative, by the equation or correction of the apparent time. For the natural days are truly unequal, though they are commonly considered as equal, and used for a measure of time; astronomers correct this inequality that they may measure the celestial motions by a more accurate time. It may be, that there is no such thing as an equable motion, whereby time may be accurately measured. All motions may be accelerated and retarded, but the flowing of absolute time is not liable to any change. The duration or perseverance of the existence of things remains the same, whether the motions are swift or slow, or none at all: and therefore this duration ought to be distinguished from what are only sensible measures thereof; and from which we deduce it, by means of the astronomical equation. The necessity of this equation, for determining the times of a phenomenon, is evinced as well from the experiments of the pendulum clock, as by eclipses of the satellites of Jupiter.

To elucidate still further the metaphysical foundations of Newton's theory of time, we shall quote the following passage from the end of his work. 'God is eternal and infinite by his essence . . . He endures always and is everywhere and, due to his existence always and everywhere, He constitutes duration and space' (*Deus est aeternus et infinitus per essentiam . . . durat semper et est ubique et existendo semper et ubique durationem et spatium constituit*).

43. In order to understand Newton's theory, we must look for its origin not only in his metaphysics as we already stated, but also in the contemporary state of natural science which he himself completely codified for the first time. Let us first consider the metaphysical origins of his theory, then. We find here the same difficulties as in Plato's system. If time is 'constituted' by the existence of God, it appears to be a manifestation of His eternity, but time cannot be identical with God, since it flows uniformly. There is no flow in divine eternity; whence then comes the flow of absolute time, which ought to take place without any relation to external things? Plato gave us, at least, the gratuitous hypothesis that the soul of the world is responsible for the fact that the image of eternity is movable. In Newton's system, this difficulty in its entirety is passed over in silence. Yet this difficulty would not exist had Newton not conceived his time without any relation to the created world. Connected to the created world, absolute time would become superfluous since its flow would be identical with the flux of change.

However, Newton was unable to accept such a consequence since, he says, time must flow uniformly, and it is possible that no uniform motion exists. We are thus in the presence of the scientific origin of his system. Aristotle did not see this difficulty since he did not doubt that absolutely uniform movement really exists in the revolutions of celestial bodies, or of the sphere of the fixed stars. The progress of science has shown that in reality this is not so. But this is why Newton saw himself authorized, even obliged, to assume the existence of an absolute and uniform time.

In our opinion, this assumption is not necessary; it explains nothing; on the contrary, it creates new difficulties. In the uniformity of time we have to deal with the equality of two things which are compared with one another. Let us assume that two celestial revolutions which seem to us to be equal are really unequal. This means that the lapse of time, represented by one revolution, is not equal to that of the second revolution, but that it is either greater or smaller. But all this has nothing to do with the reality of the flux; these are merely difficulties in our means of measurement. The temporal flux of movement is the same for uniform and non-uniform time; the problem of uniformity is that of the placement, correct or otherwise, of our reference points, as it were. If we erroneously divide time into portions, must we assume in this connection that there exists another flux outside that of visible change, wherein the reference points exist 'in themselves' beyond any error?

What do we gain by accepting this hypothesis? Only difficulties. For how does one explain the relation of this absolute flux to the flux accessible to the senses? Does the temporal flux of motions come from the participation of the absolute temporal flux? But this participation represents a new enigma!

These remarks ought to suffice for the time being. We shall return to them in the critical section of this work. Presently we shall acquaint ourselves with the system of Leibniz and his very interesting correspondence with Newton's disciple.

### *Leibniz. Time: The order of successions (non-constant possibilities). Polemic with Clarke*

44. The most important sources for a knowledge of Leibniz's theory of time are: the *Nouveaux essais sur l'entendement humain*, directed against Locke, and the correspondence with Newton's disciple Clarke, directed against Newton. The definition of time as 'an order of succession' is found in the correspondence only, but it could be foreseen from what Leibniz said in his earlier works on this subject. Already in his letter to Basnage (1693), he says that time represents 'an order in changes'. In his reply to the criticism of Bayle (1702), space is called the order of simultaneous possibilities, and time – the order of variable possibilities; however these two orders have, between them, a relation such that they can be applied not only to what exists in reality, but also to that which can be put in the place of what exists.

This opinion that space and time are not substances but rather real relations, kinds of orders, was expressed by Leibniz quite frequently. The chapter entitled 'De la durée' in the *Nouveaux essais* is opposed to Locke's opinions only in a number of secondary details. The psychological starting-point of Locke did not harmonize very well with the system of Leibniz.

A succession of perceptions rouses in us the idea of duration, but does not produce it. Our perceptions never have a sufficiently constant and regular succession so as to correspond to that of time which is a continuum uniform and simple, just as a straight line. The change of perceptions gives occasion to thinking about time, and it is measured by way of uniform changes: however, were there nothing uniform in nature, it would not be possible to determine time, just as place would not let itself be determined either, had there been no fixed or motionless body. The point is, that if one knows the rules of distorted motion, it is always possible to relate them to uniform and intelligible motion and by this means, to foresee what will happen by the various motions connected together. And in this sense, time, as uniform movement, is the measure of distorted movement.

It is probable that, while writing these lines, Leibniz already knew Newton's work because the conversation between Philalethus and Theophilus is directed to the same questions examined by Newton with respect to the uniformity of time.

The pendulum has rendered appreciable and visible the inequality of days from one noon to another: *Solem dicere falsum audet*. True, this was already known as was the fact that this inequality obeyed its own rules. As regards the annual revolution which compensates the inequality of solar days, this could well change over the succession of time. The revolution of the earth around its own axis, which is commonly attributed to the *primum mobile*, has been our best measure up to the present time, and it is the clocks and the watches that help us subdivide it. Nevertheless even this daily revolution of the earth can also undergo changes over the course of time: and if any pyramid could last a sufficient time, or if new pyramids could be made anew from it, one would be able to notice changes in the daily revolutions by maintaining there the length of pendulums, for which a known number of swings presently takes place over the course of the revolution. One would also know this change in a certain manner by comparing this with other revolutions, as with those of Jupiter's satellites since there is no indication that if a change occurred in both, it would be always proportional.

The dialogue returns again to Aristotle's definition. 'Indeed, it can be said that duration becomes known by the number of equal periodic movements, of which one begins when the other ends.'

Leibniz stresses that 'time and space agree as well with the possible periodic movements as with the existent ones.'

One can conceive the beginning of movement, though it is by no means possible to understand that of duration taken in its entire extent. Similarly, one can give boundaries to bodies, but one could not do this with regard to space. I just said, time and space indicate possibilities beyond the supposition of existences. Time and space are of the nature of eternal truths which regard equally the possible and the existent.

Space, like time, has its reality from God alone.

At the end of the chapter, 'De la durée et de l'expansion considérées ensemble', one finds an interesting remark regarding the void in time.

If there existed a void, i.e. a duration without changes, it would be impossible to determine its length. It follows that it is possible to confute anyone who would say that two bodies between which there is a void were in contact; because it would be impossible for two poles of a hollow sphere to touch one another, this being prohibited by geometry: on the other hand, one could in no way refute anyone who said that two worlds, which one of comes after the other, would be in contact as far as duration is concerned, so that one world would necessarily begin when the other ended without there being any gap whatsoever. This could not be refuted at all, I say, because this interval is indeterminable. If space were only a line and if the body were motionless, it would likewise be wholly impossible to determine the length of the void between two bodies.

What Leibniz wants to bring out in his considerations is, that the role of the mind in the formation of the idea of time is sooner active than passive. As opposed to Locke, he emphasizes that the succession of perceptions rouses in us the idea of duration, but it does not produce that idea. The change of perceptions only give us the occasion to think of time. In his 'Preface', Leibniz classifies duration among the innate ideas. When we hear that our science of time partakes in the nature of eternal truths, that that idea can be applied as well to existing things as to possible things, we already sense Kant's theory. In spite of this, we have no doubt that Leibniz's theory of time is that of a moderate realism. That which he will combat in his correspondence with Clarke is the absolute realism of Newton; and this in order to here emphasize still better the relativity of time.

45. The correspondence with Clarke continued over two years, (1715–1716), up until the death of Leibniz. It contains five letters of Leibniz and five replies of Clarke. The first two letters are of lesser importance to us. It is in the third letter that we find the known definition of time as an order of succession, and of space as an order of coexistence. Space and time are considered as wholly relative. The relativity of time is explained by the following example. Why had not God created the world one year earlier? This question has a meaning solely from the point of view of the absolute realism which attributes to time a reality beyond temporal things. But it is precisely the impossibility of giving a reasonable answer to this question that best demonstrates that the moments of time are nothing outside of things, and that they find their consistency in the successive order of things alone. Since the order is the same, the imagined state does not differ in anyway from the existing state, and it cannot be distinguished from the latter.

To Clarke, this question was not a meaningless one. Had God desired to create the world one million years earlier, the world would have existed earlier. Time is not only an order, it is also a quantity.[60] The existence of creatures is not indispensable to its existence, for time, and space as well, are 'immediate and necessary successions of God's existence, without which God would not be eternal and present everywhere.'[61]

Leibniz replies very correctly, that the idea of order did not exclude that of the magnitude. Relative things possess their magnitude just as do absolute things. An earlier creation of the world is but a chimera, *idolum tribus*, a fancy resembling that which desires to see in space more than an order of places. Let us assume that God had pushed the world slightly forward in the right direction and that this motion failed to produce any change whatever in the world. Clarke affirmed that such a translocation would be something real. However, to Leibniz, a change that does not change the world in anyway would be a fictitious change alone. To the remark of Clarke; viz. that there could be non-observed movement, Leibniz provides the following explanation, 'My reply is that motion is independent of observation, but that it is by no means independent of observability.' (Fifth letter).

Returning to the question of the anticipated creation of the world from the point of view of his observability principle, Leibniz gives the following, definitive reply: We have to agree about meaning of words. If this creation does not change anything in the visible world, it does not signify anything real. But, if it signifies a temporal, backward extension of the world, along with an increase in things, then this creation would have meaning.

Absolutism could likewise affirm that a lengthening of time also exists in the case of a nonobservable, but earlier creation of the world. But this would be, rather, a fictitious expansion, because the relations between the parts would remain the same.

Finally, Leibniz stated repeatedly that, without created, time and space would only exist in the thoughts of God as simple possibilities. Time without things created is only a simple, ideal possibility. The time that coexists with the created world – according to some expressions of Leibniz – is also an ideal thing, because of this time only moments exist and moments are not part of time.

Since time and space are only relations, they cannot supply any individuating principle (*principium individuationis*); we therefore have to look for this principle in the internal distinctions of beings themselves.

46. In an essay, written (in Latin) towards the end of his life, *Initia rerum mathematicarum metaphysica*, Leibniz expressed an idea which deserves mention because its theme was later developed by Kant.

Supposing that there exists a plurality of states of things, which do not exclude one another, then they will be designated as existing simultaneously. This is why the events of the preceding year and of the current year are not simultaneous to us, since they bring about opposing states of one and the same thing. If, of these two elements, which are not simultaneous, one implies the justification of the other, the former will be considered to be the precedent, and the latter the consequent.

Here we encounter for the first time the idea that the relation of temporal succession can be reduced to the relation of causality, or be established by the latter. Kant expressed a similar idea in *The Critique of Pure Reason* in maintaining that it is due to the causal relation that subjective succession becomes objective.

47. If we consider the totality of Leibniz's ideas regarding time, we must admit that Leibniz is the partisan of what we have called a moderate realism and even that, along with Aristotle, he is one of the most eminent representatives of this tendency. This is best borne out by his definition of time. All the same, his theory of time is sometimes called idealistic. Did not he himself say that time was an ideal thing?

This does not matter, however. We have to agree on the meaning of a word. According to our classification, the essence of idealistic theory consists in a complete negation of becoming and in an affirmation that reality is immutable, or that it is, with respect to its temporal character, completely unknowable. These two opinions were respectively represented, one by the Eleatics and the other by Kant. Nothing similar can be found in Leibniz. To him, temporal reality is not a veil that hides from us its true nature, which has nothing in common with time. When Leibniz says that time is an ideal thing, he means this word in a special sense, different from that of the idealists. 'There exist instants only, and these instants do not form any part of time.' Now this is correct, but from the mathematical point of view alone. From psychology's point of view of our real lives, the present instant is only an ideal section of a continuum. In any case, the reality of the temporal flux is not contested by it.

Now the moderate realist is fully entitled to say that time does not exist, if he means thereby that time coexists with change and does not enjoy a separate existence as Newton and Clarke affirmed. And the negation of time in this sense is essential for moderate realism which, moreover, can assume quite different forms. The point is that time does not exist separately, neither as an accident, nor as a relation or an order. It was Aristotle and Leibniz's great merit to have expressed this in a clear and decisive manner. This moderate realism also existed in the writings of other authors, but it was very often unconsciously confused with nuances of absolute realism.

In the polemic between Leibniz and Clarke, which is one of the most interesting episodes in the history of science, there is no doubt that it was Leibniz who was right. True, his definition of time is not a good one; if he wanted to define time as an order, he ought to have, at least, added that it was a continuous order, as otherwise one could not deduce from it the metric quality of time. If Aristotle's error was to have placed time's metric properties too much in the foreground, Leibniz sinned with a contrary omission, metric properties could not be deduced from his definition. However, he understood Aristotle better than did Descartes and Locke. And, although he saw the inadequacies of Aristotle's definition, Leibniz was more just with regard to him.

Clarke merely defended the opinions of his master. It sometimes seemed that the difference between Clarke and Leibniz was not very great. Leibniz likewise states that time owes its reality to God alone, but he discriminates between the time that exists with the world and time as a simple, ideal possibility, contained in the ideas of God. To Clarke, time is an immediate and necessary sequence of the existence of God, to which the created world cannot add anything. To us, the metaphysical foundation of their theories is less important than the theories themselves. In any case, the role of metaphysics is greater in the theory of Clarke than it is that of Leibniz. According to the former, time becomes an attribute of God, while in the latter it is only a relation. To Leibniz, this relation has nothing to do with God's eternity; to Clarke, time is precisely the divine eternity. What is the relation of these two doctrines to Scholasticism? Leibniz is closer to the latter. However, Scholasticism is responsible, to a certain extent, for the attitude of Newton and Clarke because, though Scholasticism did establish a radical difference between eternity and time, it nevertheless used the word, 'duration' as much for temporal duration as for the non-temporal duration of a created, or non-created, substance. We have seen that Suarez also used the word 'duration' for divine eternity. But since Descartes succeeded in convincing people that there was no difference at all between the duration of a motionless thing and time, there is nothing astonishing in the fact that eternal duration became identical with temporal duration, in Newton and Clarke.

Descartes hardly thought of divine eternity when he said that there was only a *distinctio rationis* between time and the duration of a motionless body. And Newton failed to point out that divine eternity, even if we call it 'duration', remains no less a *simul totum*, while its absolute time *aequabiliter fluit* and that, consequently, there is no possible logical transition from motionless eternity to the uniform flux. We have thereby the impression that Newton and Clarke had been duped by the ambiguity of the word, 'duration', as created by the Scholastics.

Leibniz likewise used the word 'duration' in an ambiguous sense. In his

*Initia rerum mathematicarum metaphysica*, duration is defined as a 'magnitude of time'. But in the *Nouveaux essais* he uses this word in a different sense, by designating there with both the entirety of existent time and fictitious time which relates to what is possible. We perceive this very well in the following passage: 'One can conceive the beginning of movement, although one cannot at all understand that of duration taken in its full extent.'

Finally, we must stress that neither Locke, nor Newton, nor Leibniz properly distinguish the eternity of God from the contingent eternity of the world, as Scholastics had done. The essential feature of divine eternity, according to St. Thomas,[62] is not only the fact that it has no beginning and no end (since it is not to be excluded, that the temporal world likewise has no beginning and no end), but that divine eternity is, in its entirety and at the same time, *tota simul*. If this had been remembered, divine eternity would never have been confused with the temporal flux.

### *Berkeley: The Duration of a Mind Must be Measured by the Number of Ideas*

48. Berkeley belongs likewise among the adversaries of Newton. His starting-point is, as we might guess, exclusively psychological.

> As far as I am concerned, whenever I strive to form a simple idea of time, disregarding the succession of ideas in my mind and the time which flows uniformly and in which all the things take part, I always become lost and entangled in difficulties which cannot be resolved. Although I have no idea about this myself, I nonetheless hear it said that time is divisible to infinity; I hear people who speak about time in such a way that the expressions they use suggest to me of astonishing ideas about my own existence; because such a theory places us before the absolute necessity of supposing that, either, we pass through countless centuries without any thought, or that, at every moment of life, we fall into annihilation – both of these being of an all too evidently absurdity. Since time, when isolated from the succession of ideas in our mind, is nothing, it follows that the duration of a mind that is enclosed within certain limits, or is finite, ought to be measured by the number of ideas, or successive activities, in this mind.

This point of view is very close to that of Locke. However, the problem of the uniformity of time is not taken into account. Psychological time does not entitle us to assume the existence of a universal time, common to everybody and uniform. It seems that Berkeley rejects not only Newton's absolute time, but also the common sense time. In his

*Dialogues between Hylas and Philonous*, he repeats not only the thesis that time is measured by the succession of ideas in our mind, but he also admits that it is possible for ideas to succeed one another in one mind twice as fast as in another. In so doing, Newton allows for the relativity of time in a sense which is perhaps broader than that admitted by Leibniz. Moreover, this consequence is inevitable from the exclusively psychological point of view.

As regards the question of reality, it seems to us that Berkeley, too, is the representative of a moderate realism. The relativity of time is not identical with idealism, though it can be utilized to the benefit of the latter. Yet, in the case of Berkeley, time is neither an illusion, nor a subjective form concealing the true nature of the being. We must not confuse the problem of the origin of an idea with that of its reality.

### *Hume. Time: A Mode of Appearance of Impressions, Which is Not Itself an Impression*

49. David Hume devoted to the problem of time a considerable part of his principal work, *Treatise on Human Nature*. It was the problem of the divisibility of time that most interested him. According to him, an infinite divisibility does not exist. It is obvious that our representations of space and time, our images, do not admit of infinite division. Our imagination arrives finally at a minimum beyond which any further division is impossible. The same applies all the more to our impressions.

Now, if our impressions and our images are composed of a finite number of indivisible parts, we must say the same with regard to the objects of these representations, i.e. time and space themselves. No finite space contains an infinite number of parts. Similarly time can only be composed of a finite number of moments.

The other problem treated by Hume was that of the origin of the idea of time. The idea of time, he said, is an abstract idea which we get from the succession of impressions of every kind. Just as the positions of the visible and palpable objects give us the idea of space, so also the succession of impressions and ideas provides us with the idea of time. Locke noted very well, states Hume, that there exists a natural limit to the speed of this succession which depends upon the nature and the constitution of our mind. The circular motion of a piece of glowing coal, where its successive positions unite into a continuous line, is the best proof thereof.

A motionless object which brings about only coexistent impressions cannot provide us with any idea of time. Time can only be linked to a succession which, moreover, is inherent in the impressions of all the senses. However, Hume raises the problem of whether the idea of time is derived from a separate impression, existing apart from other impressions. He opted for the latter alternative. If we hear five notes of a flute in succession, which give us the impression and the idea of time, this time is not a sixth impression which presents itself to our hearing. Thus time is only the mode of appearing of the impressions, without, itself, being one of them. But this mode can reappear as connected with other impressions, and thus it is but an arrangement of different ideas.

Hume also responded to possible objections that might be raised against him. Astoundingly, he called the minimal, indivisible parts of space and time 'mathematical points'! They are not physical points, he said, since physical points are extended in a visible manner. Now, the first objection is: the assumption that extension is composed of a finite number of mathematical points is absurd, as such points do not exist. To which Hume responded, 'This objection would be conclusive if there did not exist some intermediary between the infinite divisibility of matter and the non-existence of the mathematical points. But such an intermediary does exist. We obtain it by attributing colour and solidity to these points.'

The second objection raises the difficulties resulting from the mutual penetrability of indivisible points. Since these points are absolutely simple, external contact is impossible. But the internal contact is the penetrability, itself. Hume's answer determines, first of all, the meaning of this term. When two bodies penetrate one another, this means that the body resulting from their penetration is not larger than one of the two former ones; this is tantamount, in other words, to the destruction of one of the bodies. But if we thus understand the term, is it possible for a palpable and colored point to be destroyed by the approach of another colored, palpable point?

The third objection or, rather, a new group of objections results from the point of view of the mathematical sciences in general. According to Hume, the mathematical definitions of a point, a line or a surface remain incomprehensible unless one admits the existence of indivisible points. The foundations of mathematics are borrowed from the data of the senses, impressions, and imagination. To Hume, geometry is an empirical science the reasoning of which can be only applied to objects of the senses and imagination. The assumption that geometry is a science which is not interested in the existence of real objects seems absurd to him. He does not deny that the reasoning of mathematicians is exact, but he affirms that, at their starting points, there exists a lack of precision and, in this respect, no mathematical reasoning can exceed and deny the testimony of impressions and ideas. That is why all of the geometricians' proofs concerning infinite divisibility are devoid of value.

When all this is admitted, we also obtain the consequence that there exists neither a spatial nor a temporal void, nor that any idea of such a void can be formed. As regards time, this consequence is obvious. The point is, from whence could we arrive at an idea of time without any changeable existence? Does it come from an impression of the senses and from reflection? If only someone would show it to us! But, if it is impossible to point to any impression relative to it, we are mistaken in believing that we possess an idea of empty time. We can, however, point to the source of this false belief. We find in our mind a continuous sequence of perceptions and consequently, the idea of time always presents itself to us. When, observing a static object, we first withdraw from it and then return to it, we are inclined to carry this idea of time over to the object, as if it presented a change at every moment. When we consider that the object could really be subject to change and that it exerted an influence upon the action of the forces of nature, the origin of the idea of the duration of a motionless thing is explained.

50. Thus Hume's theory of time also represents a moderate realism, since time is merely the mode of the appearance of impressions. However, time itself is not an impression. Of course, there is a grain of truth in this affirmation. In any case, this mode must be perceived: this can either take place or not, depending on the state of our faculties of perception. If the impressions are perceived, they enter as one of the components into the totality of our apperception; but this form of composition cannot be treated as an element equal to the other sensible elements whose form it presents. With these restrictions Hume's affirmation can be accepted.

The remainder of his theory is of lesser value. Since Archytas and Aristotle, one always spoke of time as intervals separated by limit-points. Now, if Hume tells us that time is composed of a finite number of indivisible parts, he evidently has in mind what one calls intervals. But each interval has two limit-points, and intuition tells us that, between two points, there is always a new intermediary point. This is why Aristotle maintained that there exists no shortest time, as each interval is capable of division. Now to escape the danger of the infinite, Aristotle appealed to his distinction between *potentia* and *actus*. The number of intervals and points which separate them is infinite *in potentia* alone. The number of intervals and of limit-points, effectively reached, is always finite. This point of view is accepted even today.

As we have just seen, Hume is reluctant to take advantage of this distinction. He thinks obviously of the really indivisible parts, such as the shortest impressions, and the smallest parts of our visual and auditive images. But the reader is perplexed because Hume called these indivisible elements, mathematical points, instead of considering them as the shortest intervals. The idea of something indivisible now becomes highly equivocal. We should remember that mathematical sciences never accepted Aristotle's solution, *in actu* and *in potentia*; these are not terms of the vocabulary of mathematics. Modern theory set speaks without qualms of infinite sets. Nobody is disturbed by this fact, since today it is most often accepted that the mathematician is not interested in the reality of the objects of his study. However, difficulties do appear at the moment when mathematics are applied to the physical sciences. Until the advent of the most recent physical theories, this application caused no difficulties. It is not impossible however, that we shall be obliged to change our point of view under the influence of contemporary physical theories.

If we look at Hume's theory from this angle, it loses some of its apparent absurdity. Nevertheless we have to admit that he did commit some errors. The shortest intervals of concrete time (i.e. having a content) ought not to be called indivisible elements in the mathematical sense of the word, or mathematical points, since these words have been always used in a different sense. The problem formulated by Hume is not that of pure mathematical sciences, but rather that of applied mathematics. But, what is most important, Hume confused the psychological point of view with the physical one. The minimal extension of our impressions or images can vary from one person to another; it depends on the state of our attention, our fatigue or our practice. If there exist minima of physical extension, they cannot be identified with these variable, psychological minima.

Yet one circumstance can justify that attack by Hume, directed against the mathematicians. The point is that Newton and others who continued work on the infinitesimal calculus, the calculus of fluxions, spoke of an infinite number of infinitely small magnitudes and of the differences of degree between these magnitudes. Modern analysis has shown that allowing for these truly vague concepts was superfluous, and that we can do without them by defining the fundamental notions of the infinitesimal calculus, using the concept of a limit. The same attack against the mathematicians was also made by Berkeley in his *Analyst*, (1734). These criticisms are therefore justified insofar as the creators of the differential calculus had failed to establish its foundations in a completely correct manner. Today the notion of an infinitely small magnitude is rejected; but the notion of an infinite set remains, and it is almost impossible to imagine the existence of mathematical sciences without this notion. However, all this has nothing in common with the minima of temporal extension. If modern science is obliged to admit them, this will not be due to the arguments of Hume.

## THIRD PERIOD: KANT AND THE 19TH CENTURY, UP TO THE ADVENT OF THE PHILOSOPHY OF BERGSON AND OF THE THEORY OF RELATIVITY

### *The Theory of Kant. Time is only an* a priori *form of the internal sense*

51. General remarks relating to Kant's theory. – 52. Arguments of the metaphysical exposition of the Critique: *Argumentum ex antecessione, ex adhaesione, ex necessitate, ex unitate, ex infinitate. Argumentum e continuitate* according to the Latin dissertation of 1770. – 53. *Argumentum e mathematicis* of the transcendental exposition of the Critique. – 54. Consequences drawn by Kant from these expositions. – 55. The role of the *transcendental determinations of time* in *the schematism of the pure concepts of understanding*. – 56. The succession principle in time, according to the law of causality. The subjective succession becomes objective and necessary, thanks to the application of the law of causal relationship. – 57. The role of time in the 'Refutation of idealism'. – 58. The critique of this refutation. – 59. The antinomy of time in the transcendental dialectic. – 60. Critique of the antinomy.

### *Time According to German Idealism: The Critique of Renouvier and German Neo-Kantism*

61. Schopenhauer energetically defends Kant's [well known, aprioristic] time theory and even finds new arguments in favor of the ideal nature of time. – 62. Definition of time according to Hegel. – 63. The idea of time in the system of Charles Renouvier. – 64. The idea of time according to

the Marburg school. – 65. Otto Liebmann and his interpretation of the *a priori* nature of time.

*Criticism of the Theory of Newton by Ernst Mach, and the Beginning of Psychological Research [into time]*

66. Criticism of Newton. Time expresses the reciprocal dependence of things, which is incomplete. Change is the product of partial indetermination. The irreversibility of time is connected with the increase of entropy in the universe. – 67. Mach affirms the existence of time sensations. The entirety of these sensations is called 'physiological time'. The irreversibility of physiological time is connected with the increase of organic consumption and of the work of attention.

*Wundt and the Psychological Theories of Time*

68. The role of the senses of hearing and of touch in time perceptions, according to Wundt. The illusions of time, especially those of the displacement of temporal order. Wundt's explanation. – 69. The nature of time perceptions. We perceive short durations simultaneously. The role of rhythm. Each time perception forms a rhythm or a part of a rhythm. – 70. Psychological development of time representations. Temporal signs are composed of sensations and of affective states of tension and relaxation. Affective elements play a constitutive role in the formation of time perceptions. – 71. A view of the history of psychological theories, according to Wundt. (Herbart, Lipps, Czermak, Vierordt, Mach, Münsterberg, Schumann, Meumann).

FOURTH PERIOD: CURRENT THEORIES OF TIME

*Bergson's Theory of Time*

72. Time according to 'immediate data'. The second chapter of this book. True succession where in the successive states get into working order and mutually penetrate one another is only given in the consciousness. The idea of a homogeneous medium only vitiates that of time. – 73. Bergson's observations of time in physics and in astronomy. The analysis of the concept of motion. The sophisms of the Eleatic School. – 74. Two aspects

of our conscious life. Their relation to time-quality and time-quantity. – 75. The third chapter of the 'immediate data' concept of time. Defence of freedom. Observations regarding the law of the conservation of energy and psychological determinism. All of the difficulties of this problem arise from the finding that 'in duration one will find the same attributes as in extent'. – 76. Modification of the theory of time in *Matière et Mémoire*. The material universe lasts as well, but its duration differs from ours by its rhythm. The different rhythms or tensions of duration. A new conception of matter. – 77. Definition of perception in terms of duration. Perception consists in the condensation of enormous periods of an infinitely diluted existence into some, more differentiatedf moments, of an internal life. The vision of matter relieved of what the requirements of life add to it. – 78. *L'Evolution créatrice* completes the theory of time from the biological point of view. Cosmological considerations found in the last chapters. Creation and invention as the essential elements of time. The Supreme Being lasts as well.

*Husserl and the Phenomenological Analysts of the Intuition of Time*

79. The difference between the psychological and the phenomenological problems. – 80. Brentano's theory as a starting point. The 'original association' of Brentano. 'Past' and 'future' as modifying terms and unreal predicates. 'Present' as the real predicate. 81. Husserl's three objections. The original association denies the true apperception of the succession and of the change. Brentano does not distinguish between the contents and the object of the physical act. He does not explain why the temporal marks do not function in the same way as do the qualities of representations. – 82. Results of the phenomenological analysis. The *a priori* laws of time. – General remarks on Husserl's phenomenology and critical remarks. The relationship between phenomenological and cosmic time according to the *Ideen*.

*Poincaré. Equality of Two Intervals and Simultaneity of Distant Events have only a Conventional Value*

84. Conventional nature of the equality of two intervals. The definition by means of equal causes which use the same time to obtain equal effects is only approximate. – 85. The definition of the concepts of 'before' and 'after' by means of the causality relation does always contain a certain

degree of arbitrariness. In each definition or verification of the simultaneity of distant events it is convention that plays the decisive role. We always choose the simplest and most convenient methods.

*Enriques. Defense of a Natural Measurement of Time*

86. Psychological and physical time. The abstract notion of physical time comes about by unifying all of the possible, temporal scales, while physiological or psychological time represents only the scale of all perceived phenomena. Properties of temporal succession. – 87. The problem of the measurement of time. The abandonment of Newton's theory of absolute time does not involve conventionalism as a necessary consequence. The temporal scale is natural rather than conventional. – 88. The postulate of time measurement. The problem of time's independence of its place.

*Time in the Theory of Relativity*

89. Principle of relativity in classical mechanics. The Michelson-Morley experiment. – 90. Explanation of the negative result of this experiment by Lorentz and by Einstein. Three consequences of Einstein's interpretation: the relativization of length, of time intervals, and of simultaneity. The explanation of these consequences through the analysis of the Michelson-Morley experiment. – 91. The transformation laws of Lorentz. The theory of the limited relativity. – 92. Definition of simultaneity and the relativity of time measurements, which follows from it, according to the systematic study of Einstein's (1905) *Zur Elektrodynamik bewegter Körper*. – 93. The same subject according to Einstein's popular book of 1916, *The special and general theory of relativity, rendered accessible to everybody*. Reply to the vicious-circle objection. – 94. Some consequences of time relativity, according to the systematic work published in 1905. – 95. Minkowski's work of 1908. The space-time of four dimensions. This four-dimensional world presents in its properties a great kinship to Euclidean three-dimensional space.

96. Three kinds of vectorial magnitudes in four-dimensional space-time. Points of the universe. Lines of the universe. – 97. Geometric interpretation of Lorentz's transformation, as given by Minkowski. The 'luminous cone' (*Lichtkegel*) in the four-dimensional space-time. – 98. The theory of general relativity. The effect of the gravitational field on the

physical measurements of time, according to Einstein's paper of 1911. – 99. Einstein's work of 1916. All observations merely ascertain the coincidence of material points in space and in time. Four-dimensional space-time is not Euclidean. – 100. The space-time interval in the general theory. Separation of space and of time in static cases. The stationary case. – 101. The effect of the gravitational field on the running of clocks, according to the works of 1916. 102. Weyl's theory and its significance for the idea of physical time.

*The Significance of the Theory of Quanta for the Idea of Time*

103. Heisenberg's principle of indeterminacy. – 104. The parallelism between the undulatory and the corpuscular physics. – 105. Some general observations on the present state of physics.

## Part Two

## Time *Aporias* and Their Solution

### INTRODUCTION: THE PROBLEMS AND THE METHOD USED FOR THEIR SOLUTION

106. The history of doctrines of time has revealed to us a number of problems or, according to the language of Aristotle, *aporias*, which have not been resolved in a manner agreeable to all, although they have indeed been considered to be soluble. The divergence of opinions has originated less in falsehoods of reasoning, than, it seems, in the unilateral point of view; or it arose from the ambiguity of the terms used in a particular argument. Beside these two sources of error, such as insufficient observation or the incomplete analysis of the facts, and erroneous terminology, we were able to demonstrate one additional cause of the difference of opinions among the philosophers; viz., the arbitrariness of their suppositions, especially in the case where these suppositions were unverifiable hypotheses.

If such a hypothesis is indispensable, then methodological reasons should decide on their choice: however, this has not always occurred. Intuition always influences our suppositions, and this is fully justified. However, if they are divergent, we must approach them with circumspection. Many times in the history of sciences, conviction proved itself to be merely the voice of naïveté. We are determined to adopt as starting point our intuitive and evident convictions, but we cannot guarantee whether all of their consequences will be equally evident and verifiable. Let us take an example. Each of the three theories of time that we have singled out, idealism, extreme realism and moderate realism, are unverifiable in their consequences. Kant's thesis, viz., that the temporal aspect of reality would not exist had our mind not been endowed with an *a priori* form, is just as unverifiable as Newton's affirmation that time would flow even if the world did not exist. But the consequence which Aristotle's theory reaches is also unverifiable; viz., that time would no longer exist if the course of events stopped suddenly and the universe became stiff and motionless, as in the fairytale where the princess pricked her finger with a pin, and this prick led to the stoppage of time's flow and the immobility of the environment, until the young prince awoke the princess with his kiss, thereby setting suspended time in motion again.

One of these three theories is certainly true, but none of them is verifiable, and each one is based on suppositions which seem to be evident. Thus we shall have to look for still another methodological criterion by which to make the choice.

Here is another example which demonstrates the great number of errors that originate from misunderstandings as to the meaning of terms used. Some philosophers proclaim the ideality of time while referring the matter to Kant; however, if we examine their theory more closely we immediately see that their point of view has nothing in common with that of Kant and that it corresponds rather to one of moderate realism. In the latter case time does not exist as a separate being, but the expression, 'A is prior to B,' is a proposition which is true for the specific values of A and of B. On the contrary, in the philosophy of Kant, this proposition is ambiguous; it is true when it refers to phenomena, but it is stripped of all sense when applied to absolute reality, to the 'thing in itself'.

How many theses, erroneous or corrected too late, appeared because the origin of the notion of time was sought only in the life of consciousness or only in the physical world. The method which must facilitate the solution of these problems is summed up in the following: we shall try to be in agreement with the most complete experience; agreement with the laws of formal logic is self-explanatory; we shall avoid those errors caused by the ambiguity of the terms used and, finally, we shall accept a few methodological principles, universally recognized. All our suppositions, serving as our starting-point, will be evident and intuitive, but we do not guarantee the same for all of their consequences.

107. We are dividing all problems of time into a number of groups which we shall analyse successively. These problems are so interconnected that it is impossible to treat them separately, hence a division into groups is indispensible here. We havc probably not exhausted all of the possible groups, but the most important ones are here in consideration.

The first group of *aporias* contains all of the questions connected with the problem of the reality of time and with that of the *a priori* or *a posteriori* element contained in the idea of time. Kant connected these two problems in such a way that it is now inconvenient to treat them in isolation, although the first is completely independent of the second. Beside the critique of Kant, the first chapter of the second part of our work also includes the critique of Newton's absolute realism to better situate the moderate realism which we are defending.

At the end of this chapter we add some observations on the *a priori* dimension of time in Husserl's phenomenological sense.

The second group of *aporias* concerns the question of the psychological origin of the idea of time and, especially, the relation between intuitive time and physical, metric time.

This includes the critique of Bergson and discusses the problem of the so-called sensations of time, the role of memory in the formation of the idea of time, the facts of temporal illusions and, finally, the question of the relation between ideas such as the temporal order, the interval, and the instant. The chapter ends with some etymological considerations connected with the question: which of these notions of time is the oldest and the most extensive one?

The third group of *aporias* contains the problems raised by modern physics, viz. the uniformity of time and the simultaneity of distant events. While adopting the viewpoint of modern physics, we distinguish especially between physical and philosophical relativism, and we will try to demonstrate likewise that conventionalism does not provide a foundation peculiar to the theory of relativity. We endeavor to reconcile physical relativism with the classical concept of truth and, finally, we elucide certain points with respect to the logical foundations of the theory of relativity, as logic seems to us the only possible foundation and differs only slightly from that of Einstein and of Reichenbach, one of the most eminent representatives of relativism among the philosophers.

In this chapter we consider, additionally, the various attempts at temporal relations and temporal notions in general, with the help of the axiomatic method.

The last chapter is devoted to problems of continuity, of the infinity of time and of its irreversibility. The first of these problems is connected with the greatest diversity of opinions. We can distinguish here the philosophical notion of the continuum, created by Aristotle; the notion of the mathematical analysis; the intuitionistic conception represented by Bergson and the different forms of atomism of time. The results of quantum physics render unverifiable the mathematical continuity of time, but we think that the intuitive notion of continuity should nevertheless be maintained, since the atomistic theories are not verifiable, either. The problems of the infinity, and irreversibility, of time provide the occasion for considering the value of the second law of thermodynamics and of other laws of physics which determine the conditions and the direction of change. Here again, modern physics has uncovered new possibilities.

We have decisively solved none of the questions contained in this final chapter, since the state of modern science does not permit it. In our final remarks, we try to determine in what sense it might be said that time presents something irrational, and what is the role of time in the scientific conception of the universe.

## 1. THE FIRST GROUP OF THE TIME APORIAS

### *A critique of Newton's theory of absolute time*

108. The criticism of Newton's theory will not entail as many difficulties as that of Kant; nevertheless, we cannot neglect this theory because it best corresponds to human intuition; while Aristotle's point of view seems to be rather artificial, as does that of Kant. Why the average person will ask, should time disappear along with the disappearance of change, or indeed with the abolition of human consciousness? Is it not more accurate to assume, with Newton, that time would flow, even if the world did not exist?

The only argument we are able to advance against Newton consists in this, that the supposition of absolute time is superfluous, that it does not explain anything; but that it forces us, on the contrary, to make new suppositions and provide new explanations for the relation between the flow of events and the empty and homogeneous medium which exists by itself and by its own nature. *Entia non sunt multiplicanda praeter necessitatem*, this, Ockham's razor, is the only thing that we can recommend here from the point of view of scientific method. If this argument does not suffice, we have no other at our disposal.

Already in our historical section we considered the motives which inclined Newton toward his theory. Now this theory had undoubtedly a metaphysical background, but Newton could also invoke scientific reasons in its favor. The following passage indicates the metaphysical source of the theory: '*Deus est aeternus et infinitus per essentiam, durat semper et est ubique, durationem et spatium constituit*'. The sense of the term '*constituit*' is not clear. There are two possible interpretations: either time is identical with divine eternity, or else it is a product, a manifestation, of this eternity. If it is identical with it, we ought to attribute to divine eternity everything that we attribute to time. Consequently, eternity would flow just as does time, which seems to us absurd. If time were a manifestation of this eternity, different from its source, we would obtain a paraphrase of the statement of Plato; viz., that time is a moving image of eternity. However, in such a case, we do not understand why this image is moving if its model, eternity, is immobile.

Newton, however, had a scientific argument in favor of his theory. Without the supposition of the uniformity of time, it would have been impossible to formulate the laws of physics; for instance, the law of inertia. But, no measured time is uniform; at least, we have no guarantee at all of the absolute uniformity of these measured times. Hence, we have to suppose that a time does exist which flows uniformly of its own nature.

This supposition is not necessary even if it were true that the meaning of the laws of classical physics could not be preserved without it. We know today that these laws do not present the only possible, or even the most probable, form of the order of nature. The theory of relativity shows that these laws can assume a much more general form, to which the supposition of the uniformity of temporal flow is not indispensable.

However, the point of view of moderate realism does not demand necessarily that we give up the uniformity of time. The point is that the uniformity of time does not express a property inherent in the flow of events. We say that events flow uniformly if two periods which seem equal to us are really equal. For Aristotle, this difficulty did not exist since he had no doubt that the revolution of the sphere of fixed-stars gave us equal periods; however, modern astronomy does not share his certainty. If astronomy introduces corrections, the problem is reduced to the fact that the reference points which divide the flow into equal parts are, as it were, well arranged. If they divide time poorly, must we thereby admit that there exists a second flow, outside that of visible change, wherein the reference points, invisible to us, exist in themselves and divide time without any error? We can always relate the correction to the visible flow. However, let us suppose that there exists an absolute flow outside time which is capable of being measured. How can we explain the participation of the flow of events in the flow of absolute time? Here, the theory of absolute time only increases the number of difficulties, instead of dispelling them, since one has either to admit that events flow without any relation to the flow of absolute time which, thus, becomes superfluous, or else the latter exerts a dynamic influence upon real events, which becomes a gratuitous and rather adventurous hypothesis.

Finally, one can guess that in his theory, Newton relied on intuition, on the rooted conviction of commonsense; viz., that time would flow even if the world did not exist at all. This idea, which seems to be independent of experience, corresponds fairly well to what Kant and the phenomenologists of the Husserlian school call the *a priori* dimension of time. The second argument of the 'Transcendental Aesthetics', '*argumentum ex adhaesione*', appeals to the invincible force with which the idea of time forces itself upon our mind, because of which it is impossible to suppose that the temporal place for phenomena disappears if these phenomena are moved aside in thought. However, the manner in which Kant made use of that intuitive conviction shows that it can be interpreted in a different way. Its source lies in the Scholastics, although they did not make much of it; according to them man possesses, beside the idea of time, also the idea of '*tempus imaginarium*', of a possibility of time, which is sometimes mistaken for real time; which is why we find it difficult to imagine, e.g., that time did not exist before the creation of the universe.

109. The fact that this tendency of the human mind to objectify the idea of empty time can be interpreted differently best proves that the inference made in favor of the theory of absolute time is not necessary. In any case, this tendency provided the strongest support for the theory of absolute time. The point is that we should not forget that Newton was not the only defender of absolute time. This tendency toward objectifying empty time has always existed. In our most recent period, Anton Marty supported Newton's theory, with a slight modification, in his work *Raum und Zeit* (1916). The existence of time is, for him, *a priori* certain. Time is incapable of any action ('*ist nicht wirkungsfähig*'), it does not possess dynamic properties; hence, in this respect, it is not real. However, *mutatis mutandis*, empty time 'subsists' ('*ist etwas Subsistierendes*'). It is a

continuum that conditions each being, but it is not itself conditioned; it is the only, absolutely unconditioned continuum.

Marty admits that he has no empirical proof for the existence of time, but this existence is not merely a hypothesis for him because it is evident *a priori*, and necessary. Even eternal truths exist in time, their alleged non-temporal character consists in that they exist in all times. Only non-being is not temporal.[63]

Is it really necessary to admit the existence of empty time by virtue of its *a priori* evidence, and to create for this existence the new category of non-real subsistence? We do not think so; we do not see any *a priori* necessity to do so, and as an hypothesis this existence seems to be superfluous. The words 'time flows' have no other meaning for us than that of the following phrase: 'there exist events that take place' or, even better, 'the thing A exists prior to the thing B, etc.' All of the expressions containing temporal terms can be transformed into expressions which relate to events and to things.

There exists a point of view according to which all expressions concerning relations, properties and processes, are merely the expressions relating to things. It is only the names of things that are, from this point of view, proper nouns, all others are apparent nouns which are used only for abbreviating certain expressions or for making them more appropriate. This point of view is represented in contemporary Polish philosophy by Kotarbiński; it was accepted by Brentano, and even in Leibniz's *Nouveaux essays* we find, in Book II, Chapter XXIII, the following passage:

> It is also known that it is abstractions that give rise to most difficulties when one wants to examine them in detail, just as it is known to those who are acquainted with the subleties of the Scholastics, from which the things that are most thorny drop at one go if one will banish abstract beings and resolve to speak ordinarily and using concrete terms only, and admit no other terms in the demonstrations of science than those which represent substantial subjects.

Now, all difficulties disappear indeed if we begin by considering the word 'time' as an 'apparent noun', an abbreviation replacing more complicated expressions which relate to things, to 'substantial subjects'. However, moderate realism's view of time is not necessarily obliged to accept this point of view as the only possible one. It can also admit Bergson's philosophy of continuity, according to which the only concrete reality is represented by events, by motions, and, on the contrary, every division of the world into independent beings is an artificial division, at least if matter is concerned.

This point of view of Bergson was accepted by the English philosopher and mathematician, Whitehead, in his 'geometry of events'.[64]

However, like Bergson, Whitehead could not do without the idea of an object. The relation of objects to events is expressed by Whitehead in two ways; according to him, both of them are correct. One can say that 'objects are the characters of events', or else that 'the event is the situation of the object'. The term, 'object', is taken in the widest sense of the word either as a quality, or as the conjunction of object-qualities, or finally as a scientific object: atom, electron, etc.

The choice between these two orientations goes beyond the limits imposed on us by our subject. Yet we would like nevertheless to observe that, if one considers the universe from the point of view of its macro-structure, then 'reism' (for this is the name given by Kotarbiński to the point of view 'of substantial subjects' of which Leibniz speaks) cannot be discarded; but, if we descend to the micro-structure of the universe, the parallelism between corpuscular physics and wave-physics shows us that the philosophy of continuity, the philosophy of events, ought to be considered as admissible as that of 'substantial subjects'.

The reists consider inertia to be the most essential feature of physical bodies, and since the field also possesses an inertia, they therefore conclude that even a wave field is not an event lacking a substrate, but rather a thing. Let us, however, not forget that this inertia is only a secondary product of energy; it is not the inertia which, in the sense of classical physics, could be attributed to motionless bodies.

These observations suffice, in any case, to justify our point of view, which attributes no separate existence to empty time. Following the criticism of the theory of Newton and his adherents, we have still to criticize Kant's idealistic theory.

### *Criticism of Kant's theory*

110. Let us now consider Kant's arguments one after another. The first '*argumentum ex antecessione*', according to Deussen, was the only argument in the Latin dissertation of 1770 if we leave aside the observation of continuity. Kant began with the words: '*idea temporis non oritur, sed supponitur a sensibus*'. Alas, we share the opinion of those who see in this argument only a vicious circle; Kant supposes in advance that which he has to prove. He says that the order of succession or of simultaneity cannot be drawn from experience if the representation of the time does not serve *a priori* as its foundation. However, this is the

dogmatic affirmation of the thesis that should have been proved!

In the Latin dissertation we find only the same dogmatic affirmation.

*Quae enim in sensus incurrunt, utrum simul sint, an post se invicem, nonnisi per ideam temporis repraesentari potest, necque successio gignit conceptum tempris, sed ad illum provocat.*

Consequently, all that strikes our senses, be it as succession or as simultaneity, can be represented only by means of the idea of time; succession does not itself create the idea of time, but has recourse to it; viz., it causes us to make use of the idea.

Everyone will agree that the idea of time is not found ready-made in the data of the senses, but where do we find the proof that it cannot be acquired by way of abstraction or by some other effort of our mind? Still, according to Kant, if this idea is drawn from experience, it is only because the idea was put into experience earlier on.

Kant's opinion is not herewith excluded. Perhaps we do draw from experience only what we have previously put into it. But we lack the proof of this. We can see in the dissertation that Kant was irritated with the empiricists, who tried to draw the idea of time from the succession of ideas in our mind. He claims that, if I understand the significance of the term, 'after', it is thanks to the fact that I already possess the idea of time.

However, this observation does not prove that the temporal character of this 'after' is of our own invention. Kant appeals to the fact that time cannot be found in sensations, that it cannot strike our senses since it is no real object. All this is true; but all this does not yet prove Kant's thesis. If time is neither an accident, nor a relation in the common meaning of this word, it can at any rate possess a '*fundamentum in re*'. But this '*fundamentum in re*' has nothing in common with our time, as Kant might have said.

This is possible, but Kant has not proved it with his argument. Everything he said could be summed up in these words: *idea temporis non oritur, sed supponitur a sensibus*. We do not see here any proof.

Our requirements for the exactitude of logical proofs are today too strict for us to take Kant's idle talk on this subject as an exact and rigorous proof.

The second argument, '*ex adhaesione*' provoked many misunderstandings. The objection raised with respect to Kant was that, if we abstract the entire contents of our consciousness, then consciousness itself will be thereby suppressed. However, we think that Kant's idea has a different meaning. Let us take an event for example. We may assume that, instead of this event, another event could be produced in the same time, or else that no event whatever occurred in that time. But the free place for an event in this time could not be annihilated. Let us apply the same reasoning to all events, and we shall understand what Kant meant to express.

Commonsense is prone to the belief that, even in the case of the non-existence of the world, we would be able to distinguish between a before and an after in the nothingness. For the Scholastics, this was the '*tempus imaginarium*'. It is precisely this '*tempus imaginarium*' that is the cause of our difficulty in representing the idea that, before 'the creation of the world', time did not flow or, to put it otherwise, that there was an absolute beginning of time.

Leibniz explains in the *Nouveaux Essais*, that time and space denote possibilities beyond the supposition of existences. 'Time and space are of the nature of eternal truths which concern equally the possible and the existent.'

Now this is precisely the idea Kant wanted to express in his second argument when he said that we can suppress by thought all phenomena, but that we could not exclude time, itself.

There is no doubt that we possess a deeply-rooted conviction, one might say an intuition, that, should the world not exist at all, time would flow and could not be annihilated. But is it possible to draw from this the consequence that it is an *a priori* form of our mind, and that it does not possess any reality? This intuition can be explained in another way. Of course, it can be used to the advantage of Kant's theory, but on the other hand, it can also be made to agree well with Newton's absolutism, as Clarke has done. Finally, it is not impossible to reconcile it with moderate realism by explaining it in an empirical way. We have no certainty that some intuitions of our mind are not erroneous.

The third argument '*ex necessitate*', has no value at all insofar as it brings out something new here. The time axioms which Kant enumerated, to wit, that time has only one dimension, and others like it, are merely analytical judgements. Whatever the origin of an idea, when once it is fixed as to its contents, it can become a source of analytical, and hence of apodictic judgements.

Now, all the chronometric axioms are merely judgements of this sort, and this has long been established.

The last arguments of the metaphysical exposition are the argument of infinity and that of continuity, (that is, if one wants to complete their list by Kant's observations contained in the Latin dissertation). Now, there is no doubt that these two properties cannot be extracted directly from experience. What is given is always finite intervals; and continuity, taken at least in the mathematical sense of the word, is only a construction of our mind. True, one can draw from experience a naïve notion of continuity as something wherein every thing is mingled to such an extent that neither gaps, nor leaps can be seen, but it is not excluded that such continuity is of spatial origin.

In any case, from the fact that some elements of a notion are not drawn directly from experience, it does not follow that the entire notion is *a priori*.

111. We distinguish between constructive and reconstructive notions. Only the latter are but a copy of immediate data, while the former bear the stamp of our own creation. We might be inclined to believe that all the notions relating to the universe ought to be merely reconstructive, and that it is only mathematical notions that could be the product of our unlimited creation. However, the history of science in general teaches us that this is not so. There always exists a series of notions of the natural sciences which contain constructive elements alongside the reconstructive ones. We are obliged to complete empirical elements with elements that are not directly given. It is because of this that our scientific notions are subject to endless corrections and to evolutions which will, perhaps, never end. As often as our experience discovers new properties of objects, we are obliged to recast and modify our theoretical constructions, in order to better adapt them to the new investigations.

Now the scientific notion of time does possess, in our view, reconstructive or empirical elements, as well as constructive ones. The latter can be named *a priori* in a sense which might not be identical with that of Kant, but which nevertheless, resembles his meaning considerably. Now, the infinity of time is not an empirical element; it is the result of our own creation, and perhaps scientific continuity is only the result of our own elaboration.

For Kant, however, all the essential elements of the notion of time are of an *a priori* origin; for example, the notions of the temporal order and of change.

'The concept of change is only possible by means of, and within, the representation of time.' Kant does not contest that empirical elements play a role in the apperception of change,[65] but he cannot say anything about this role, except that 'the object affects our mind in a certain way.'

As far as this affection is concerned we know only that it becomes, for our sensibility, the opportunity for applying the *a priori* form of time. In his *First Critique*, Kant tells us that this form 'is found ready in the mind' (*liegt im Gemüthe bereit*); in the Latin dissertation, it is represented as an activity of our mind, useful for co-ordinating the sensations according to a law, but which is nothing innate.

'Conceptus temporis tantum modo lege mentis interna nititur, neque est intuitus quidam connatus, adeoque nonnisi sensuum ope actus ille animi, sua sensa coordinantis, eliciatur'. Another passage of this 'Dissertation' even states: '*tempus non est objectivum aliquid et reale, nec substantia, nec accidens, nec relatio, sed objectiva conditio per naturam mentis humanae necessaria, quaelibet sensibilia certa lege coordinandi, et intuitus purus.*'

But neither is this form of interpretation satisfactory. If pure intuition is only the subjective condition of the coordination of sensations according to a law, it would follow that, outside of that subjective law 'necessary for human nature', no order would exist, or rather that each order of succession and simultaneity would be but our work.

To the question of whether there exists change outside ourselves, Kant answered in one of his letters, that reality in itself is neither variable nor invariable.[66] From the point of view of this agnosticism, the temporal order and change become something extremely enigmatic.

Nor can we give a strict and rigorous proof of Kant's error. At the very most, we could indicate that the attitude of our senses is passive and receptive with respect to change and to the temporal order. The strongest objection we can raise against Kant is that his system does not account for the difference between present, past and future. No order, no succession, no continuity will be temporal, as long as it is not connected with the difference between present and past.

Now, we find, in Kant, not a trace of explanation as to whence this difference comes. On this account a deep silence reigns throughout all his critical works. And this has nothing astonishing to it, since it is impossible to doubt the empirical origin of this difference. Since this difference is the most essential element of the notion of time, every affirmation that this notion is not of empirical origin becomes a manifest absurdity.

Even if Kant were right in his arguments, his doctrine would be absurd since Kant left out the most essential element of time without which no *a priori* form can be so named.

At the very most, Kant should say that some elements of the notion of time, not the most essential ones however, are *a priori* ones.

Moreover, in the metaphysical exposition of the concept of time, Kant emphasizes, the following point; viz., that time is not a discursive concept or a general concept, and that different times are merely parts of the same time. This property of the notion of time was already known to the empiricists; for instance, to Locke. Now this point does not tell us anything about the origin of this idea. Doubtless, it does indicate the exceptional role of our notion among other notions, but nothing more. In any case Kant's affirmation, that this representation has only one object, is explained by him to a lesser extent than in any other system. For, whence comes the certainty that the pure *a priori* intuitions of all living beings refer to one object alone?

112. The 'Transcendental Exposition' of the concept of time carries us over to another ground. The *a priori* form of time must explain to us the possibility of the synthetic *a priori* judgements of arithmetic and of rational mechanics.

A full criticism of this point in Kant's theory would require a long discussion with respect to the relation between the mathematical sciences and logic. It would involve a change in the subject of our research. hence it is impossible for us to examine this problem thoroughly here. Moreover, we have devoted a special dissertation to this subject.[67]

In this respect we share with Frege, Couturat, Russell and Whitehead a negative opinion of Kant's theory.

It is not necessary, however, to proceed as far as these authors did in order to see that Kant was wrong. According to them, there exists no difference between logic and mathematics, since all of the primitive terms and the axioms of arithmetic (we need not speak here of all the mathematical sciences), can be reduced to primitive terms and axioms of logic. Thus the idea of Leibniz has been realized, since it was he who saw in the mathematical sciences only the continuation and extension of logic.

This opinion is not shared by all mathematicians, however. We find among them a formalist school, the most eminent representative of which is Hilbert, as well as an intuitionist school, represented by the Dutch mathematician L.E.J. Brouwer. The formalists defend the traditional point of view; viz., that the object of mathematics differs from that of logic, but even they have no doubt that logic provides the only foundation of arithmetic and that the mathematician need not appeal to intuition to verify a mathematical formula. Only Brouwer's intuitionism has some common features with Kant's intuitionism.

Yet one ought to distinguish between two questions, in Kant's opinion, of the relation between mathematical sciences and logic, viz. the synthetic, *a priori* nature of the judgements of arithmetic, and the relation between the concept of a number and the *a priori* form of time. These two questions can be considered independently from one another. We know for instance, that the famous French mathematician, Poincaré, saw in mathematical induction something similar to the synthetic *a priori* judgements of Kant, without, however, looking for the origin of this synthesis and of the idea of the number in the *a priori* form of time. He harbored no doubt that time was an empirical idea, and that the principle of mathematical induction, which is for Poincaré a synthetic *a priori* judgement, has nothing in common with the idea of time.

It is but the affirmation of the power of the mind, which alone is capable of conceiving the indefinite repetition of the same act, as soon as this act becomes possible for the first time. Mind has a direct intuition of this power.

However, the neo-intuitionism of Brouwer seems to be, in this respect, closer to Kant's point of view, since Brouwer finds the origin of the concept of number in the *a priori* form of time, as we see it in his paper, 'Intuitionism and formalism'.[68]

Brouwer's point of view is not identical with that of Kant, however, and neither the one nor the other is clear to us.

According to Brouwer, each moment of our life is divided into two qualitatively different parts, viz. that of the present and that of the past, but these two parts never cease being united, and thus the two-one-ness relation is formed. The intuition of this relation serves as a foundation for mathematical intuition creating, as it does, not only the numbers one and two, but all of the ordinal, finite numbers insofar as one of the elements of the two-one relation can be envisaged as a new relation of two to one.[69]

But if Brouwer is right in that the ordinal numbers have their origin in this 'two-oneness' relation, we do not see clearly what role is played here by the *a priori* character of time. Nothing will be changed *if we assume* that the origin of the idea of time is *a posteriori*. In any case, Kant's point of view is totally different. In the *Prolegomena* Kant says that, 'arithmetic creates its concepts of numbers by adding successively the units in time'. However, in the chapter of the *Critique*, devoted to the 'Schematism of the Pure Concepts of the Understanding,' we read:

The number is nothing other than the unity of the synthesis, operated in the diversity of a homogeneous intuition in general, by the very fact that I produce time itself in the apprehension of the intuition.

Now these two passages have occasioned the most varied interpretations. The former passage suggested the idea that the concept of number is linked to time – a very feeble argument if one takes into account that every activity of our mind, not only that of counting, takes place in time. The other passage seems to indicate that the *a priori* form supplies the elements of the synthesis of our mind, and that, through the synthesis of these elements, both number and time are produced simultaneously.

However this passage is even less understandable than the preceding one. The point is that time is a continuous magnitude, while arithmetical numbers and their units form only a discrete set; how, then, can both of them be produced by a common act of synthesis? One could answer that the synthesis in time is relative to the segments of time, the number of which corresponds to the number of units synthesized.

This answer does not dispense with all of the difficulties either, since the act which is the production of time ought to be an action outside of time, a non-temporal act, because time does not exist before its production. On the contrary, the synthesis of units in a number takes place in time, as does every arithmetical addition.

Thus, the identification of these two activities always remains mysterious. A defender of Kant's theory could have said that the act, as such, is never temporal, and that it is its product which assumes a temporal form. This would be an unverifiable metaphysical affirmation.

Our most important objection consists in that we do not see any relation between this theory of the ordinal numbers and Kant's thesis that arithmetical judgements are *a priori* synthetic. Even if all orders were of temporal origin, we nonetheless protest against the thesis that the theory of cardinal numbers presupposes the concept of order. Kant did not distinguish between the arithmetic of cardinal numbers and that of ordinal numbers; he did not see that the idea of the cardinal number is simpler than that of the ordinal number, and that the counting operation already presupposes the knowledge of cardinal numbers.

The detailed criticism by Couturat, Frege and others detected in Kant's thesis still other errors. Kant did not see, in arithmetic, the difference between an axiom and a theorem. The fact that the definition of the number is synthetic led him to the affirmation that the judgements of arithmetic are likewise synthetic.

This source of error was discovered for the first time by Richard Manno.[70] Each judgement, the truth of which is the consequence of the definition, is of necessity analytic, even if this definition is synthetic.

113. Our criticism of the transcendental argument is very incomplete; we have limited ourselves to dogmatic affirmations, but this has been

solely in order not to deviate excessively from our principal subject. We must again point out that, in Kant's arguments, logical problems are confused with psychological ones, and that is why the controversy around the method of the *Critique* has extended up to the most recent decades. It is true that Kant distinguished between '*quaestio iuris*' and '*quaestio facti*', that he separated the psychological deduction of ideas from the metaphysical and transcendental deduction, but nevertheless, in spite of Kant's intention, psychological elements stole into the *Critique*. This is especially true in the case of his theory of time. We are familiar with Schopenhauer's comparison. He said that the *a priori* forms of space and of time are like the colored spectacles on the eyes of a man. Just as a man who wears green spectacles sees everything in green, so a man endowed with an *a priori* form of time and space sees everything in a form that is alien from reality itself.

The philosophers who interpreted the *a priori* of Kant in a strictly logical and methodological sense became indignant with this comparison which stresses above all the psychological sense of the *a priori*. However, in reality, this comparison is not erroneous. Is it possible to interpret the theory of Kant, in a different way if he assures us that the thing-in-itself has nothing to do with space and time, that reality itself is neither variable nor invariable, and that the temporal order and every change exist only for our mind? The adherents of the logical interpretation forget that they themselves are obliged to close their eyes in order not to perceive the many passages of the *Critique* which give evidence against them. If it is true that the *a priori* of a judgement emphasizes in particular its logical character of apodictic, necessary and universal truth, then the *a priori* of pure intuition would be completely incomprehensible were it deprived of any psychological meaning.

Schopenhauer did commit one error: he identified the *a priori* of time with the ideality of time. But this is precisely Kant's own mistake. The arguments aim at demonstrating that the idea of time is independent of the senses and of experience in general, but, in the continuation of the *Critique*, Kant is persuaded that he proved that time and space, because of their *a priori* origin, possess no absolute reality.

It has been justly noticed that ideality is not a necessary consequence of the *a priori*. From the fact that our mind possesses the ability to know something independently of experience, it does not follow that the object of this knowledge is our own invention and possesses no absolute reality.

114. However, prior to passing our decisive judgement on Kant's theory, we must again consider the 'Explanation' and the 'General remarks in respect of the Transcendental Aesthetic', which Kant added to his proofs of the *a priori* of time.

This is what Kant says in the 'Explanation':

I have met, on the part of perspicacious men, so unanimous an objection against this theory which attributes an empirical existence to time, while fighting against its absolute and transcendental reality, that I concluded that this objection should present itself naturally to the mind of any reader who is not accustomed to these considerations. The objection is formulated as follows. Real changes do exist (which is proved by the succession of our own representations, even if one would deny the external phenomena, as well as their changes). Now, changes are only possible within time, and thus time is something real.

To this objection Kant replies:

I admit the argument in its entirety. Time is, of course, something real, viz. the real form of internal intuition. Hence it possesses a subjective reality with respect to internal experience, which means that I have, indeed, the representation of time and of my determinations in time. Therefore, it must really be considered not as an object, but as a mode of representation of myself as an object. Yet, were I able to submit myself to intuition, or could another being act upon me in this way, without this condition of sensitivity, then the same determinations which we understand as changes would give us a knowledge in which one would no longer find the representation of time, nor, consequently, that of a change.

Thus, the empirical reality of time remains a condition of all of our experiences. It is only absolute reality that cannot be attributed to it, according to what has been put forward above.

Obviously, the explanation does not explain anything, it is even unreliable, since Kant pretends not to understand what it is all about. His opponents will never consent to the distinction between the absolute and the empirical reality of time, solely because another vision of the world is not excluded.

However, Kant continues his explanation when looking for the reason why his opponents treat the ideality of time in a different way than that of space. They suppose, says Kant, that the reality of external objects is incapable of any rigorous demonstration, while the reality of the object of our inner sense (of myself and of my state) is immediately clear to our consciousness. The external objects could be a simple resemblance only (*Schein*) but the object of our inner sense is, according to our opinion, incontestably something real. 'However, they do not believe that these two kinds of objects, without our being permitted to combat their reality in so far as representations, belong always only to a phenomenon with two faces: one, when the object is considered in itself

. . . and the other in which one takes into account the form of the intuition of this object, which one should look for not in the object itself but in the subject to which the object does appear and which nevertheless does belong with a lesser reality and necessity to the phenomenon of this object.'

Most important to us is the end of this explanation, where Kant recognizes the participation of an empirical element in the apperception of change and motion.

The fact that the 'Transcendental Aesthetic' contains no more than two elements – space and time – results obviously from the fact that all other ideas belonging to sensibility, even the idea of motion which combines the two elements, presupposes something empirical. Indeed, motion presupposes the perception of something moving. Now in space, considered in itself, there is nothing movable, hence it must be that the movable is something which is only found in space by experience, and therefore something given empirically. Therefore the Transcendental Aesthetic could not reckon among these *a priori* data the idea of change, as it is not time itself that changes, but something that is in time. This therefore supposes the perception of a certain existence and, of the succession of its determinations, consequently the experience.

The latter proposition provides food for thought. No wonder that what moves in space and changes can be given by experience only. However, the last-mentioned proposition contains something more; it tells us, or seems to express, that experience gives us not only the perception of a certain existence, but also the perception of the succession of its determinations. This would be an enormous concession to realism, but we must not forget everything Kant said beforehand, viz. that 'the idea of change is only possible by way, and in the representation, of time', and that 'the succession would not fall by itself under perception if the representation of time did not serve it, *a priori*, as foundation.'

All the same, we do not see clearly, in the succession, the difference between what belongs to empirical data and what belongs to the form *a priori*. One would be inclined to believe that succession exists as much in empirical determinations as in our intuition of a one-dimensional, homogeneous medium. However, according to Kant, the latter (*a priori*) conditions the former (empirical data), or, in other words, the apperception of the succession is only the product of the co-operation of two elements, that of an *a posteriori* agent and that of an *a priori* one. The role of the latter agent is so great that, according to Kant, without it the time outside ourselves 'is nothing'. Hence, change and motion are given, thanks to the co-operation of the form [of intuition], without which, the reality itself, 'is neither variable nor invariable.'

115. All this finds its confirmation in the 'General Remarks on the Transcendental Aesthetic', in which three points are of especial interest to us: Kant's relativism, the paradox of the inner sense, and the relationship between Kant's theory and natural theology. In the second Remark we read:

> To support this theory of the ideality of the external, as well as of the internal, sense, and therefore of the ideality of all of objects of the senses, as simple phenomena, one can make use advantageously of the following important remark: that everything which, in our consciousness, belongs to intuition, (hence it is not here a question either of the feeling of pleasure or of pain, nor of volition, none of which are knowledge at all), comprises only simple relationships, relations of places in a single intuition (extension), relation of the change of places (motion), and the laws which determine this change (motor forces). But that which is present in the place, or that which happens in the things themselves outside the change of place, is not hereby given.
>
> Now, simple relationships do not permit us, however, to become acquainted with the thing itself. . . . The same applies precisely to internal intuition . . . which cannot be anything other than the manner in which mind is affected by its own activity.

The internal intuition would be an intellectual and not only a sensory intuition if everything 'diverse that is in the subject were given to us spontaneously (*selbsttätig*). But the case is different here. Mind intuits itself, not as it would represent itself immediately and spontaneously to itself, but according to the manner in which it is affected internally, consequently such as it appears to itself, and not such as it is.'

The last-mentioned words will become more understandable to us if we also take into consideration the Fourth Remark of the Transcendental Aesthetic, where Kant speaks of the relation of his theory to natural theology. If time and space were conditions of every existence in general, they would have to be so, likewise, for the existence of God. In order to avoid this absurdity, we have, according to Kant, to assume his point of view as the only possible one. Our mode of intuition 'is called sensible, because it is not original that is to say, it is not such that by it alone is given the very existence of the object of intuition (a mode which, as far as we can estimate, can only be found in the Supreme Being), but because it depends upon the existence of the object and, consequently, is possible only insofar as the subject's capacity of representation is affected by it.'

Our sensible intuition is called derived intuition (*intuitus derivativus*), as opposed to the intuition of God, which is original (*intuitus originarius*). This is the same intuition that Kant also called intellectual intuition (*intellektuelle Anschauung*), 'it can never belong to a being dependent as to its existence and its intuition, which determines its existence relative to given objects.'

Kant admits that the only aim of this remark is to elucidate, but not to prove his theory. In any case, we shall stop at this point, since the following problem is posed: is Kant's point of view indeed the only possible one for preserving the belief in a non-temporal Being? And yet Aristotle and the Scholastics taught that the nature of God was non-temporal, without admitting the idealistic theory of the universe. If God lives in the eternal present, does it follow that the temporal order of the world cannot possess any absolute reality? It seems that we do not need to affirm at all that time is an indispensable condition of every existence, nor that it is the form of our mind. Changes may have an absolute nature, and there can also exist a component of reality, perhaps the most essential one, which is eternal and immutable.

For the Scholastics each substance, and especially our 'I', is a being outside time, since its determination in time is only incomplete and external. On the contrary, according to Kant, the transcendental 'I' is not only outside time, but it is also the maker of the temporal aspect of the world, a thing difficult to understand if we consider that it is not the creator of its existence.

It is almost impossible to understand that time, as our *a priori* form, possesses nothing corresponding to it in absolute reality. In the Latin dissertation of 1770, time is called '*aeternitas phenomenon*', just as space is called: '*omnipraesentia phenomenon*'. However, if the thing that corresponds to time in absolute reality is nothing but eternity, what is the meaning of motion and change? Much more understandable is Plato's expression, since time, as the image of eternity, becomes mobile, and, being mobile, it ceases to be eternity.

In the *Critique*, moreover, Kant says expressly that 'if one disregards the subjective conditions of sensory intuition, time is nothing'. He does no longer look for the correlate of time in the immovable eternity, and already in his letter of 1772 to Markus Herz, he says that phenomena, as things-in-themselves, are neither variable nor invariable. Thus he rejects the opinion that everything is immutable, like the thing-in-itself, and admits that the determination of the thing which corresponds to time in absolute reality, surpasses the abilities of logical thought.

The same idea emanates from what Kant says, in his *Critique of Judgement*, with respect to the conciliation of mechanism with teleology; for the human mind, attached to temporal form and to discursive thought, this conciliation is impossible. However, this is possible for an 'intuitive understanding', to which everything does not presents itself as an accidental junction of its parts, as also to our 'discursive understanding'. In the case of such an intuitive understanding, mechanism and finality would coincide, and perhaps the temporal becoming – though Kant does not say anything about it – would appear in its true light. The idea of this intuitive understanding was developed later by Schelling, but with a modification that the temporal relations, as we know them, acquired in Schelling an objective significance.

However, if one accepts Kant's idea, expressed in his letter to Markus Herz in 1772, viz. that the correlate of the temporal flow of phenomena in absolute reality is not an immutability, but that it is something which exceeds our abilities of logical thinking,[71] one does not see what profit natural theology should gain from Kant's doctrine. What is essential for theology, consists in distinguishing between the non-temporal existence of God and the temporal existence of His creation.

Now, the realism of Aristotle and of the Schools introduced this distinction in a much simpler way, by stating that the substantial being is exempt from the temporal flux. However, Kant only added a complication to this in distinguishing between the eternal being, the temporal aspect of the world which exists only for a living being, and the correlate of this temporal aspect of the world whose nature is completely inextricable for logical thought. We could object that he damaged natural theology since, by taking away from the world its temporal form, he brought creation nearer to God and, thereby, made the difference between God and creation less substantial.

Were this consequence the unavoidable result of these proofs, we would nevertheless have to accept it. However, we have seen that the proofs were not unshakeable; on the contrary, all of them rested on misunderstandings, and not one possesses an absolute value.

116. The 'General Remarks' contain still other, highly interesting things. Kant stops here at the fact that the internal sense does not find itself in a better situation than the external sense by disposing of its temporal form; even in the internal sense one has to distinguish between the temporal phenomenon and the noumenon outside of time. Kant calls this the paradox of the internal sense '*das paradox des inneren Sinnes*'.[72]

This doctrine has brought about protest which is fully justified. Its defenders are not lacking either. For instance, Schopenhauer appeals to the case of dreams which, on the occasion of a brief objective sensation, unfold long stories to render this sensation rational. In his article 'Versuch über Geistersehen'[73] he even cites cases of telepathy as empirical proofs of Kant's doctrine, and the same ideas are developed later by du Prel in his *Oneirokritikon*. Yet this interpretation seems to us nonetheless highly problematical, at any rate it is neither necessary nor the only one possible. The above-mentioned facts merely prove that estimation of duration in dreams is almost always wrong. Now, such errors can be very well reconciled with the absolute reality of time.

The distinction between two realities in our internal life, one of which provides us with the raw material of internal sensations and the other apprehending these sensations in temporal form, seems to us to be tenuous. How can we prove that these sensations, prior to being apprehended by the *a priori* form of internal sense, were deprived of any temporal duration? The analogy with the external senses should be, according to Kant, so complete that even in the internal life our science apprehends only relations whose elements are not cognizable. Thus the primacy of internal experience before external experience is abolished, as regards its evidence. 'With the stroke of the pen, and without any discussion, Kant has removed the firm foundation of modern philosophy, the rampart erected by Descartes against scepticism'.[74]

Kant expresses himself with great circumspection in saying that the pure intuition of time is only the 'manner through which mind is affected by its own activity', and that he does not want to exclude the possibility of intuiting himself without this form. However, even if this possibility were realizable, the reality of our terrestrial life would not thereby be abolished. Thus we can accept the Kant's supposition, viz. that a being different from us, God for instance, or we ourselves, when placed under other conditions, would have another aspect to our psychic life. Yet, for us, this would be a new reality, one in lieu of the preceding one. And if God has his own point of view, whence comes the certainty that He does not participate in our point of view and does not view the world from both sides? The degradation of our point of view to an illusion or appearance is by no means founded.

However, Kant does not want to identify the phenomenon with appearance or illusion. We reply that, if this distinction between the phenomenon and the appearance is justified for the external senses, it is not justified for the internal sense, at least when one takes the word 'appearance' in one of the two senses in which Kant uses it. The point is that we can demonstrate that Kant sometimes opposes appearance to phenomenon, and sometimes appearance to the truth, or the absolute reality. In order not to repeat things already discussed in the philosophic literature, we refer the reader to Vaihinger's Commentary.

It suffices moreover to compare the third remark of the Transcendental Aesthetic with the introduction to the Transcendental Dialectic, in order to see these different meanings.[75]

An objection could be raised against us that, if Kant sometimes opposes appearance to the truth, we must not forget that, to him, the word 'truth' does not have the same sense as in realistic philosophy. Now, we can prove that Kant also uses the word 'truth' in two different senses. According to one of these two senses, truth consists in conformity to the laws of the mind.[76]

However, beside this definition of truth, so characteristic of Kant's philosophy, we find in Kant a second definition coinciding with the traditional one, where truth is merely 'the conformity of our concepts with their objects.'[77]

We can attempt to reconcile these two definitions of truth in Kant,[78] but we affirm that Kant cannot do without this second definition taken in the most literal sense, because otherwise it would be impossible to find any sense at all in the expression, 'if I could intuit my own self without that condition of sensibility, which is time, these same determinations which we represent as changes would give us a knowledge in which we would not find the representation of time'. If this new knowledge were to be true, this would be a truth according to the second definition, conformity with the object, and not agreement with the laws.

Hence, if we accept the second definition of the appearance as opposition to truth, and the second definition of truth as conformity with the object, completely independent of our cognitive abilities, we are entitled to say, in the very words of Kant's, although in spite of his intention, that the ideality of time makes illusory all of our internal experience.

117. We have still to consider Kant's relativism as expressed in one of the 'Remarks'. There is no question here of any skepticism: Kant wants only to say that all of our science deals merely with relations. Kant likewise applies this thesis to our internal sense. The point is that, if external senses provide us only with simple representations of relationships, we must not forget that the representations of the external senses constitute the proper material with which we enrich our internal sense. However, Kant remarked that the external senses do not give everything

to the internal sense since, in saying that everything which in our consciousness belongs to intuition contains only simple relationships, he observes parenthetically that 'it is neither a question of the emotions of pleasure and of pain, nor one of volition, which themselves are by no means items of knowledge.'

In the second argument in support of the thesis that the internal sense only gives us relationships, he points out that time contains only relations of succession, of simultaneity, and the relation of 'simultaneity to succession (permanence).'

'Now, simple relations do not enable us to acquire the knowledge of a thing in itself; we only learn in what manner our mind is affected by its own activity.'

This psychological relativism of Kant does not show itself with sufficient clarity if one takes into account that the emotions and volition, which 'are no cognitions', play an exceptional role. We could object here that sensations are not cognitions either, but that they can become objects of cognition to the same extent as emotions and volition. At any rate, it is noteworthy that some handbooks of psychology indeed quote the law of relativity 'as a general law of our sensations, but this law is not formulated identically by all psychologists (Wundt, Ulrici, Bain, Höffding).

Some of these forms recall very well the idea of Kant; for example, Höffding gives the following form to this law: the objects of our sensations are not absolute objects but relations between the objects, while objects themselves are only something unknown. This law is sometimes formulated in a different way, viz. that, in our sensations, the relationship between an object and our mind is expressed as a subject.

We have no need to criticize this law which can be correct in some of these forms. What is of interest to us is solely whether this law can be regarded as a confirmation of the ideal nature of time. Of course, the relationship may possess properties which the elements of the relationship do not possess. However, in order that this law might confirm the ideal nature of time, we must have an absolute certainty that the elements of the relationship are completely deprived of any temporal character. Now, such a certainty is given us neither by the above-mentioned arguments of Kant, nor by the superficial image of the law of relativity.

Thus we see that the 'General Remarks' do not improve the situation in any respect and that, in some details, they even provoke astonishment before the levity with which Kant treats problems of prime importance. Above all, the paradox of the internal sense provoked much protest, and

Cousin rightly stresses that: 'the inconceivable negligence with which this claim is set up and as if hidden in a corner of the Transcendental Aesthetic, has thus far screened it from attention, while it deserved profound examination because it contains enormous consequences; it is the unnoticed root of the errors of Kant and of those of his successors.'

118. Our criticism was directed, in the first place, against the ideal nature of time, while the object of Kant's arguments was to demonstrate the *a priori* of time, whereof the ideal nature of time was considered by him as an immediate consequence. However, it was noticed very early that these two problems – the problem of the reality of time and that of its origin – are not linked together as closely as Kant supposed.

As early as 1786, a German empiricist, Christian Selle (1748–1800) objected to Kant, in an article written in French and published under the title, *De la réalité et de l'idéalité des objets de nos connaissances*, that ideality does not follow from the *a priori*; that time and space can be forms of our sensuality, but that the thing in itself can also possess these forms. This problem was later the subject of a long controversy between Trendelenburg and Kuno Fischer, in which Kuno Fischer defended Kant's point of view, whereas Trendelenburg pointed to a third possibility overlooked by Kant. Much as the problem had at first been formulated by Trendelenburg in a rather defective manner, in principle we must admit that he was right; with this correction, there are in truth four possibilities, as Vaihinger observed, viz.:

(1) time is real, and its knowledge is *a posteriori*,
(2) time is real, and its knowledge is *a priori*,
(3) time is ideal, and its knowledge is *a posteriori*,
(4) time is ideal, and its knowledge is *a priori*.

The first case is the point of view of empiricism and of ordinary realism, as represented by Aristotle and Newton (we make no distinction here at all between extreme and moderate realism). The second case is that indicated by Trendelenburg. The last one is the point of view of Kant.

However, who represented the third point of view in the history of our problem? Vaihinger introduces this division only for theories of space, and he has quoted Berkeley as being representative of such a theory of space. However, as for time, it seems to us that it is the Eleatics. They repudiated the reality of time just as Kant did, but there is no doubt that they accepted the empirical origin of the notion of time.

We reject Kant's point of view and that of the Eleatics. There remain only the realistic points of view. Thus, the problem is what value should be attributed to the possibility raised by Trendelenburg and by numerous other German and non-German philosophers, among whom we also find some Italian philosophers such as Spaventa. In a thesis published in 1880 under the title *Kant e l'Empirismo*, the latter wrote, that 'the error consists in the belief that subjectivity in fact excludes objectivity and *vice versa*'.

It is improbable, we think, for the idea of time to be of a completely *a priori* origin, since such a solution to the problem would require the admission of a pre-established harmony between reality and the faculties of our mind. It seems to us that the influence of experience, as far as time is concerned, goes farther 'than awakening our power of cognition and setting it in action.' There needs be accepted intermediary solutions which attribute an empirical origin to at least some features of the idea of time, while assigning to some others an origin independent of experience. Obviously, we also have to distinguish between extreme realism (Newton) and moderate realism (Aristotle and Leibniz), so that, even if we give up the point of view of Kant and the Eleatics, there will result, according to these distinctions, six possible points of view; viz.:

(1) extreme realism of time and its origin: *a posteriori*,
(2) extreme realism of time and mixed origin (*a priori* and *a posteriori*),
(3) extreme realism of time and *a priori* origin,
(4) moderate realism of time and *a posteriori* origin,
(5) moderate realism of time and mixed origin,
(6) moderate realism of time and *a priori* origin.

As we rejected the exaggerated realism of Newton, none of the first three points of view are of any significance to us. Of the three latter ones, according to what we have said before, the 5th case is the soundest, viz. that of moderate realism, which claims that the idea of time is of mixed origin. Such was, for instance, the point of view of Aristotle who found in the idea of time subjective elements beside objective ones, and who therefore stated that time would not exist without the human soul. We have already explained this thesis of Aristotle in the historical section of our work. Here, we want to consider something else.

We can indeed state that all the elements of the idea of time are of empirical origin, but that the idea of time is nonetheless the product of our mental activity, because our mind's attitude in the formation of this idea cannot be a solely passive and receptive one. Now, some philosophers assume the notion to be *a priori*, simply because the synthetic activity of our mind is indispensable in such a case; viz. that of the synthesis of the understanding, of reproduction and of recognition. Otherwise, in spite of the succession of impressions in our mind, we would have no consciousness of this succession. The succession of sensations is not the sensation of the succession, as Husserl said. Because of this synthetic activity, which is indispensable to the creation of the notion of temporal succession, the notion of time can be considered as *a priori*, although all these elements are borrowed from experience. However, in our opinion, the active role of our consciousness does not suffice to call the notion an *a priori* one, because this role exists in each synthetic notion, be it constructive or reconstructive, with the possible exception of analytic notions.

If we assume that the idea is of mixed origin, this means that some of the components are not drawn from experience. These are the infinity of time and its continuity, understood in the mathematical sense of the word. On the other hand, succession and simultaneity are of empirical origin, although to perceive them, the active role of our consciousness is indispensable. But above all, the distinction between present and past is of purely empirical origin. The elements of nature, which do not present a faithful copy of what is given, can have a conventional character. Our conception of the *a priori* does not exclude such an interpretation. And even the significance of the reproductive elements of a notion can be changed. Indeed, we must not forget that the field of our experience is very limited. Our experience is unable to reach either the macrocosm or the microcosm: that is why our interpolation or extrapolation of the direct data may always be unreliable. We do not rule out possible modifications of our notion of time.

119. We reject the idea of the *a priori* as a blind and mysterious force which weighs heavily upon us, just as did the *fatum* of the Ancients. Such a conception of the *a priori* has nothing in common with science or with the philosophy of science. It is the great merit of Husserl to have modified the sense of Kant's *a priori*, by limiting it to the consciousness of what is necessary and essential in the structure of an object. However, when penetrating the essence of objects, Husserl fell into the errors of the old metaphysics. Discrimination as to what is essential to, or accidental in, an object may change with the progress of science and, according to our opinion, it depends solely upon the definition of an object, which can also vary with experience. The point is that all *a priori* truths undergo a limitation of their application. Phenomenological investigations can in no way

settle the problem of the reality of an object, or of its origin. The progress of sciences gives us the respective warning. It is from this point of view that the phenomenological axiomatization of time should be considered.

## 2. THE SECOND GROUP OF TIME *APORIAS*

120. The research referred to in the preceding chapter has led us to the result that the most important components of the idea of time are of empirical origin and that, if there are components which are not a perfect copy of immediate data, even they enter into the idea as, at least, the probable correlates of what exists independently of us. Every uninformed man is inclined to look for the origin of the idea of time in the facts of the experience of our senses as well as in the facts of internal observation, seeing that all the phenomena of the physical, as well as the psychical, universe take place in time. However, the history of philosophical doctrines has shown that some philosophers attributed to only one of these two groups a fundamental role as far as the genesis of the idea of time is concerned. That is why, among the realistic theories some possess a rather psychological color, while others have a naturalistic one. The former group comprises the doctrines of the neo-Platonists, of St. Augustine, of Locke, Berkeley, and especially of Bergson. Since the latter group is the more exclusive and the most original, we shall start our criticism with this group before proceeding any farther. It is our opinion that the role of the direct data of consciousness in the formation of the idea of time is exaggerated.

What Bergson attributes to time is our psychical life. It is beyond doubt that, when the states of consciousness succeed one another, they penetrate each other, organize themselves, and are added to one another dynamically like the successive notes of a melody. But anyone may ask why he should limit the sense of the word 'succession' solely to this special succession, the example of which is given to us by our psychical life; why should we limit the sense of the word 'duration' only to the duration of states which merge into one another, without the ones having any tendency of externalizing themselves with respect to others? Bergson's answer is that any other representation of duration and of succession is only a phantom of space, only an image wherein the idea of space occurs surreptitiously. If this is the case, it proves that, even in concrete duration there exists something which assimilates it to space, in spite of the differences between homogeneity and the succession of heterogeneous elements which merge into one another. Even in the succession of states which do merge there exists an order, and it matters little whether this is, or is not, the order of states externalizing themselves the ones relative to the others.

However, let us suppose that Bergson is right in saying that one ought to restrict the sense of the word 'succession' to the succession in which penetration, interdependence and organization do exist simultaneously. What are we to do with the succession wherein this dynamic relation does not exist? According to Bergson, it does not deserve this name; it is not a true succession. Bergson expressed this opinion with respect to the succession in the physical world where 'each of the successive states exists separately, and their multiplicity is endowed with reality only for a consciousness which is capable of preserving them.'

However, in *Matière et Mémoire*, Bergson changed his mind; he sees in the physical world a duration resembling our own, wherein succession exists with dynamic organization, the only difference being that the tension and the 'rhythm' of this duration are not identical everywhere, and that the foundation of the organization of these durations in the organic world is not memory. We accept these ideas with much sympathy, as we do with everything Bergson says in the chapter 'Duration and Tension', in the same work. We think that this chapter possesses a much greater value than anything else Bergson has said about past and future.

Now let us assume that successions are everywhere organized dynamically, sometimes by the action of physical forces, as in the inorganic world, sometimes by the advent of memory, as in living beings; it would nonetheless be impossible to dispense with the term 'succession' in the primitive sense of the word. The point is that, both in the states of nature and in the states of organic and spiritual life, there can well exist more or less isolated successions of facts. What happens to the organized succession when an animal dies?

Bergson is right throughout, except with respect to the terminology he employs, since succession always remains succession, and the succession linked to the organization of the successive states is a succession connected with something that is not indispensable to the existence of the succession itself. The same should be said with respect to the significance of the term 'duration', with this difference, that the vicissitudes of its meanings are much more varied. The term 'duration' has been always ambiguous. In the widest sense, it can have the same meaning as time, for instance when one says that a phenomenon, an event, lasts a certain number of minutes. However in other expressions, duration means the permanence of a being when something varies. It is in this sense that the Scholastics contrasted the temporal flow of movement with the duration of the substance which, according to their understanding, was to a certain extent outside of time. Kant distinguishes three kinds of temporal relationships: simultaneity, succession and duration taken in the sense of permanence. However, the latter relationship can be expressed, according to Kant, by means of the two preceding ones, since permanence is a kind of simultaneity of something with a succession.

Now in Bergson, the term 'duration' has acquired a sense almost opposite to the preceding one, because it does not signify the always identical permanence of a being, but, on the contrary, the temporal flow of states which merge together in such a way that it is impossible to clearly separate what remains unaltered from what varies. He shares the opinion of the Scholastics; viz., that the being which is always identical with itself has no duration, and this is why he says, in connection with atoms, that they remain in an eternal present[79] since time has no hold on them. Time slides over atoms without changing anything in them, and the latter preserve no trace of the time elapsed. However, in *Matière et Mémoire*, Bergson changes his opinion. He develops – both in this work and, to a greater extent, in his later works – a new conception of matter, according to which each separation or distinction of the substance and its attributes is only an artificial one. We attribute considerable importance to this idea of Bergson; it has already had repercussions in Whitehead's theory, and the progress of quantum physics seems even to confirm it to a certain degree.

But we do not proceed as far as Bergson does. If we are inclined to attribute to matter a duration similar to ours, this does not mean that nature can be considered 'as a neutralized, and consequently, a latent consciousness.'[80] This would be a gratuitous metaphysical hypothesis, without any value for positive science, and we do not want to affirm anything more than what follows from the actual state of science. Now the state of science causes us to think that there do not exist, whether in the physical or in the psychical world, absolutely permanent substances, as the ancient atomists supposed, and that the successive states of the world are added dynamically to the preceding states in such a way that it is impossible, as Heisenberg says, 'to attribute to an electron or an atom a determined point of space as a function of time.'[81]

If we thus make such a case for Bergson's idea, we wish to treat it simply as a scientific hypothesis subject to revision with the progress of sciences.

This is also why we do not completely reject the idea of permanent duration. It roughly retains its scientific value, as there is no doubt whatever that some properties of bodies seem to be preserved while others vary. However, if there are things in physical bodies which behaves tolerably like substance, e.g. in their mass or electric charge, we do not approve of the scholastic idea that everything that seems to be permanent ought to be considered as exempt from the flow of time. This permanent duration can be measured if it is accompanied by the flow of change and, in so far as it is measured, it does exist in time. Therefore Descartes very well observed with *considerable reason* that we measure permanent duration using the same units of time as in the case of successive duration. And, if we agree with Bergson that every permanence in the world of phenomena is only approximate, or that it is impossible to separate it from its variable states, then the argument of Descartes acquires a new strength. Thus we can consider Bergson's theory as an elaborate continuation of the theory of Descartes.

We shall return again to this point in Bergson's theory, and examine now his other ideas which seem doubtful to us. Everything he says about motion in the *Données Immédiates* should be corrected by what he says in *Matière et Mémoire*. In the former work, Bergson thinks that 'motion, insofar as it is a passage from one point to another, is a mental synthesis, a psychical process', and that this synthesis 'is endowed with reality only in our consciousness.'

On the other hand, in *Matière et Mémoire* we learn that motion is not solely a mental synthesis in our consciousness, but rather an organization outside of our consciouness, which is not 'without some analogy with the continuity of our own consciousness'. We think, however, that this correction should be purified of any panpsychic element, which is solely a gratuitous hypothesis. If motion is not merely a series of definite positions, if the successive states are organized, we can speak here only of an organization of physical forces.

121. But the most important objection we raise against Bergson consists in that he wanted to eliminate from our consciousness every quantitative element of time. Above all, it is not true that each measure of time can be reduced to counting simultaneities, although the establishment of simultaneities is the foundation of all measurements. What we do count is not the ascertained simultaneities but the intervals which separate them. Bergson strives to avoid at all costs the word 'intervals', but it is in vain. He tells us that, if we count the beats of a pendulum, we

consider each sensation as being equal to the preceding one, but they are really not equal because of their reciprocal penetration. However, the one does not exclude the other; besides a continuous, qualitative change, there exists something that is repeated. And should our intuition fail to sense this something, how could we succeed in counting the beats of the pendulum? Bergson explains that this is only because we see that the external cause always remains the same. Is it true that we perceive the repetition of a note or of an auditory sensation thanks to the perception of the identity of an external cause? According to *Matière et Mémoire* we know that, even in the physical world, repetition is not absolute, and all the same, we do measure physical time. Evidently, the one does not exclude the other.

In opposition to what Bergson affirms, we should state that intuition provides us with one of the most solid foundations for the measurement of time. It is intuition that grasps the equality of two intervals, which, moreover, Bergson himself ascertains by opposing the relativity of the physical time to the real duration of our own life. By opposing time-quality to time-quantity, he is right insofar as our consciousness makes use of the qualitative data of tension, expectation, and surprise for the estimation of the time intervals. In any case, however, consciousness can estimate the equality of two intervals, and it perceives them through the repetition of sensations, although this repetition is not an absolute one. Hence, neither the order nor the interval of a duration are foreign to our intuition and to the immediate data. But how is that so, it will be asked. Do we not need the effort of thought and of abstraction in order to discover this order and these intervals in this data? Certainly we do! However, to discover that the successive states of our real duration organize, penetrate and merge into one another, Bergson was also obliged to use a great abstraction and thought. It does not suffice that our *I* becomes absorbed in its own sensations and allows itself to live only; it is necessary that the immediate data be observed and analysed internally. Bergson's analysis is perhaps even more thorough and more scrupulous than that which only discovers in consciousness order and intervals of succession.

If Bergson says that the immediate data may be vitiated by analysis, he probably means this: we must not substitute the products of analysis for the immediate data, since, in so doing, we atomize and immobilize them. Instead of a living continuity we thus obtain immobility and atomism, alien to the immediate data. But if we abstain from this substitution, if one does not realize the products of the analysis, the latter cannot vitiate anything. On the contrary, it is then indispensable, since it explains to us what the immediate data do contain.

Hence it is not analysis as such that is dangerous, but rather a propensity of our mind to substantialize, so to speak, the products of our analysis and to substitute these products for the immediate data. However, we do not want to do this. By stating that order is not foreign to the immediate data, we do not dispute the fact that this is an order of states which organize themselves and penetrate one another. If we state that the equality of two intervals is not alien to intuition, we do not deny that we sometimes need an effort of attention to perceive this equality.

However, Bergson's greatest worry is to refrain from introducing into the immediate data the phantom of a homogeneous space. It seems to us that, by interpreting the work of the analysis as we are doing, we are not left open to this objection. But if, in spite of everything, there exists an analogy between the temporal order and the spatial one, this is neither our fault nor that of the analysis. This analogy does exist and Bergson can do nothing about it. These two orders have widely differing properties, but an order is present here and there. Finally, we repeat the same objection which we raised against Plotinus, whose theory much resembles that of Bergson. Both speak of the continuity of our mental life. Now what is the meaning of this term 'continuity' if not that of an ordered, though non-spatial, extension.

Kant refers to our 'I' as to the numerical identity of an absolute being which is the source of all of the *a priori* syntheses. Now, Bergson sees here only continuity. Hence it is impossible to eliminate from this the notion of order and the metric properties. Should this continuity mean something else, we would have to stop here. However, neither Plotinus nor Bergson have done this. Plotinus confined himself to the remark that time need not be measured in order to exist. Quite so, but we ought to distinguish between measurement effectively performed and measurability as such. Now, Bergson's statement that our real duration represents a time-quality rather than a time-quantity has the same value. If the continuity of our duration possesses quite special qualitative properties, this is no proof that the idea of a measure cannot be applied to it. The difference between a distinct multiplicity and an indistinct, or qualitative, one has no value at all. Any continuous area, even a spatial one, can be termed an indistinct multiplicity if one likes, since it has neither leaps nor gaps.

Our temporal duration, insofar as it is continuous, is measurable;

*it would be absolutely impossible for our analysis to apply the notion of measurement here if our intuition failed to give here any foundation for it.* Immediate data can be invalidated by analysis only if we substituted the products of analysis, as a new reality, for the direct data. However, there is no need to do this.

122. By opposing physical time to our own duration, Bergson observed, by the symbol *t*, physics does not denote a duration, but rather a relation between two durations, a certain number of intervals, and that the intervals, themselves, do not enter into the calculations of the physicists; on the contrary, consciousness perceives these intervals because they are lived duration. This observation is absolutely accurate. Indeed Bergson should have said more, that it is consciousness that appraises the equality of two intervals, which forms the basis of each measurement. However, this appraisal is not absolute. If physical time is relative, then our estimates are not infallible. They deserve to be trusted only up to a certain extent. Thus, were our intuition the only basis of our measurement of time, then time itself would become very subjective. In his last book dealing with time, which was written in connection with the theory of Einstein in 1922 (*Durée et Simultanéité*), Bergson admits that the idea of duration did not entail, as a necessary consequence, the hypothesis of a universal time, 'but that it was very naturally in keeping with this belief').[82] If it was in keeping with this belief, it was only owing to the fact that Bergson had widened the foundation of his time doctrine, viz. by basing it not only upon the immediate data of consciousness, but also upon the data of the biological theory of the *élan vital*. However, even this extended foundation does not vindicate the idea of universal time if we recall all that Bergson has said with respect to the different rhythms of time in living beings in his *Matière et Mémoire*. In order to demonstrate more clearly that this psychological and biological foundation is insufficient to support the idea of a universal time, we shall take the liberty of quoting here the idea of a famous Russian scientist of German extraction, Karl Ernst Baer (1792–1876), expressed in his article of 1860 under the title, *Quelle conception du monde organique est vraie*?' It is likely that Bergson was aware of these ideas when he wrote his *Matière et Mémoire*. These ideas are so interesting that they deserve to be known precisely.

The starting-point of these ideas is a fiction, but a highly instructive fiction.

The velocity of free motions, of sensations and of the psychic life appears to vary, in various animals, as the speed of the heart-beats. Since, in the case of rabbits, the heart-beat is four times that of the bull, the rabbit has, during the same period of time, four times the number of sensations and is able to carry out four times as many voluntary acts of motion as the bull; in a word, the rabbit lives four times as fast as the bull. Speaking more generally, we can say that, during the same astronomical time, the internal life of different animals and that of humans takes place with different velocities and, therefore, the subjective appraisal of time is different for every animal. It is only due to our subjective measurement of time that we are able to consider an organism, e.g. a plant, to be something stable and immutable as regards its form and size, at least within certain limits. The point is that we can see it a hundred times every minute without noticing any external change. However, if we imagine that the pulse beat, the ability of perception and the whole of the external and spiritual life are considerably accelerated or retarded, then everything will assume a completely different appearance; phenomena will have an improbable and almost supernatural form from the point of view of our present physiological structure which, however, will become necessary and natural as soon as our structure changes. Let us assume that our life, from childhood until old age, has shrunk to its one-thousandth part, to the duration of one month, and that the pulse beats one thousand times faster than in reality: we shall then be able to follow, more easily than the flight of a butterfly at present, the movement of a rifle-bullet from one point to another. The point is that the movement of a projectile during one second would correspond to one thousand heart-beats and, consequently, when measured on the scale of our sensibility, it would become 1000 times slower. Had our life shrunk again by the factor of 1000 to the duration of 40 minutes, the flowers and grasses would seem to us as stable and immutable as do the mountains and rocks, the crumbling of which we can discover, but not observe directly. During our entire life of forty minutes, we would not be able to see the geological transformation of the earth's crust. The movement of animals would be imperceptible to us because of their slowness: at most, we would discover them just as we now discover the movements of the stars. Should the contraction of time become still greater, light might take on the nature of an acoustic phenomenon. Instead of seeing it, we would hear it, and all of the phenomena which we presently call sound or voices, would no longer affect our ears.

However, let us imagine something opposite, viz. that our life, instead of shrinking, would begin to increase enormously; what new aspect of reality would open up before us! If the pulse-beats and the perceptive ability would become slower by the factor of 1000, so that our life would last about 80,000 years, and if over one year we would live, through as much as we do now during one-third of a day, we would be able to see, within four hours, the melting of the winter ice, the growth of vegetation and its wilting. Many a phenomenon, because of the high velocity of its evolution with respect to the pulse-beats, would be completely imperceptible. Fungi would shoot forth rapidly before us, as if they were fountains. Day would follow night just as a bright minute follows a dark one, and the sun would traverse the dome of the heavens as if it were a projectile.

If our life would slow down again by the factor of 1000, so that we could make, within one year, about 190 distinct perceptions, the difference between day and night would disappear completely, the trace of the sun would assume the form of a luminous arc, and all of the forms that seem to be ordered and solid would dissolve in that course of events, absorbed by the violent tempest of the phenomena.

Although these considerations are supported by a fictitious supposition, they nonetheless indicate the relativity and subjectivity of our appraisal of time. Otto Liebmann, a German neo-Kantian, quotes this passage with true rapture, and finds in it a confirmation of Kant's doctrine. However, we do not sanction this inference, since the relativism and subjectivism mentioned here are not identical with Kant's idealism. In all of the Baer's reasonings, the absolute reality of the temporal nature of the changes is an indispensable supposition. It is only our appraisal of the length of time that is subjective, insofar as it is conditioned by the physiological process of pulsation. Now, no real psychologist will dispute that this appraisal can be subjective, much as the reality of the temporal flow is indubitable. However, Liebmann's reasoning is as follows: since neither the subjective, psychological time nor the time of the physicists is absolute, it is Kant who is right, and not Newton!

However, beside Kant's and Newton's theories, there exists still a third point of view. If Newton is not right, this is by no means tantamount to Kant's being right. We accept Liebmann's affirmation; viz., that if we disregard all of the physical and psychical phenomena, there remains only the idea of an objective possibility of succession, which is nothing real. Now, Liebmann believes that this objective possibility is a δυναμει ον, [potency of being], which is connected with the organization of our intellect and which corresponds very well to what Kant has called the *a priori* of the time. In our opinion this inference is not necessary. The reader will remember that the reasoning of Liebmann corresponds to the second argument of Kant (*argumentum ex adhaesione*). Now, already in the criticism of this argument, we noted that this possibility of time can be reconciled with each of the three theories of time, which we distinguished. According to realism, this δυναμει ον is sooner inherent in the nature of being, as the possibility of its self-realization through change and, has nothing to do with 'the organization of our mind.' Our mind also possesses the faculty of realizing this possibility through change and, moreover, of perceiving it. However, the faculty of perceiving the change has nothing to do with the realization of the temporal flux through change. Only the appraisal of this flow, or that which Bergson called the tension of the duration, is subjective.

By introducing the term, 'rhythm of the duration', in his *Matière et Mémoire*, Bergson drew much nearer to Baer's considerations. True, their points of view are not identical. But both of them represent to us the same moderate realism. And we see that the point of view of the 'immediate data' is given up once and for all by Bergson. In one of the following chapters we shall try to study thoroughly the sense of the term, '*tension de la durée*', (rhythm of the duration), as applied to the physical world. As far as the question is one of applying these terms to consciousness and to living beings, this concept is well explained by Baer's examples and by what Bergson says in his *Matière et Mémoire*.

123. Before proceeding further, we should like to give some explanations regarding the terms 'absolute' and 'relative' in order to eliminate all possible misunderstandings. Each of these terms has more than one meaning. We have called Newton's time absolute, because it flows in itself and by its own nature. However, there exists another sense of this word, which can be reconciled with the moderate realism of time, and even with idealism. This other meaning is connected with the idea of the uniformity and measurement of time. Even if we accept the point of view of Kant, or Aristotle, we can still assert that time flows uniformly, and that the length of this flow is independent of our subjective appraisals which are discussed, e.g., by Baer and by Bergson in *Matière et Mémoire*. In this sense, the term 'absolute' is applied to the term 'relative' and has nothing to do with the problem of reality. By identifying the flow of time with the flow of events, we can nevertheless protest against the possibility of a 'contraction' of time in our own appraisals, a contraction that would be brought about by physiological conditions, as Baer believes, or by the 'tension of the duration' according to Bergson.

It is true, however, that moderate realism very often entails relativism. This relativism is quite visible in Leibniz's polemic with Clarke, concerning the meaning of the proposition that God could have created the world one year earlier. Perhaps this realism was already foreshadowed by Aristotle when he said that the number attributed to the movement would not exist without soul. However, the term 'relative' also has several meanings. One of them leaves intact the meaning of uniformity. The other, introduced by the theory of relativity, can even cause the term 'uniformity' to be deprived of any sense whatsoever. The point is that, in saying that the appraisal and the measurement of the length of time are relative, we can nevertheless assert that the relation of the intervals is always the same, that equal parts always remain equal and, consequently, that the uniformity is not affected. That is the relativism of Leibniz and of classical physics. But the relativism of relativistic physics goes further. For parts which to one observer are equal are not equal to another.

Now, Bergson defended the absolute nature of our appraisals of duration in his 'immediate data'. It was only physical measurements that were relative according to him. However, in *Matière et Mémoire* it seems to us that he also introduced a relativism for the appraisal of the intuition, since the same duration can be long or short, depending upon the tension and the rhythm of the duration of a living being. But this relativism did not abolish the uniformity of time and this is why Bergson was able to defend the idea of universal time in his *Durée et Simultanéité*, although this idea had not been a necessary consequence of what he asserted in *Matière et Mémoire*. Bergson, himself, says the following in *Durée et Simultanéité*: 'nothing proves rigorously that we would rediscover the same duration when we change the surroundings'; different durations, i.e. durations endowed with different rhythms, could co-exist. We once put forward a hypothesis of this kind, with regard to living species. We distinguished between durations endowed with a tension which is more or less high, characteristic of the various degrees of consciousness which would be spaced out over the length of the animal kingdom; nevertheless we did not perceive then and we do not now see any reason for extending this hypothesis of a multiplicity of durations to the material universe. We have left open the question of knowing whether the universe was divisible or not into worlds, independent of one another; our world, with that special impulse (*élan particulier*) which life manifests in it, sufficed for us. However, were it necessary to settle the question, we would opt, in the contemporary state of our knowledge, for the hypothesis of a material Time, one and universal.[83]

Let us summarize once more all of our objections to Bergson.

1. He has set limits to, and changed, the meanings of the terms 'succession' and 'duration'. For him, true succession is only the organic succession (for the sake of brevity, we give this name to the succession linked with the penetration, interdependence and organization of successive states). Similarly, the term 'duration' – which, when used in the sense other than time, signifies the permanence of a being – has acquired with Bergson a limited sense of flow, corresponding to organic succession, wherein one need not distinguish between the thing and its states, since 'the thing and the state are only snapshots taken artificially during the transition'. The new notions are not without value; on the contrary, we attribute to them great importance, but we do not approve of the elimination of the broader, universally accepted, senses of these two words; nor do we approve of the motives behind this elimination, since:

2. The motive of this restriction of meanings is insufficient. If we

introduce really spatial elements into our representation of time, this only proves that there exists a difficulty in representing time in the form of concrete images. Besides, even the Bergsonian duration is not without an analogy in space, since space too forms a continuum, which implies an order and measure.

3. It is not true that, in measuring time, 'one counts simultaneities.' What we do count are the intervals created by the positing of simultaneities. Bergson tries in vain to eliminate, at all cost, the idea of the interval from the measurement of psychic time.

4. If it is true that physics has no hold over the length of the intervals themselves and that it only explains the relations between these intervals, our intuition does not find itself in a much better situation. The point is that, if intuition does explain sufficiently well the equality of two intervals, provided that these are not too long, the appraisal of the length of an interval, itself, is highly subjective. Even if we disregard the varying conditions of waking and sleeping, of fatigue and rest, of the variation of attention, of diseases and mental disorders, appraisal of intervals likewise depends on what Bergson has quite rightly called the 'tension of the duration' of a living being.

Finally, the last objection of a general nature concerns the relation between analysis and intuition. Even intuition benefits from analysis and, consequently, it is not true that analysis falsifies the direct data. These would only be vitiated if we substituted the results of the analysis for the direct data. The point is that only then do we introduce immobility and atomization, in the place of mobile continuity.

The final chapter of the section, 'Immediate Data' introduced, as a new property of the time-quality, or of the concrete duration, the freedom and creation of unforeseeable things. It is this property that, along with irreversibility, plays the most important role in *Creative Evolution*. We would observe here that, even had Bergson been right as far as freedom was concerned, it would still have been impossible for us to introduce it as a component into the idea of time. This freedom is not found everywhere; we do not perceive it in the inorganic world and, since all events take place in time, freedom is not essential for the temporal flow. However, one detail ought to be mentioned at this juncture. The reader will remember Mach's statement that, if everything were absolutely determined in the world, the world would be immovable. If anything does happen, it is solely thanks to the indetermination which exists in the midst of the Universe. These ideas were not well justified in

Mach, and they even seem to us to be in contradiction with what he says elsewhere with respect to physics; viz., that change and its directions are determined by the difference between energy levels, and by the law of entropy.

In any case, Bergson's ideas concerning freedom deserve comparison with those of Mach, and we shall return to this subject when discussing the value of the ideas of quantum physics for the notion of time.

Bergson's ideas concerning freedom are likewise insufficiently justified, although some of his observations are quite right, e.g. that in the life of the mind nothing is repeated in the literal sense of this word. It is not out of the question that the same be true with regard to the states of the physical world, if we look at this from the point of view of modern theories. However, Bergson has failed to provide the proof that the uniqueness of a fact completely excludes its determination. Similarly, the argument that the concepts of necessity, of foresight and of determination, cannot be applied to what is engaged in becoming, is not entirely convincing; had it been right, it would also be valid for the process of becoming in the physical world, especially if we consider that Bergson was forced later on to extend his theory of time to the physical world.

Finally, we must not forget that determinism was supported not only by means of the law of causality, but also by means of the law of logic according to which every proposition which has a meaning is either true or false. This argument, which has not played a large role in modern philosophy, was known however to the Stoics and it is absolutely correct in our opinion. Thus, the indeterminist must not forget that, when defending freedom, he also defends the existence of propositions which are neither true nor false. Each proposition that refers to an action depending solely upon my will is neither true nor false, if indeterminism is right. However, it is questionable whether all indeterminists realize this.

124. Having stressed the weak sides of Bergson's theory, we will now enumerate its merits. One of these merits is that, having broken with the idea of time as a medium or a homogeneous form introduced by Kant, he represented the flow of time as a flow of temporal phenomena, thus rendering time accessible to scientific research, a merit which he moreover shares with empirical psychology. An analogous step forward was made by the theory of relativity which, having broken with Newton's theory, contributed to the knowledge of the nature of temporal phenomena from the point of view of physical sciences.

The second merit of Bergson consists in his emphasizing the fact that the answer to the question of the nature of time is a response to the question, 'what is reality when taken in its most general and most essential properties?' since the flow of time and duration present a common and most profound property, not only of the entire universe, but also of the entirety of reality, of being as such. According to Bergson, even God should be considered *sub specie durationis*. However, leaving aside the last-mentioned detail, which has perhaps the least value of Bergson's ideas, let us not forget the throng of new ideas introduced by him, which tempt us to carry out more profound investigations. Such are the ideas regarding the difference of rhythm, or of the tension of duration in living beings, perhaps suggested by Baer's considerations; the idea of the relationship of the mind to the body from the point of view of time; the definition of perception as a process which immobilizes or condenses the enormous periods of duration diluted into a few, very differentiated moments in a more intense life; finally the idea that the discrimination between a thing and its states is only an artificial one. It is really a great pity that Bergson, instead of developing these ideas in his later works, devoted his efforts to the defence of the idea of a singular, universal time. For indeed it is precisely the theory of Quanta that, we believe, has cast new light on the philosophy of continuity. For this reason we will devote a separate chapter to the problem of continuity.

Finally, we would be inclined to cite, as a merit of Bergson, that he cleared up the origin of the idea of time. But has he really done this? We do not think so. All he did in this domain is to stress that the successive states organize themselves, and add themselves, dynamically to one another. However, does this suffice to explain the origin of the idea of time? Of course, not. It has been even affirmed that Bergson 'did not take his theory from psychology'.[84] Now this is an exaggeration. There is no doubt that the theory was constructed from the point of view of psychology, but it is true that there exist psychological problems which Bergson failed to mention and that even those problems which he considered were represented in a unilateral manner and in such a way that Bergson himself was obliged to correct his opinions in his later works. He not only wanted to eliminate all metrical properties from the intuition of time, but he even tried to impart to the terms of order and continuity a qualitative meaning altogether unique.

125. In order to obtain a clear answer to all questions attached to the problem, both of the origin of the idea of time and of the connection

between psychological and physical time, we shall take into consideration the following points: (1) Do we owe the idea of time to the observation of motion, as the neo-Scholastics assert, or is the source of this idea to be found in the direct data of consciousness, according to Bergson's claim? Can we say that one of these two times, the psychological and the physical time, is original, while the other one is only secondary? (2) Is it possible to speak of the sensations of time and even of the existence of a sense of time? (3) What is the role played by memory in the formation of the idea of time? (4) How should we appraise the existence of the illusions of time? (5) What role do the ideas of order, of moment, and of the interval play in our idea of time itself?

We have already prepared the reader for our answer; viz., that neither Bergson nor the neo-Scholastics are right. If a child understands the difference between past and present, can we say that this happens only thanks to one group of facts, and to the exclusion of others? We believe that it is all of the child's experience that cooperates herein and that the role of sense-data is just as important here as that of memory. The difference between what is really given to our senses and that which the child remembers and expects, opens the child's eyes to the difference between that which is present and that which is not present. It has been pointed out that children often confuse the difference between 'tomorrow' and 'yesterday'. What is important to them is the difference between presence and absence. Probably after a long series of experiences, children grasp the difference between what exists at present but is not present, being either hidden or distant, and that which is neither presently available nor presently existing, being past and never able to return. Later, they see better the difference between what will take place and what is already past; between absence in the past and absence in the future.

The origin of the first idea of order, tied to the use and understanding of the terms 'before' and 'after', is still more difficult to discover. However, one thing is certain to us, that the phenomena of motion alone do not give us the opportunity of knowing order. In this respect, Locke's criticism was totally justified.

In any case, we can venture the response that the phenomena which have played the most important role in the life of persons, societies and nations have likewise most contributed to the formation of these ideas. Perhaps these phenomena were not identical for everybody and everywhere. We must distinguish between the origin of this idea in a child, in the life of an adult, and in that of a nation. The order of the succession of befores and afters could have been discovered in each group of phenomena, and the role of memory was here as inseparable from direct sense-data, as it was central in the formation of the general idea of present and past.

This so important role played by our memory in the formation of the idea of time in general, or of the temporal relations – such as succession, duration, order – has brought about many misunderstandings. The point is that, if memory is indispensable in the perception of temporal relationships, it is nothing as far as their significance is concerned. Let us assume that we live without memory; probably we would never succeed, under such conditions, in forming an idea of time, but our life would nevertheless take place in time. It is only by way of misunderstandings that memory can enter the definition of time. Even if the psychical states are added dynamically to one another, as Bergson assumes, time canot be identified with memory, since the past is not identical with the recollection of this past.

It is for this reason that the distinction between psychological and physical time, inasmuch as it is based on memory, has not played any role in the formation of the idea of time, neither in the life of nations nor in the life of an individual. This distinction so conceived is only the artificial product of all of those scientists who had confused the meaning of an idea with the subjective conditions of its formation. It is justified solely because it indicates that our duration represents a special case of the general duration.

If duration in general denotes the permanence of a being during the temporal flow of states which themselves change ceaselessly, then our own duration differs from this permanent duration, as Bergson pointed out, because the permanent being of our 'I', and of its states intermingle in such a way that any distinction between 'I', on the one hand, and its states, on the other, becomes artificial. However, we cannot identify this duration with memory. Memory never does embrace the entire duration; it is always only fragmentary; it only resuscitates, from case to case, the fragments of our past. And even if it embraced our entire past, this past life in our memory would be completely different from the past itself. The recollections of our youth are not identical to the life of youth.

Now what sense can we attribute to the distinction between physical and psychological time? Does it suffice to say that the former is time-length, while the latter time-quality? This assertion remains in contradiction however with what Bergson observed elsewhere; viz., that it is our consciousness, our intuition, which ascertains the equality of two intervals and which could not be deceived as regards their duration, even if by some chance, the intervals of the clocks were to contract or to dilate. Consequently the only difference consists in that the measurement of time in the natural sciences relies upon rational principles, on chronometers constructed according to these principles, whereas the measurement of psychological time is based upon the appraisal of our intuition. Such appraisal is not without error, but it is this error that, to a certain extent, guarantees that the measurement of time is not absolutely conventional, as Poincaré supposed. We remind the reader here of Enriques' considerations, of which we completely approve. It is above all the rhythmical processes that give our intuition occasion to recognize the equality or inequality of two intervals.

Now does this perception of the equality of time intervals occur before or after the perception of order? It is difficult to give a conclusive answer to this question. However, it seems that it is easier to establish the difference between 'before' and 'after', than it is to appraise the distance between these by comparing it with another distance. Measurement implies order, as Enriques has rightly established, but order does not imply measurement.

We can constrast the time of intuition, or the time of direct data, with the metrical time of the natural sciences, with this correction: intuition also measures time, but it does so following what it senses directly, while science measures time relying upon a few rational principles. Finally, we must add that the measurement of time is not the sole object of the physical science of time. Physical science can also specify and correct the direct data which concern the temporal order. In his work *Philosophie der Raum-Zeit-Lehre*, Reichenbach distinguishes the qualitative properties of time, such as the topological, from the metric properties; thereby imitating the distinction between the topological and the metric properties of space. We can accept this terminology without being afraid of misunderstandings. Yet, we must observe that the topological properties of time, just as the metric ones, rely sometimes on intuition, sometimes on rational principles which correct and control intuition.

If the Scholastics considered physical time to be the primordial one, and psychological time secondary, they wanted therewith to express the fact that the data of introspection can exist only after that of the senses. In this respect they are right. Direct perception precedes reflexion, and the object of our first knowledge is the external world. Nevertheless, we find that this opposition of psychological time to physical time is dangerous because of the ambiguity of the term 'psychological'. When speaking of physical time, the Scholastics thought only of the time created by local movement. Now, is the time of the auditory sensations physical or psychological? Bergson illustrated the relations of internal duration by the notes of a melody. Several psychologists distinguish, in each perception, the act of perception, the contents of this act, and its object. The object is physical, but the contents are always mental. Now it is not always easy to distinguish the physical object from its psychical contents. We wish to leave out all these subtleties since they would take us too far away from our subject and, instead of opposing psychological time to physical time, we prefer to set intuitive time in opposition to metric time, based as it is on rational principles, or sometimes on necessary conventions. In so doing we are obliged, however, to admit that intuition also measures time and that the data of intuition can be of fundamental value for ascertaining the equality of two intervals.

126. The second point that we will consider is the problem of the sensations of time, and even of the sense of time. Mach was the first to speak of the sense and sensations of time. Hume, who distinguished between 'impressions', and ideas as copies of these impressions, acknowledged that we have no temporal impressions, that time is only the manner in which impressions appear, and that the idea of time comes to us from the succession of our impressions. This explanation is not clear. If each idea is only a copy of the impression, whence comes the idea if the impression does not exist? Mach stressed the fact that the recognition of the same rhythm in different series of tones is a matter neither of intelligence, nor of reflection, but one of sensation, the sensations in question spanning only very short durations of time. Wundt is not pleased with the introduction of the term, 'sense of time', but he nevertheless speaks of direct temporal ideas. He criticizes those psychologists who do not distinguish between the direct apperception of time and all the indirect appraisals, in which different associations and the secondary influences play an auxiliary role. He thus blames both the psychologists who do not recognize the existence of direct ideas of time and those who see it everywhere yet do not understand that the succession of ideas is not identical with the idea of succession. Now, there is not the slightest doubt that temporal relations are not readily perceived everywhere with the same ease; according to Wundt and to most psychologists, it is especially the sense of hearing that deserves to be called the sense of time and,

following that, the sense of touch as the second 'mechanical sense'. And, among the auditory sense data, mental images of rhythm play an exceptional role, since they are intimately linked, according to Wundt, to certain complex sentiments and emotions.

These temporal images, including those which Wundt calls the direct ones (*unmittelbare Zeitvorstellungen*) are not identical, however, with sensations in the proper sense of the word. They are rather units of a superior order, based on perceptive elements which correspond to what the German psychologist, Ehrenfels has called '*Gestaltqualität*' (quality of form, of the whole). If we consider attentively everything Hume and Mach meant to say, we come to the impression that they were thinking of something similar to Wundt's notion. Mach speaks explicitly of the identity of the temporal aspect, or of the temporal form (*Zeitgestalt*), in the various tonal complexes. However, even if we accept this manner of solving the difficulty, there still remains another problem. For indeed, some psychologists (and it appears that even Ehrenfels belongs to their ranks) suppose that the preceding notes should still exist in consciousness when other notes succeed them, in the absence of which we would not understand the total action of the melody. Others, on the other hand, suppose that the preceding notes may act on consciousness, without being present in it.

Schumann is of this opinion, and H. Delacroix, in his article '*La Perception du temps*', (which forms a chapter of the Second Volume of the *Traité de Psychologie* of Georges Dumas), appears to share Schumann's opinion. 'To form the whole, means to act as a whole, and not to be present simultaneously. In a general manner, what Ehrenfels and Meinong call formal sentiments, sentiments of relation, all this does not indicate that the elements are present simultaneously, but merely that upon these elements, a unit of a higher order is built. And this criticism includes something just, since there comes necessarily a moment where the melody is complicated, where the elements escape our present apperception, either remaining or not in our memory, and where the form that they exhibit is given as it were separately, surviving the ephemeral appearance of the elements. This only proves, however, that memory and perception are closely linked, as we will see through the study of direct memory, and that one glides imperceptibly from the one to the other.'[85]

We cannot exclude the possibility that a part of truth is to be found in one camp and part in the other. But Wundt would probably not be satisfied with an intermediate solution such as this since he wanted to retain a difference between the immediate idea of time (*unmittelbarte zeitvorstellung*) and the mediate or indirect idea (*mittelbare Zeitvorstellung*). Now, the difference between the one and the other, according to Wundt, consists in that: in the former, the totality is supposed to be given simultaneously, while the indirect idea is based on reproductions and associations.[86] Schumann criticizes the existence of Wundt's direct perceptions, for the former admits the existence of both past and memory where Wundt would speak only of the present and of consciousness. In theory, Schumann seems to be right, for how can we see simultaneously that which, by its essence, is not simultaneous. But is the theoretical present identical with the present of consciousness? This is why Delacroix sees, and not without reason to our minds, an element of truth in Wundt's statement: 'It is no less true that the observation is in agreement with Wundt's experiments wherein he seeks to demonstrate the complexity and fullness of the present, the reciprocal implication, the simultaneous inclusion and exclusion of the moments of duration. It is unquestionable, on the one hand, that the different elements of the present are not on the same plane, otherwise the present would seem static to us and, on the other hand, that the present does not contain merely a single element given with the character of the present, while all the rest is pure memory. This character of the past present, of the immediate past, of the future in the making, is the very paradox of time. For James and for Bergson, duration is precisely this mutual penetration of the states of consciousness.'[87]

This passage is rather disheartening because we see that experimental psychology has not cleared up any of the difficulties created by the theories of time. In any case, we must admit as certain that it is absolutely impossible to separate perception from remembrance, since every perception has a duration, and each moment of perception superimposes itself upon the residue of the preceding moment. It is in this way that the problem of time perception, as well as the problem of the distinction between psychological and physical time, brings us to the same problem of memory.

127. The reader will recall that, having done with our criticism of Bergson, we further resolved to deal with five points, viz.: the scientific value of the distinction between psychological and physical time; the problem of temporal sensations and perceptions; the role of memory; temporal illusions and the mutual relations of such concepts as order, duration, continuity, etc. Thus we shall now consider a third point. We have already seen that it was impossible to discuss the first two points

without taking into account the role of memory, however. It remains for us to complete that which we already stated. The discussion of the first point gave us the result that memory is indispensable in order for the idea of time to be formed, but that memory cannot itself enter into the definition of time, even if we admit Bergson's thesis; viz., that it is our consciousness which instructs us that the successive states are added dynamically one to the other. This observation is correct, and it is this dynamic or organic succession that makes impossible the distinct separation between that which Wundt called the direct, immediate idea of time and its indirect idea. But Bergson's observation does not justify the identification of the past with memory, which he seems to do in his theory of time-quality.

What are the causes of this error? No doubt it is the equivocal nature of the words 'present' and 'memory' that is responsible. The word 'present' can refer as well to that which is the real present, as to that which is present only thanks to recollection. From this fact, Bergson concludes that there exists no precise moment at which the present becomes the past; the consequence being, that any distinction between present and past is only conventional and depends solely upon the effort of attention. Who would deny that he allows himself to be carried away by the ambiguities of the terms used?[88]

The term 'memory' can also mean several different things. In the proper meaning of this word, we understand by memory that faculty which reproduces [a representation] after an appreciable interval and after the original excitation had been blotted out of the memory. But there exists also another form of memory, 'the direct memory', which means 'a privileged, but very short, persistence of impressions'. Delacroix devotes to this an interesting chapter in the *Traité* of Dumas. 'These direct recollections resemble consecutive images, they possess a particularly lively character of freshness and vividness. We still hear the names and the words spoken resound. The deferred memory possesses this feature only exceptionally.'[89] Now, Bergson did not discriminate at all between direct and deferred memory. Whenever he speaks of the continuity of states of consciousness, we have the impression that it is simply a question of the unity of consciousness, of the unity of our 'I', which maintains itself over the duration of our entire lives. Thus consecutive images, direct memory, deferred memory and the unity of consciousness – all these figure under the same title of time-quality. Of course there is something in common among these properties of our internal life, but the omission of the order [of succession] and of the interval of time which likewise belong, we believe, to the direct data of intuition, renders inexact and superficial Bergson's entire representation of the internal life.

Among the time theories of the German philosophers, there is one which recalls that of Bergson, especially because it too stresses strongly the role of the 'I' for psychological time. This is the theory of Baumann. Having explained the main theories of time in modern philosophy up to Hume, Baumann gives us, at the end of the Second Volume of his work, his personal views on the subject. He distinguishes between psychological time on the one hand, and psychological–astronomical time, on the other.

Psychological time is conceived as a relation between the successive internal phenomena and the image of the 'I' which coexists with them. The sequence, or succession, of phenomena becomes time only if it is referred to the permanent duration of the 'I'. We therefore see that the difference between Bergson's theory and that of Baumann is quite large. The resemblance of the two theories is confined to this detail, that the unifying principle, which connects all the successive elements one to another, is recognized as indispensable. This psychological time, lacks regularity and infinity, at least *a parte ante*.

The time of everyday life is psychological–astronomical time, which differs from the preceding one in that the series of our representations is considered as being caused in us by movements in the external world. Finally, astronomical time tries to bring the preceding time to a greater degree of precision, thanks to mathematical calculations based upon the observation of celestial motions.

However, the special role of the psychological time is not clear in Baumann. Sometimes we have the impression that what he wants to say is that, without the unity of our consciousness, the idea of psychological time would have been impossible. Moreover, we should point out that without this unity, no idea whatever would be possible, not merely that of psychological time.

The condition of the formation of the idea is not identical with the constitutive note [*sic*] of that idea. However, if Baumann wanted to introduce the 'I' as the constitutive element of the idea of psychological time, he ought to have given us an explanation of the meaning of his terms. Of the three temporal relations: simultaneity, succession and duration (in the sense of Kant), the second one is the most important for the idea of time. It seems to us that this opinion is shared by most philosophers. The permanence of a being is indispensable for its duration. Now, in matters of duration, it is psychological time that provides us with an example. It is not impossible however, that there exist in the physical world as well, something permanent that lasts – always the same – in the flow of time.

A similar remark must be applied to Bergson's system, as we have already done several times. What he describes as time is not even duration (in the classical sense of the word), but a special type of duration, wherein the successive states are added dynamically to one another, and wherein unity is maintained only via successive states. Also, in this case, it is not out of the question, (and Bergson accepts this in his later works) that such a duration also exists outside of ourselves. Now, if the 'I' plays a role in this duration (and, with the 'I', also something of what Bergson calls memory), all this can enter into the description of this special duration, but not into that of time in general. And even in this special duration, the role of memory ought to be specified in order to avoid causing misunderstandings such as those to which Bergson himself fell victim.

128. The fourth point to be considered is that of temporal illusions. Wundt devotes a large place to them in his *Physiological Psychology*. This is a very interesting problem, but for us it has only secondary importance because, in our opinion, we cannot draw from it the slightest consequence of importance to the questions of the reality of time or to the origin of the idea of time. At any rate, it is a great pity that Bergson left this problem out of his first work, since it would have shown him that basing the idea of time solely upon the direct data of consciousness was a more difficult task than he thought it to be. Wundt distinguishes between three groups of illusions. The first refers to the appraisals of duration in direct perceptions of time, (as Wundt understands these); the second concerns the transpositions which take place in momentary sensations; and the third group has, as its object, the positional shifts occurring (*Zeitverschiebungen*) within a continuous series of sensations. This list is not a complete one; we should add to it the illusions linked to the memory of time, e.g. the law of temporal perspective which Spinoza already discussed, but Wundt has left all of this out, since these illusions are connected with the indirect temporal ideas based on reproductions and associations. Further on in his work, he turns to the discussion of the reproduction of temporal ideas, but he there takes into account only the results of laboratory experiments whereof it is a question of the comparison, for example, of two time intervals one of which has already disappeared from direct consciousness. We can say therefore that the problem of temporal

illusions has not been treated in a complete manner, even by this great, systematic mind.

We have put forward the opinion that the fact of temporal illusions can neither play a decisive role in the problem of the reality of time, nor in that of time's origin. However, we may have some doubts about this. We fought the idealism of time in Kant, by supporting claims with the facts of internal experience. How can we reconcile this evidence with the fact of internal illusions? Now, we are not attributing this evidence to the convictions or judgements which relate to our abilities and our past. My conviction that I understand, a mathematical proof for example, that I am able to reproduce it, might be wrong. The same applies to our appraisals of duration and of the positions of certain events of our past. Only the judgements which relate to the direct contents of our consciousness are obvious; those judgements whose contents coincide with the contents of our consciousness. This obvious quality comprises, nevertheless, the distinction between present and past, since the present is only given to us as an incessant flow, as a present which passes.

As regards the problem of the psychological idea of time, the fact of the existence of an illusion does not give a decisive answer either. At best we could say that illusions indicate the need to verify the internal data with the results of our reasoning, in order to arrive at an idea of time which would have a practical value. Perhaps this need became indispensable to man from the very first days of the activity of his intelligence, when he first realized the difference between his own duration, during sleep, and that of the external world over the same time. The differences between wakefulness and sleep, days and nights, undoubtedly forced man to create an idea of time that could account for all of these experiences simultaneously.

129. The last point upon which we shall focus our attention is the relation between notions such as: order [of succession], the interval, the moment. Lalande's *Vocabulaire de la Philosophie* gives us three meanings for the term 'Time', distinguished by the letters A, B, C. Meaning A shows us time as a period which proceeds from a prior event to a posterior one. 'Time' is used in this sense in the expressions: the harvest-time, the time of vintage, often in the plural to designate a period whose limits are only vaguely defined; e.g., heroic times, modern times.

Meaning B denotes the continual change (usually considered to be continuous as well), by which the present becomes the past. This is the meaning of time in Bergson, and also in Fouillé, according to Lalande's,

*Vocabulaire*. Finally, meaning C denotes the indefinite medium in which a sequence of events would unfold, but which, in itself, would be given entirely and indivisibly, to thought (whether it exists by itself, as Newton and Clarke believed, or in thought only, according to Leibniz and, especially, to Kant).

To these explanations are added critical remarks dealing with the problem: 'which of these ideas is the proper and primitive idea' of time. There are, in addition, some observations of other French philosophers, members and correspondents of the French Philosophical Society, who express their personal restrictions as regards the sense accepted in the *Vocabulaire*. Now, according to the *Vocabulaire*, meaning A, which is placed first for this reason, 'is characterized as being by far the most frequent one, either in Greek or in French, not only in everyday language, but also among the philosophers'. . . . 'The use of the word "time" in grammar, music, horsemanship, military training, surgery is likewise connected with this acceptation. We can also add here the time 't', of the mathematicians, considered either as 'the lower limit of a period which is getting shorter and shorter, or as the moment which starts or ends that period.' Meaning A, owing solely to the history of its use and to the breadth therefore, maintains pre-eminence.'

If we compare our classifications with the three meanings of the *Vocabulaire* we shall see that meaning C comprises the theories of absolute realism and of idealism, which we reject. The *Vocabulaire* underlines very clearly what both theories have in common, since both of them conceive of time as an empty and homogenous medium. But [unlike the *Vocabulaire*'s article] we exclude Leibniz from the company of Kant and Newton, and we have attempted to show in our historical section, that Leibniz used the term 'ideal' in a special sense with respect to time; his sense was completely different from that of Kant, and has nothing in common with the idea of an empty medium.

On the contrary, meanings A and B seem to correspond to the moderate realism which we espouse. We say 'seem', because the relation between meanings A and B is not entirely clear. The *Vocabulaire* does not take into account that meaning A can be completed by meaning B, or interpreted according to meaning C, for if we say that time signifies the 'period of a prior event to a posterior event', the question remains open as to whether this is the interval within the flow of events or the interval which is the medium beneath the others. Moderate realism requires the former conception, viz. the clear unification of meaning A with meaning

B, and not with meaning C. Perhaps the authors of the *Vocabulaire* would not oppose this, for they agree that it is only meaning C that radically changes meaning A:

'meaning C can be considered as a widening to infinity of meaning A, but this widening does radically change its character, since the main point of meaning A consists in defining a period, whether short or long, but still distinct, and which, being in opposition to other periods, determines the place of an event with regard to the others. When this period is no longer a reference point but rather a universal container, it is no longer able to serve in this capacity, and only responds to the need of considering all moments and all periods as co-ordinated in a uniform manner.

One thing does worry us, however. When we read the above passage, we get the impression that the authors are opposing meaning C to meaning A, since C introduces not the idea of a medium but the idea of that medium's infinity. In the name 'indefinite medium' which we find in meaning C, it is the word 'indefinite'that seems to be stressed, and not merely the word 'medium'. If this were so, the classification would be incorrect and would confuse the two principles of the division. The problem of the infinity or the finitude of time should be treated separately.

However, the most important of the remarks is that of Drouin, who protests against meaning B, pointing out at the same time that the *Vocabulaire* has left out another meaning of the word time:

Meaning B, whatever Fouillé might say on this subject, is not accepted universally as becoming is not time. What ought to figure here as the second meaning and what would naturally be the passage of meaning A into meaning C, is the general concept of time, the totality of relations, time, the unique order of succession, to which all meaning A's times refer. This idea is not confused with that of an indefinite medium. In my opinion, the filiation of meanings is as follows: A's times, defined by their limits, may begin at a common limit, whence the comparison between equal and unequal times, the division of the same time, the addition of successive times, all of the operations which envelop the univocal relation before–after and lead to the clear establishment of the general idea of the time (B). Now this abstraction is made real by Newton (C) and, perhaps, by Spinoza, but not by Leibniz, to whom the ideal unity of time (just as that of number or of space) is simply the existence of the same [common] reason for continuing. By positing time as intuition, Kant oscillates between meanings B and C.

This note of Drouin contains observations which are correct, yet not completely so. We do not understand why meaning B in the *Vocabulaire* does not at all deserve to be called time. To say that time is a ceaseless flow, by which present becomes past, is this not tantamount to grasping one of the most characteristic features of the notion of time? Obviously this notion does not exclude the complement [of meaning B] by way of the idea of order and of intervals, which are implicitly contained in the idea of the flow.

We have already noted that analysis could not discover in direct data, either order or distances, (periods, 'intervals'), if the data failed to contain any basis for these. Yet Drouin represents this in such a way that the idea of order would exclude from the time the character of the flow. Now, this rather astonishes us.

Yet, on the other side, his great merit consists in his having pointed to a gap in the classification of the *Vocabulaire*; viz., that time also represents a set of relations. In this respect we find a surprise in the *Vocabulaire* since, when introducing right away the metric notion of the period, it omits the notion of the relation of order, which is much simpler and which is implied by the idea of the period. We here remind the reader of the remarks of Enriques on the temporal order and on the measurement of time. We do not want to decide which of these ideas occupies first place from the psychological point of view, as it is impossible to decide. Probably the one and the other were observed at the same time, but we do not doubt that the first thing noticed by the primitive man was the qualitative difference between the present and the past, given as something mobile. In this flow primitive man also saw that some object did endure without undergoing change, and that he himself endured as well, in spite of changes in his internal states. Yet it is hardly probable that he was aware of the subtle difference between duration in the classical sense of the word (as permanence), and duration in the sense of Bergson (as the continuation of being, through states that merge with one another). Although primitive man doubtless experienced this latter duration, if he had formed the idea of time, it was not his own, intuitive duration that interested him above all.

However, to do justice to the authors of the *Vocabulaire*, we have to admit that their definition respects the notion of the 'before–after' order, since they speak of the period which proceeds 'from a prior event to a posterior event'; the moment is introduced here as a limit or as one of the extreme points of the period. In so doing, they avoided complications, for if we define time 'as a set of relations,' the following problem arises: what are the elements of these relations. Are they the relations which exist between the moments or between the intervals (periods)? Is it possible to define the interval as an ordered set of moments?

When the notion of the temporal succession 'before–after' has been introduced, we find before us two paths: either we introduce the notion of the moment and define the interval of time as an ordered set of instants, or else we start with the idea of an interval by introducing the idea of a moment as a limit-point of the interval or as a point of division of that interval. However, all these options are no longer psychological questions. Each of these two paths can be tolerated as a theoretical construction and each of them agrees with the everyday meaning of the word 'time', since the relation of the succession can be applied as well to intervals as to moments. We shall be obliged to return to this problem in the chapter devoted to the idea of the continuity of time, wherein these different points of view have led to different conceptions of the continuum.

The authors of the *Vocabulaire* try, however, to demonstrate that their conception is more natural and that it is justified by the etymological sense of 'tempus': 'root *tem*, whence the Greek τέμνω, properly called: division, section, whence, particularly, division of time, epoch . . .' This etymology would be in agreement with the hypothesis of the sociologists, for whom the idea of time was formed by the regular return of religious festivals, (tied to agricultural labor and to astronomical rhythms), these festivals dividing social life into periods of a special and homogeneous character.'

The *Vocabulaire* observes that there exists also another explanation of the word 'time'. Bréal and Bailly state (*Dict. étym. latin*, sub V) that the primitive sense of *tempus* would be temperature, heat. 'This word', they say, 'is of the same origin as *tepor*, from which it differs only in gender and by the presence of a nasal sound.' Hence '*temps*' [from 'tempus'] (good or bad) in general. Finally, we have come to the abstract idea of duration.

However, the *Vocabulaire* observes that this etymology is not quite satisfactory from the point of view of semantics, and that, even if it is correct, meaning A preserves its pre-eminence. In the reference, L. Robin adduces also an observation regarding the origin of the name of the god Κρόνος, which had been confused, as early as the 6th century B.C., with the term χρόνος, time.

Now, it is generally accepted that Κρόνος comes from Κραίνειν: thus Cronos would be the god who achieves, who conducts things to their term. This confusion would favor the derivation adduced in the second place: Cronos became Chronos, since time is what makes things ripen, and then, by extension, what leads beings to their maturity and to their end.

Yet Lalande remarks correctly that this assimilation agrees still better with the idea of duration, of the completed period, than with that of temperature. Κράινειν, in the intransitive sense of the word, means also to draw to a close, to end. We shall add from our side that if Κρόνος (Saturn, of Roman mythology) had not been identical with time χρόνος, then the etymology of the name of the god does not explain the origin of the Greek word χρόνος. If the origin of the term χρόνος and *tempus* is not absolutely certain, the same uncertainty applies also to the time-words in other languages. The German '*Zeit*' corresponds to the English '*tide*', but the origin of both is uncertain. Similarly, the origin of the Slavonic terms '*čas*' and '*vreme*' lacks certainty. It is possible that the Polish word '*czas*' has the same root as the German '*Zeit*'. However, in Old-Polish, there also existed another word for time, viz. '*Vreme*', which appears at present in Russian and in the South-Slavonic languages. However, its origin is highly problematical. It recalls quite vividly another Slavonic word '*breme*', the meaning of which is, however, 'burden'. It is difficult to suppose any relationship between time and burden. For that reason one finds no relation between these two terms in Brückner's *Etymological Dictionary of the Polish Language*.

The origin of the term 'time' was probably not identical in the case of different nations. At any rate, the observation found in the *Vocabulaire*, viz. that the *original sense of the word 'time' means interval or period*, may be true, for the Graeco-Latin languages at least. The conception of duration as an ordered set of moments is rather the product of mathematical analysis.

## 3. THE THIRD GROUP OF TIME *APORIAS*

130. To the third group of aporias there belong the problem of the uniformity of the temporal flow and all those questions created by the theory of relativity. Their common characteristic feature consists in that they issue logically from the fields of natural sciences, of physics and of chemistry. We could thereby conclude that we are here dealing with questions tied to the measurement of time. However, metrical time and physical time are not identical notions; physics is also interested in the qualitative properties of time. We have said already that Reichenbach introduced the term 'topological properties of time' to denote all of the properties which are not metrical, by analogy to the topological and metrical properties of space. This term does not correspond to the etymological meaning of the word τόπος, place; but it gives rise to no misunderstandings. Of course,

it would be better to contrast chronology with chronometry, but unfortunately, the former term has already acquired too specialized a meaning in the historical sciences.

In geometrical research we designate by the term 'topology or *analysis situs*' the science of all the properties of geometrical forms which do not disappear if we subject them to continuous and bi-univocal transformations; i.e. if we deform the figures by dilating, contracting or curving them, without however tearing them apart or gluing them together. Consequently, the circle, the ellipse, the triangle are equivalent from the topological point of view, but a segment is a form which differs from these closed lines, as it is impossible to transform it into a closed line in a continuous and bi-univocal manner without sticking the ends together. With respect to time, *cycling* would be, for example, a topological property differing from the open form of common time.

Since time has only one dimension, its topology and its metrics are simpler than the corresponding parts of geometry. To the topological problems of time there belong, according to Reichenbach, the order of succession and simultaneity which results from the mutual comparison of temporal series in different places. Among the metrical problems we should include the choice of the unit of measurement, the question of the equality of two time intervals which leads to the idea of the uniformity of time and, finally, the question of simultaneity which, as the theory of relativity has shown, is not only a topological, but also a metrical question.

We stated in the historical section of our work that the consideration of these problems led to two opposed orientations. One of them considers the ideas of the uniformity of time and of simultaneity as completely relative and conventional, while the other defends their intuitive and absolute character. The viewpoints of Poincaré and Enriques have very well illustrated the divergence between these two opinions. We are not sure whether Enriques changed his opinion following the publication of his valuable book, *Les problèmes de la science*, but this is presently immaterial to us. In any case, his book defends a view contrary to that of Poincaré. He has also shown that the fact of giving up Newton's absolute realism does not necessarily entail the relativity of the uniformity of time. Similarly, it is possible to preserve the traditional concept of simultaneity. The moderate realist can select either the conventional point of view or the intuitionist one. We must admit that intuitionism found itself in a more difficult position after the advent of the theory of relativity.

However, we do not exclude it, and indeed shall try to assume an intermediary position which takes into account the results of modern physics while not neglecting the data of intuition.

131. Prior to doing this, we shall analyse the point of view of Reichenbach, whose book *Philosophie der Raum-Zeit-Lehre* is one of the most interesting publications of the 'relativistic' literature. In this work, the point of view of Poincaré predominates in its most radical form. Reichenbach was in 1920 already well-known, owing to his publication *Relativitätstheorie und Erkenntnis apriori*, in which he subjected the philosophy of Kant to criticism from the point of view of the theory of relativity. According to Kant, it is our mind that creates the system of principles, thanks to which the cognition of nature becomes possible. These principles cannot be challenged by experience since experience is realizable only with the aid of these principles. But, Reichenbach remarks, Kant did not prove that this system of principles might never be in contradiction with experience.

Now, the theory of relativity discovered a case which limits the choice of principles. Kant did not foresee that there can exist different coordinate systems of the facts of experience. We find ourselves with the necessity of reshaping our principles of knowledge. But this modification of the constitutive principles of knowledge should be carried out in such a way as to leave the preceding principles true, as a special case. Reichenbach called this manner of modifying the principles, 'a method of gradual widening'. Kant was right in stating that scientific experiment would not be possible without the constitutive principles, but he had no reason to state that these principles ought always to be evident and intuitive.

In our work of 1923, '*The Axiomatic Method and Natural Sciences*', published in Polish, we accepted this principle of Reichenbach, having changed only its name, for it is not a question here of widening the principles, but of modifying them. Now there is no doubt that, if the progress of science forces us to modify the constitutive principles of science, this modification should be carried out in such a way that the preceding principles, admitted by virtue of their evidence, ought to be included in the new principles, at least as a special case; otherwise it would be difficult to comprehend how evidence itself could bring us to a point of evolution where we see the necessity of forsaking this faithful companion in order to find another one. Moreover, the progress of the mathematical and physical sciences tells us that this principle is true. Classical mechanics is contained, as a special case, in relativistic mechanics; similarly, the special theory of relativity is contained, as a limit case, at least approximately true, in the general theory of relativity, etc. Now, it is not impossible for a similar method to be applied to the constitutive principles of all sciences. These principles are indispensable, but they need not necessarily possess an apodictic certainty.

We accepted Reichenbach's result because, already in one of our earlier works '*Causality and the Functional Relation*', published in 1912, while analysing the sense of the *a priori* in Kant's works, we recognized that the most important component of this sense is to 'be constitutive' for a scientific experiment.

Yet Reichenbach admitted, in his work *Relativitätstheorie und Erkenntnis a priori*, still another idea borrowed from Schlick; viz., that scientific knowledge is based on the one-to-one correspondence relation between facts and symbols. The entire meaning of truth is reduced to this relation of correspondence. Nevertheless, Reichenbach introduced some corrections into Schlick's theory, since this correspondence differs from that which takes place between the elements of two sets. The elements of sets, prior to being coordinated, must be well defined. But in human knowledge, one side of this relation is not yet defined prior to the coordination. We cannot consider the objects of our perception as the well-defined elements of reality. Their content is often too complicated to allow them to be considered as determinate elements. Perception does not even supply us a criterion for recognizing whether a datum belongs to a certain set or not. Illusions are the best proof thereof.

In knowledge, we find ourselves faced with a particular case, viz. that the elements of one or two sets are only determined by the fact of their coordination. Thus, knowledge requires the admission of 'principles of coordination'. Kant's system was precisely an attempt to enumerate such principles of coordination. On the other hand, Schlick, while defining the truth of knowledge by the one-to-one coordination of facts and symbols, did not consider the need to introduce the general principles of this coordination. Kant might, perhaps, have enumerated them rather badly, but there is no doubt that they exist.

However, in the later considerations, Reichenbach arrived at the result that these principles of coordination can be replaced by suitable definitions. It is probably for this reason that in his work of 1924, *Axiomatik der relativistischen Raum-Zeit-Lehre*, wherein he proposes to axiomatize the relativistic geometry of space and time, we find no further mention of coordination principles ('*Zuordnungsaxiome*') beside the laws of nature, such as the 'principles of connection' (*Verknüpfungsaxiome*). In this work we see only axioms and definitions with the remark that all the definitions of the natural sciences, as opposed to the mathematical definitions, are only definitions of coordination. A mathematician defines a notion by means of other notions, while a physicist has to make notions correspond to facts, and that is why these definitions become definitions of coordination.

We will not analyse this work of 1924, since all that is most important for us can be found in the work of 1928, *Philosophie der Raum-Zeit-Lehre*. The system of coordination is introduced in the latter by means of definitions. These definitions do not yet express any knowledge; they are neither true nor false, they are completely arbitrary and conventional. It is thus that Reichenbach moved away from Kant's philosophy and drew nearer to Poincaré's point of view.

Reichenbach is convinced that, had this point of view not been introduced, the theory of relativity would be incomprehensible. Einstein's definition of simultaneity is merely a conventional definition. It can be more or less useful, but we cannot speak of its truth. It is abandoned in the General Theory. Similarly, the equality of two time intervals can be introduced only by way of a definition. The entire philosophical value of the theory of relativity is reduced, according to Reichenbach, to the fact that it posits definitions of coordination, where hitherto we saw knowledge, which can be either true or false. However, it is not only in chronometry, but also in the topology of time, that definitions play a decisive role.

132. The theory of physical time is based, in Reichenbach's opinion, upon five definitions; three definitions are tied to metric properties, and two to topological properties. These definitions do not yet give us any knowledge, but since the relations which exist between the notions introduced do not depend upon us, they become foundations for a knowledge which can be true or false. In the foreground we find the metric definitions. We will leave aside the first definition relating to the choice of the unit of measurement, and shall deal with the second one, which refers to the problem of the equality of two time intervals. This problem is tied to the question of time's uniformity or, in other words, the question of 'isochronism'.

Reichenbach distinguishes between two ways of measuring time: one is based on periodic processes, while the other reduces the measurement of time to the measurement of space by the profitable utilization of either the law of inertia, or the movement of light. When we measure by means of periodical processes, it is not even necessary for the movement to be uniform. The movement of a pendulum is not uniform, but this is no obstacle when it is a question of counting periods repeat themselves. In some cases, the course of a period is uniform, as is for example, the movement of the earth. However in all of these observations we make use of a supposition which concerns real mechanisms; viz., that some processes require the same time for traversing the same path, or the same period.

How are we to verify this supposition? 'We have only one answer; this supposition can never be verified.' We are unable to move back a time segment and place it against the preceding segment for the purpose of comparing the two segments with one another. However, if two clocks run one beside the other, we can observe when the beginnings or the ends of their periods coincide. But, even in this case, we cannot have any certainty that the periods will always have the same duration in the future.

But why is this question insoluble? Do not the laws of physics force us to believe in the equality of time periods? Of course, those laws which appear in school textbooks oblige us to do so. But if we ask where those laws do come from, we shall see that they were obtained by means of observation, by means of clocks constructed in accordance with the supposition of the equality of periods. The whole proof is a vicious circle. Had we used another supposition in our measurement, we would have obtained other laws, and these laws would oblige us to view this new manner of construction as correct. That is why measurement, conducted by means of non-periodic processes, cannot help us at all. Of course, the law of inertia offers us one measure of time; however we could express this law otherwise; viz., by choosing another measure of time. In that case a body not subjected to external forces would move with a retarded motion, and a body left to itself above the ground would fall with a uniform movement, and all this would not entail any contradiction.[90]

All of these difficulties can be removed only by the definition of coordination (*Zuordnungsdefinition*). The equality of two successive time intervals cannot be recognized, but only defined. Any definition is equally admissable. 'Yet physics decides in favor of a definition which is determined by special properties.' It possesses three independent ways of defining uniform time: (1) definition by means of natural watches driven by the internal forces of the watch, e.g. by the spring; the atomic clock also belongs to this category; (2) definition by means of the laws of mechanics; to this group belong: the definition by the law of inertia, by the movements of the earth, and by the pendulum, for which the propellant force is the attraction of the earth. These clocks are not natural; they require that we take into account the relation of the given body to another body. (3) Finally, definition by means of the propagation of light.

Now, it is an empirical fact that these three definitions lead us to record the time elapsed in the same manner. The natural processes prefer a certain temporal flow. If we choose, for our definition, this flux distinguished by nature, it is solely because this choice simplifies the description of natural processes. 'But this simplicity has nothing to do with truth.' This is a descriptive simplicity which makes no claim to truth. 'On the contrary, that this discerned time flux does exist has a right to be called truth.' However, this is the result of experience, and by no means a necessary truth. Moreover, we should note that this distinction is merely approximate, being valid only for fields without gravitation. All of these clocks require the introduction of certain corrections. But even with these corrections we do not obtain a true time, 'only, astronomers, using the laws of mechanics in their calculations, thus obtain a time which is assumed in physics *ex definitione*, if the laws of physics are to be valid. By changing the definition we could obtain other laws and another time.'[91] Here, we see Poincaré's ideas repeated literally.

To make this point of view more convincing, Reichenbach calls upon certain analogies with space measurement, and upon the difference between universal and differential forces. Under the name 'universal force', he understands a force which acts in the same way upon every material and against which we have no insulating walls. The forces lacking this property are called by him, 'differential forces'. It is impossible to ascertain the existence of universal forces. Solid bodies are defined by the author as bodies which are not subject to differential forces, or as bodies for which we are able to eliminate the effect of differential forces by using corrections. But no body is solid with respect to universal forces. All of the forces considered in physics are differential. Nevertheless, the assumption of universal forces makes it easier for us to comprehend the conditions under which chronometric statements are true.

The natural clock (in the author's sense) can be defined as a closed periodic system. However, the concept of closed is not definite if we accept the existence of universal forces. If we had called the time of the revolution of the Earth variable, by saying, e.g., that one revolution is twice as great as the preceding one, and so forth and so on, then such a definition would give rise to the emergence of a force in the equations of physics. This force would be responsible for the change of the time of revolutions. But we could discover that this force would act upon all clocks in the same manner, and that it would act on the motions of bodies free from the effect of all the other forces.

Now, we admit by definition that this force equals zero. Hence we can completely disregard this force in the definition of a closed system. Our definition thus decides on the zero point relative to which we measure all other forces. Without this zero point, the magnitude of a force is indeterminate. The idea is that force is something responsible for changes; but the change of space and time segments can be recognized only if we have previously introduced the definition of equality. It is for this reason that, for time as well, the definition of equality is tied to the question of the field of forces. The definition of equality in the comparison of times serves at once as a foundation for the measurement of forces and, inversely, this definition of equality can be given also because we decide something with respect to the manner in which forces are measured.

In reality, when we speak of a closed system, free from differential forces, we cannot be sure that even these forces themselves do not act. It is for this reason that the concept of closedness must be treated as an approximate one. A system can be closed for some forces, but not for others; it can be closed once, but not closed at another time, or only less closed.

133. In a similar manner, Reichenbach solves the problem of simultaneity, by accepting the third definition of coordination. The fact that uniformity has been established only by means of a definition was already, according to Reichenbach, recognized by Mach, and Einstein established the same for simultaneity. But Einstein required this definition for a physical theory, and this is why the epistemological character of this discovery was not separated and distinguished from the physical results. Reichenbach does not want to follow Einstein's route by directly introducing the principle of the constancy of the velocity of light, but he takes the epistemological problem as his starting-point in the following way.

At first, we must distinguish between the simultaneity of events in the same place and simultaneity in distant places. For simultaneity in the same place, there is no need for comparison, for in this case the coincidence of the events is such that we can speak of the identity of place and time. On the contrary, in order to compare distant events, a definition is indispensable. Without this definition everything would be arbitrary and, moreover, the conviction that, in establishing simultaneity, we are dealing with knowledge and not with a convention would draw us into a vicious circle.

The comparison of distant events is always ascertained by means of signals, it being supposed that we know the distance between the two places and the velocity of the signal. But how are we to measure this velocity? When knowing the time $t_1$ of the emission of the signal from point A, and the time $t_2$ of its arrival at B, we would have to divide the distance AB by the time $t_2 - t_1$. However, in order for this value to have any meaning for us, it would be necessary to know beforehand that the clocks are running equally, i.e. that they indicate the same hour at the same time. We therefore see that, in order to measure the velocity of the signal, we already have to know the simultaneity of the distant events. It seems that this difficulty could be passed over. When measuring the velocity of light, Fizeau used a clock at point A only; he sent out the light ray which was reflected at point B and returned to A. But this measurement rests upon a non-verified supposition; viz., that the velocities of light along AB and BA are equal; and to check this supposition it is necessary to know the time $t_2$ of the arrival of the signal at B. Thus, it always remains true that the measurement of velocity in one direction presupposes the knowledge of simultaneity.

Thus the vicious circle is unavoidable. In order to establish the simultaneity of the distant events, it is necessary for us to know the velocity of the signal, but to measure this velocity we have to know the simultaneity of the distant events. The apparition of this vicious circle proves that simultaneity is not based upon knowledge, but upon the definition of coordination; for the vicious circle proves that knowledge of simultaneity is impossible.[92]

Now Reichenbach demonstrates the second distinctive feature of the definition, i.e. its arbitrary quality. It is entirely conventional which figure we place on the clock-dial at B, at the moment when the signal B arrives at this point. Let us assume that the signal started from A at $t_1 = 12.00$, and that it returned at time $t_3 = 12.06$. We can therefore allow that the signal was at B at $t_2 = 12.03$. Fizeau's measurement could also have been based upon this value.

There is no sense in saying that this would be false; there is nothing here to know, we have to give a definition. And, with this definition, will be given at once the velocity of the light signal and the [principle of] simultaneity. This is why this convention will never lead us into a contradiction. If we establish this simultaneity and desire later to calculate, on the basis of measurements, which events are simultaneous, we shall obtain at the end always the simultaneity which we previously introduced by definition.

'It is following this consideration that we now see how we must understand Einstein's definition of simultaneity,

$$t_2 = \frac{t_1 + t_3}{2}$$

(1) or

$$t_2 = t_1 + \tfrac{1}{2}(t_3 - t_1)$$

which attributes to the moment of arrival of the ray at B the mean of the time of its emission and that of the return of the signal to A. This definition is essential for the Special Theory of Relativity, but it is not necessary from the epistemological point of view. The point is that Einstein's definition is only a definition. If we choose a more general formula:

$$(2) \qquad t_2 = t_1 + \varepsilon(t_3 - t_1)$$

$$(3) \qquad 0 < \varepsilon < 1$$

it could likewise have led us to our target. We could not call this formula false. If the limited formula admits $\varepsilon = \frac{1}{2}$, this is only because such a formula expresses the phenomenon in a much simpler way, and it is only this simplicity of description which has significance. Our choice is tied only to condition (3), according to which $t_2$ should be placed between $t_1$ and $t_3$; this follows from the topological properties of time, which must be respected by the metrical ones.'[93]

These considerations demonstrate to us that, in the problem of simultaneity, it is the definition that is concerned. This definition can be compared to that of the meter, by way of the earth's circumference. In our definition, light corresponds to the circumference of the earth, while the emission of the signal from A to B and back, with the establishment of time at B, corresponds to the operation of the division of the meridian. Reichenbach insists that, in this definition just as in every definition of coordination, there is an ambiguity; for it depends solely on us to know what should be understood by simultaneity. We establish that the simultaneity of two events in distant places denotes that, in the temporal series of these two places, these two events will have the same numerical sign. However, it is necessary to indicate which points of these two parallel series should possess these equal signs. The same occurs in the definition of a unit of length. We decide what should be understood by the words 'unit of length', but later we must indicate a real length which should serve us as this unit.

In subsequent passages, Reichenbach criticizes all the attempts at saving the absolutist conception of simultaneity in physics. These attempts can be divided into three groups: The first ones indicate the possibility of a velocity exceeding that of the light, but science has so far failed to discover this. Others are based upon the construction of special mechanisms which use electrical forces; e.g., for establishing absolute simultaneity. However, a more critical consideration shows that they are unable to achieve their target. Some of these attempts contain the false conception of a rigid body, while forgetting that there are no absolutely rigid bodies in the theory of relativity. The third group of attempts is based on the conception of the transport of clocks; however, this conception forgets that, in the theory of relativity, the clocks' indications are not independent of the velocity of the transport; and that, even if this independence existed, the question of simultaneity would be also a question of definition, and not of knowledge.

This entire chapter ends with the general remark that there exists neither an absolute uniformity, nor an absolute simultaneity, if the term 'absolute' means that the given time is the only true time. We can only say that physical mechanisms give preference to the simplest definition. 'We are taught by experience that the definition of uniformity is distinguished by the clocks and by the law of intertia; this distinction is maintained in Special Relativity, but it disappears in the General Theory.' If it is a question of simultaneity, the existence of an infinite velocity or the transport of the clocks could distinguish one solution among the others; however, this is a purely empirical question, and experience has solved it negatively.

134. The second group of definitions refers to the topological properties of time. There are two such properties: one for the temporal order in the same place, the other for comparing the temporal orders in different places. If we have to deal with two sharply separated events, the direct consciousness of the observer tells him which of them is before and which is after. 'However, in this chapter we do not want to make use anywhere of the impression of the observer'. Reichenbach finds an objective criterion in the causal order, in conformity with the teaching of Leibniz and Kant. 'If $E_2$ is the effect of $E_1$, we say that $E_2$ is after $E_1$'. Such is the topological definition of the temporal order. $E_2$ may be a partial effect only, just as $E_1$ may be only a part of the cause. Reichenbach wonders whether this definition is not a vicious circle. Indeed, how do we recognize the causal relationship if we do not yet know the temporal order? But, says the author, the causal relationship expresses not only the connection between two phenomena, but at the same time it determines the sense thereof; it is an asymmetrical relationship. Changes in the cause always entail changes in the effects, but not the other way around. If, e.g., I send a light signal from A to B, and, in the path of the beam, at A, I place a small piece of red glass, the light at B will be coloured; on the other hand, a glass placed at B will not change the colour of the light at A.

Reichenbach warns us that we must not identify the asymmetry of the temporal order with the asymmetry of a series of numbers; e.g., the asymmetry of time has an altogether special sense which we know from our internal life and which, at the same time, characterizes all of reality. Attempts were made to clear up this unique and specific sense of the time direction (*Einsinnigkeit*) by calling upon the law of entropy (the Second Law of Thermodynamics). Reichenbach finds that these explanations are insufficient, but that, for the theory of relativity, this is immaterial.

More important, however, is the question of whether an order thus defined can be established without contradiction. The idea of the eternal return seems to create difficulties. Reichenbach does not believe that this idea could correspond to reality, but, if anything similar were to take place, it would be necessary to limit the adopted definition to series within the periods, and the beginning of the period could be chosen arbitrarily.

Let us suppose, however, that only a few temporal series would return and cross one another. This would be tantamount to accepting their encounter with their own doubles. This hypothesis ought to be rejected, as it would render impossible the conception of our 'I' as an identical individual in the temporal flow. Not only would it be impossible to establish the temporal order in a univocal manner, but the identity of the individual, which develops within time, would be violated.

The last problem of the topology of time considered by Reichenbach is the comparison of the series of time in distant places. To this end the signals are indispensable. We thus have to adopt the axiom that, between two points of space, there are always communicating signals. By virtue of the first topological definition, we can treat the events of each series as ordered ones with respect to the others. If we signate by $E$ the event of the signal emission from the point $P$, and by $E'$ the event of its arrival at the point $P'$, then these events, $E$ and $E'$ are already ordered by the very fact of their communication by means of a signal, and $E'$ should necessarily be treated as posterior to $E$.

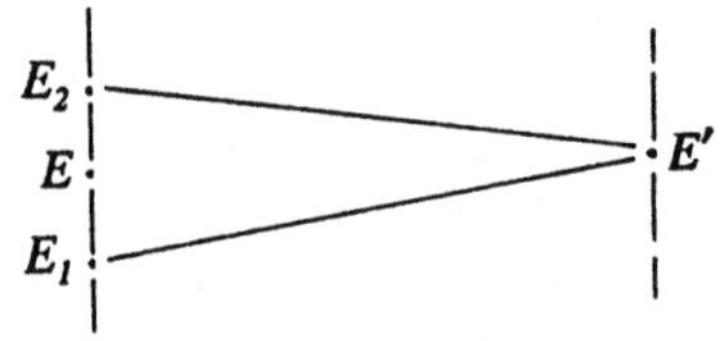


However, from the point $P$, signals of different velocity can be sent simultaneously. One of them is the fastest messenger. This signal is called by

Reichenbach the first signal (*Erstsignal*). Reichenbach has introduced this conception in order not to speak, from the beginning, about light and its velocity. Let us now suppose that this first signal, which is emitted from point $P$ as $E_1$ and arrives at point $P'$ as $E'$, is instantaneously reflected and turns to point $P$ as the event $E_2$. We now want to establish the order of the phenomenon $E'$ with respect to $E_1$ and $E_2$, knowing as we do from the first definition that:

$E'$ is after, relative to $E_1$
$E'$ is prior, relative to $E_2$

Let us choose, on the temporal line $E_1E_2$, any point $E$ between $E_1$ and $E_2$, and let us ask what is the position of $E'$ with respect to $E$. The first definition does not say anything about it. This event has not yet been determined as to its temporal order (*zeitfolgeunbestimmt*).

However, we can distinguish between two cases. The former case represented in the diagram makes an entire segment $E_1E_2$, (except its limit points), correspond to the event $E'$. Each event of this segment is indeterminate with respect to $E'$ as regards its temporal order. The second case would occur if we had to deal with signals of an infinite velocity. In such a case, segment $E_1E_2$ would be contracted at one point in relation to $E'$ as simultaneous. This is the case of the classical theory of time, while the former case corresponds to the relativistic theory. The fact that the velocity of light represents the maximum velocity in nature decides in favour of the relativist theory. Light is the first signal of which Reichenbach speaks at the beginning of these considerations. We thus have a free choice of one of the points on the segment $E_1E_2$ as simultaneous with $E'$. Hence the topological definition of simultaneity is: Any two points can be simultaneous provided that they belong to indeterminate points, as regards their temporal order.

This definition is so general that it does not exclude the classical case. In that case there exists only one indeterminate point in this group. But in general we have no univocal prescription, and it is in this that the relativity of simultaneity consists. Simultaneity is reduced to the indeterminacy of the temporal order because it means that some points are eliminated from the group of causal relations (*Ausgeschalteheit des Wirkungszusammenhangs*).

We can see that the intuitive conception of simultaneity is not absolutely erroneous; it is only its result that is erroneous; viz., that simultaneity ought always to be determined unequivocally. It is in this way that Reichenbach wants to reconcile the two opposed views. He imputes two errors to the relativists.

One of these consists in that the relativity of simultaneity is motivated by the state of movement of a system. However, the movement of the reference system is not solely responsible for this relativity, it is sooner the existence of a finite, maximum velocity for the transmission of causal effects [which is at stake]. Nor is it necessary that the metrical definition accept $\varepsilon = ½$. Still worse is the fact that some call upon the subjectivity of the observer. Now, the relativity of simultaneity has nothing in common with the subjectivity of sensory perceptions or with the monadological character of human consciousness. The appeal to an observer can only signify an appeal to different logically possible, metric methods.

The second error consists in a prejudice of the relativists; viz., that the absolutist theory could never be established. It rested solely upon the view that the causal structure of the world was erroneously considered as evident *a priori*, instead of leaving this decision up to experience.

135. We shall now express our critical remarks with respect to Reichenbach's point of view and shall develop, at the same time, our own views concerning the questions discussed. Our point of view is that of physical relativism, but it is not completely identical with that of Reichenbach, although we have to admit that we have made considerable use of his works, as well as those of Schlick and Cassirer. Indeed, all these German authors have the merit of giving a deep philosophical interpretation of the problems discussed, and of contributing to the elucidation of certain difficulties. In order to eliminate all of these misunderstandings, we shall emphasize three important points in our orientation:

(1) Physical relativism is not identical with philosophical relativism and should not be interpreted in favor of the latter, as was done by a German philosopher, Joseph Petzoldt, who believed that the theory of relativity was only a confirmation of the doctrine of the sophist, Protagoras; viz., that man is the measure of all things and that only what seems to be true for each man is true.

(2) Some properties of objects are only relative, i.e. they only exist under a certain condition, because they express the relation of the objects in question to another object. If it is a question of time and space, the coincidences of the spatial and temporal points preserve their absolute character, but the length of their distances and their equality or inequality are merely relative properties which depend upon the conditions of measurement. Propositions which leave out both these conditions and [their accompanying] metrical suppositions ought to be regarded as

*elliptical* [sic]; when completed in a suitable manner, they become true or false like every proposition. Giuseppe Peano, the great Italian logician and mathematician, very correctly observed that the error committed most frequently in the history of science, even in the mathematical sciences, is the error '*a dicto secundum quid ad dictum simpliciter*'. The novelty discovered by the Theory of Relativity consists solely in having found the same error there, where nobody suspected it. Just as a geometrical body can have various projections, so a physical process can possess different spatial and temporal properties, in accordance with a system relative to which these properties were established. This does not mean that the same object A could possess two contradictory properties at the same time; for, in saying that a man lived 50 years according to one system, and 40 years from the point of view of another system, I speak in the first case of the relation of this man's age to a group of objects A, and in the other case of the relation of his age to a group of objects B. Thus, the objects of these two propositions are not identical; each of these propositions has, for its object, another relation, because such is the sense of the explanation concerned in the case of relative properties. The same applies to simultaneity, with this complementary explanation that the meaning of this term has changed slightly, because simultaneity is considered by physicists as a relation which characterizes objects from the dynamic and causal point of view, [that is] from the point of view of their reciprocal behavior.

(3) The third point of view that we wish to stress is that this interpretation is not absolutely certain. We cannot give a strict and rigorous proof that the classical conception is absolutely false. Thus the return to the classical point of view is not absurd.

None of these three theses is new. We might say that they have become the common knowledge of all the defenders of modern physics. The last point was stressed by Reichenbach, and still more so by Schlick, who said:

> there exists no logical necessity to present the Theory of Relativity as uniquely true, as opposed to the absolutist theory; but we can demonstrate only that a theory is simpler than another one, or that it contains fewer arbitrary elements than the other one and, consequently, that it is more satisfying.

Most important is, doubtless, the second point, since it shows that in order to understand the Theory of Relativity, it is not necessary to distance ourselves from the classical idea of the truth construed as *adequatio rei et intellectus*. Now, this idea deserves to be defended at all cost. However, let us analyse Reichenbach's point of view more closely.

136. The first thing we do not like is that he observed the topological properties of time after the metrical properties, while [it is] the latter [which] presuppose the former. Moreover, the idea of basing the objective meaning of succession and simultaneity upon a causal relationship seems to us to be successful. As far as succession is concerned, it is not new. It already existed in the writings of Leibniz and of Kant; Reichenbach[94] has only taken it one step further in the same direction by defining simultaneous events as those which are eliminated from a causal connection. Thanks to this [definition], the two concepts obtain a more consistent foundation, and the relativistic definition of simultaneity is not as isolated and artificial as before. The relativization now becomes completely justified; it results from the nonexistence of momentary causal actions. It also helps us to understand the fact that the topological definition of simultaneity must be completed by a metrical definition.

The reduction of objectivity from the temporal to the causal order seems to entail some difficulties. Yet Hugo Bergmann,[95] who wants to make the same use of the causal relation in his work, *Kampf um das Kausalprinzip*, points out a small difference between his position and that of Reichenbach. According to Reichenbach, the asymmetry of the causal relation has nothing to do with time, and time gets its asymmetry from the causal relation. On the other hand, Bergmann thinks that the asymmetry of the causal relation stems from the asymmetry of the temporal order and that, without the latter, the structure of the causal relation would be completely symmetrical. Thus, the cause–effect relationship does not introduce asymmetry into the temporal order, but rather it borrows it from that order and only gives to it an objective character.

The explanation of all of these difficulties would require of us a special study of the nature of the causal relation, which we are unable to do here. We beg to observe, however, that the point of view of Bergmann seems less satisfactory to us. The point is, in what way could the causal order render objective the asymmetry of the temporal order if the causal order is by essence symmetrical? What meaning has the term 'objectivity' here? To Bergmann, who is an adherent of Kant's philosophy, this term is probably equivalent to the necessity and the universality which excludes all exceptions. Thus the symmetrical relation between cause and effect would be the sole guarantor of the asymmetry of the temporal order. In other words, we do not understand how one relation could give to another relation that which it does not itself possess.

The point of view of Reichenbach seems to us to be more natural. The relation between condition and something conditioned appears to be asymmetrical even if we disregard the asymmetry of the temporal order. Thus we can assume an intermediary position, by stating that the asymmetry of the temporal order and the asymmetry of the causal order are inseparably connected with one another. Let us, however, not forget that the attempt was already made to reduce the asymmetry of time to the asymmetry which results from the Second Law of Thermodynamics. This idea, expressed by Mach, was defended in the interesting paper by Zilsel, '*Über die Asymmetrie der Kausalität und die Einsinnigkeit der Zeit*'[96] We shall return to this problem in the following chapter.

137. Our objections are more serious with regard to Reichenbach's metrical definitions of isochronism and simultaneity. Like Poincaré, Reichenbach stresses their conventional character in an exaggerated manner. He speaks of convenience and simplicity where we would be tempted to hear the word 'truth'. Reichenbach is right in stressing the role of definitions in the construction of physical theories, but he is wrong when he affirms that the definition never expresses knowledge. The role of definitions in mathematical sciences cannot be confused with the role of definitions in the natural sciences, where each definition must be accompanied by a judgement of existence. The opinion of the majority of philosophers is here at variance with what Reichenbach says.

The following is what we read in the *Logique de Port-Royal* about the difference between the definition of things and the definition of words.

> The definitions of words are arbitrary, while the definitions of things are not arbitrary . . . The definitions of words are incontestable, since they are arbitrary. Nobody can contest that a man gave to a certain word the meaning of which he speaks. However, one can often contest the definitions of things since . . . they can be erroneous. . . . The definitions of things are really affirmations which can be contested by anyone who sees obscurity in them and, consequently, these definitions require proof.

Has the voice of this ancient book lost its value?

Of course, both Reichenbach and Poincaré were right in saying that it would be a vicious circle if we justified the equality of two beats of a clock by appealing to physical laws, as these laws were formulated thanks to the use of chronometers. It is likewise true that the appeal to our intuition is insufficient, since the testimony of intuition has only an approximate value, contained within very restricted limits. The same should be said about the causality principle invoked for this purpose: 'causes which are nearly identical take roughly the same time to produce roughly the same effects.' This definition is also an approximate one.

However, if we reach the result that some properties are only relative, this is not identical with convention and arbitrariness. Relativism and conventionalism are two different things. When Reichenbach shows that the three definitions of the uniformity of time do not lead to any conflict, and that the processes of nature distinguish a temporal flow from amongst the other flows, is it pure convention to give preference in our constructions to what is distinguished by nature? Reichenbach's answer is that we give preference to a flow by reason of simplicity and practice, and that it is a question here of descriptive simplicity and not of truth. However, the description is also expressed in propositions which have a claim to truth. The definitions of things cannot be completely arbitrary. Their change would entail change in the laws of nature and the appearance of systematic errors relating to the universal forces. If we assume these as being equal to zero, it is true that we give preference to that which is simpler, but why should we close our eyes to our propensity for catching the true state of things.

138. We must repeat, *mutatis mutandis*, the same objections as regards the metrical definition of simultaneity, which completes the topological definition of this relation. If we accept the value $\varepsilon = ½$, it is not only for the purpose of giving preference to what is simple and more practical, but for satisfying our need of seeking truth. And the same need forces us to forsake this value in the General Theory. However, Reichenbach raises the objection that to look for knowledge in a definition is to fall into a vicious circle. Now, what circle are we speaking of in this case: is it a '*circulus in probando*' or a '*circulus in definiendo*'? The simple postulate, viz. that a definition ought to be real, cannot give rise to any of these two circles. The sole motivation of this definition consists in indicating the facts which give us the possibility of checking whether the definition is correct or otherwise. Such is the opinion of Einstein who states that 'the concept exists for a physicist only at the moment when he possesses the possibility of convincing him, in a concrete case, whether the idea is correct or not.'

It can happen, however, that several notions hang together in such a way that it becomes impossible to check the correctness of one of them, without assuming the correctness of the other one, and inversely. Now it is not out of the question that we are facing such a case in the metrical definition of simultaneity. 'In order to establish the simultaneity of distant events, it is necessary for us to know the velocity of the signal, but, to measure this velocity, we must know the simultaneity of distant events.'

The fact that Einstein really was afraid of such a vicious circle objection can be established in a passage found in his popular article, in which he says that the speed of light is introduced in his definition solely as a convention, and not as a hypothesis. Hence this passage speaks in favor of Reichenbach's point of view; but this is but an appearance because, further on, Einstein is obliged to agree that this convention should inevitably be transformed into a physical hypothesis.

However, the difficulty has not disappeared. Reichenbach can respond that, if from the outset we want to transform Einstein's convention into a hypothesis and, thereby, into knowledge, this is so much the worse for us, as the vicious circle then becomes obvious. How can we measure velocity when we have not yet any means for verifying simultaneity and for measuring time?

The difficulty should be put aside as follows. Einstein's postulate that each definition of a physical notion must be connected with the possibility of testing whether it is true or not, does not yet entail any relativism. This postulate ought to be admitted also in classical physics. Let us assume that simultaneity preserves its intuitive meaning even in the case of distant events; this is what we do in classical physics. As long as the speed of light has not been accepted as a universal constant, independent of the system of reference; or, as long as the reference systems are immobile with respect to one another, Einstein's definition will merely confirm the intuitive meaning of simultaneity. No relativism will appear. However, let us assume that the speed of light – measured originally by the supposition that simultaneity preserves its intuitive meaning for distant events – remains unchanged for all of the systems of Galileo (and we are entitled to do this by virtue of certain astronomical observations), the *relativism* [sic] of simultaneity will appear immediately. This means that the supposition that simultaneity preserves its intuitive sense even for distant events, when completed by the principle of the constancy of the speed of light, entails its own negation, hence it is false according to a law of logic, which was formulated for the first time by Clavius.[97] Let us signate by '$p$' any proposition, its negation by '$\sim p$', and let us introduce the sign '$\supset$' for logical implication; then the law of Clavius will be symbolically formulated as follows:

$$(p \supset \sim p) \supset \sim p$$

i.e., that each proposition entails its own negation is false.

Our representation is the only correct means for explaining the foundations of the relativism of time. This explanation was supplied neither by Einstein nor by Reichenbach. If the reader remembers the development of ideas of Einstein in his popular work, he will find therein a perfect accord with our explanation.[98] What Einstein has called the definition of simultaneity in his work, *Zur Elektrodynamik bewegter Körper*,[99] is only a precept for verifying the definition thereof, which does not yet say whether this verification will always confirm the presupposed synchronism or not. It is in the following section[100] that he demonstrates that, if we introduce systems endowed with a uniform motion, and if we respect the principle of the constancy of light, then the synchronism will be abolished.

Reichenbach is wrong to consider all the real definitions as devoid of cognitive value. This radically conventional point of view agrees neither with the general tendency of the human mind towards the search for truth, nor with the tendency of Einstein's theory of relativity. To what use is the precept of verifying each definition, if it is true that the definition has nothing to do with knowledge?

Conventionalism is right solely insofar as we can construct several scientific theories consistent with the same experience, and that we never have absolute certainty as to whether the chosen theory is the best one. The proof of the uniqueness of a theory is excluded by the very fact that the extent of known facts always increases. These new facts oblige us to change definitions, hence the assertion that the definition does not express any knowledge is untenable. Perhaps Reichenbach went too far in spite of himself. In his earlier writings he was more cautious. His paper, levelled against Dingler, '*Erwiderung auf H. Dinglers Kritik an der Relativitätstheorie*' (*Physik Ztschr.* 22, 1921, paragraph 379) is excellent. Dingler protested against the idea that each empirical theory can be either verified or abolished. He was right if what he had in mind was a mathematical proof of truth or falsehood. In opposition to this, Reichenbach pointed to the existence of methodological criteria which oblige us to give preference to one theory over a certain number of other possible theories. Beside noncontradiction and simplicity, he cited the interpretation of empirical data by virtue of the normal, inductive procedure of physics.

Dingler blamed the relativists for falling into a vicious circle, since they seek to demonstrate the falsity of Euclidean geometry by using experiments, forgetting that the apparatuses used in these experiments were constructed on the supposition that solid bodies obey the laws of

Euclidean geometry. In his reply, Reichenbach shows however, that this vicious circle is non- existent. There exists only the remarkable fact that the relativists made use of Clavius's law in the reasonings. Reichenbach observed that this use was pointed out to him by the German logician Kurt Grelling. We presuppose the validity of Euclidean geometry; but this supposition, combined with other premises from the General Theory, such as the identity of inertial and of gravitational mass, and the validity of the Special Theory (in a field without gravitation), entails as a consequence the negation of the validity of Euclidean geometry. Hence we are entitled to conclude that this supposition was false according to Clavius's principle

$$(p \supset \sim p) \supset \sim p.$$

Of course, this principle alone does not suffice; in order to arrive at the negation of the supposition, we must always use other premises. In the above-mentioned case we have, beside this premise '$p$', also two other ones.

Now it is a great pity that Reichenbach did not turn to these ideas in his main work *Philosophie der Raum-Zeit-Lehre*, and that he failed to notice that the same reasoning was repeated in the Special Theory with respect to time. We suppose the synchronism of clocks for all of space in the classical sense,[101] and the speed of light in agreement with the experiments of classical physics. Einstein's precept for the verification of synchronism: '$t_B - t_A = t'_A - t''_B$' confirms only the classical supposition. Thereupon, we introduce new premises: (1) that the speed of light is not affected by uniform motions of luminous sources: the observations of the Dutch astronomer de Sitter here play the same role as the experiments of Eötvös on the identity between gravitational mass and inertial mass; (2) that the new principle of relativity for inertial systems ought to contain the principle of classical physics as a limiting case. It is then that there appears, by virtue of Einstein's precept, the negation of synchronism and isochronism, assumed in the first premise.

The analogy with the negation of Euclidean space is complete. Not one of the premises is conventional; regardless of whether it is called a definition or not, every one of them expresses the results of experiments, although each contains much more than are given us by experiments. In any case, Einstein's precept that each statement relating to simultaneity should be verified by experiment, introduces from the outset a warning against the intuitive time for which every verification seemed superfluous.

Here again we see an analogy with the space of the General Theory, since the propositions of geometry merely express the relations of bodies which also must be the result of experiment. Similar to geometry, chronometry expresses the dynamic relations in the universe. No wonder then that the topological properties have also changed their meaning. Einstein's precept presupposes that communicating signals always arise between two points in space. Since the existence of one signal expresses at the same time the existence of a dynamic relation, it is no wonder that the relativity of simultaneity is connected with the possibility of causal relations between facts.

139. It is also advisable to consider the importance of physics' discovery that temporal relations, considered as dynamic relations, become relative. Now, in the astronomical considerations of the theory of relativity, time has again become independent of space, and it has covered the role it possessed in classical physics. Einstein cites the fact that the relative velocity of fixed stars is not great in relation to the velocity of light, that we can therefore consider matter as remaining at rest or, at least, that we can consider its mean density as varying slowly and almost imperceptibly. Consequently, the discord between times of different systems becomes so small that it can be neglected or considered as mutually compensating errors, and that as a result, we obtain a universal time which is approximately equal to the time of classical physics.

However, Einstein's cosmology is not the only one. The Dutch astronomer de Sitter created a distinct cosmology which differs from that of Einstein. Whereas, for Einstein, the size of the universe depends upon the amount of matter, in de Sitter's cosmology the structure of the universe is independent of matter; it would remain the same even if the universe were empty. Moreover, in the case of Einstein, the universe is almost static; on the other hand, the universe of de Sitter is not static. In de Sitter's universe there are systematic movements. If, in that world, we introduce a body and an observer, the body would withdraw from the observer at a velocity depending on its distance from the observer and on the radius of the world. Now, astronomical observations confirm the existence of systematic movements, but do not agree that the mean density of matter equals zero, whence we conclude that neither the former nor the latter cosmology is true.

Now how shall we extricate ourselves from this difficulty? The solution of this problem was indicated by G. Lemaître of Louvain. The static universe of Einstein and the empty cosmos of de Sitter seemed to be the only possible solutions, since it was supposed that the universe was in a state of equilibrium. If observation does not confirm either solution, this proves that the universe is not in a state of equilibrium. The radius of the universe cannot be invariable; the cosmos expands or contracts. Astronomical research allows us to suppose that, in its present state, the universe expands.

But what becomes of cosmic time? Is the flow of time uniform when the universe expands? This is a question to which we failed to get any answer. However, a German physicist, Arthur Haas, observed that the affirmation that the cosmos expands is equivalent to affirming that its temporal flux accelerates, and that there are no means of distinguishing between the spatial expansion of the universe and its temporal acceleration. The uniformity of time would lose its meaning even for cosmic time.

140 Let us now consider the point of view of Bergson, who gave us one of the most interesting attempts at defending universal intuitive time in his work of 1922, *Durée et Simultanéité*. This philosopher, in whose system time plays a central role, could not remain indifferent to the scientific system which seems to throw new light on this notion. We might have hoped that he would address his discussion to Einstein's theory; which he actually did in the above-mentioned work. In the foreword to this work, Bergson states:

> We wish to know to what extent our conception of duration was compatible with Einstein's views on time. Our admiration for this physicist, the conviction that he not only gave us a new physics but also some novel ways of thinking, and the idea that science and philosophy are different disciplines made to complement one another – all this inspired us with the desire, and even imposed on us the duty, of proceeding to a confrontation.

Later on Bergson explains that he wants to do more than to defend his philosophy. This philosophy did not entail 'as a necessary consequence, the hypothesis of a universal time, but it did harmonize quite naturally with that belief. It was thus something of the ideas of everyone that we wanted to verify against Einstein's theory.' Bergson dwells on the paradoxes of the theory of relativity, on multiple times; he analyses the significance of each term entering into the Lorentz transformations, one after another, trying to find what concrete object of reality corresponds to each of them.

This investigation gave him a rather unexpected result; not only do Einstein's theses no longer appear to contradict the natural human belief in a unique and universal time, but they even confirm this belief.

The paradoxical aspect of Einstein's theory comes, according to Bergson, from the confusion of two different conceptions of relativity, one of which is abstract and incomplete, one-sided, while the other is vivid, finished and two-sided. In dispelling this confusion, we cause paradox to fall. It is in this way that Bergson attempts to obtain a double result: on the one hand, he wants not only to confirm, but also 'to complete what he said in the past about duration', and on the other hand, he wants to contribute 'to the clarification of the theory of relativity in the eyes of the philosophers'. Let us see whether he was successful.

As we just said, he distinguishes between two interpretations of relativity, the so-called semi-relativity or 'unilateral' relativity and the complete or bilateral relativity. We find ourselves in the first stage of interpretation if we still believe in immovable ether and in absolute rest, and if we nevertheless try to understand the negative result of Michelson's experiment. We thus see that what Bergson calls unilateral relativity is simple Lorentz's point of view. An absolutely immobile system here plays a privileged role: beside it there exists an infinite number of systems in motion, each of which possesses its own local time. The second stage, that of bilateral relativity, corresponds to Einstein's point of view, wherein we realize that there exists no privileged system and that everything that observer $S$ says about the simultaneity and the length of time intervals in system $S'$, the observer at $S'$ is entitled to say about the simultaneity and the length of the time intervals in system $S$.

The need to distinguish between these two stages is particularly important when we want to pass from physics to philosophy. It very often happens that, having recognized the relativity of motion, we return in imagination to the first stage. It is of little use for us to say that there exists only the reciprocal movement of $S$ and $S'$ with respect to each other, we do not study this reciprocity without assuming one of the two terms as a system of reference. Now, once a system has been thus immobilized, it becomes provisionally an absolute reference point, a substitute for ether. 'In short, absolute rest, driven away by the understanding, is restored by imagination.' And it is here that, according to Bergson, is found the source of our errors.

Prior to showing in what way the author arrived at this conviction, it is fitting to note that he recognizes, without any restriction, the relativity of physical movements. What is really given, is a reciprocity of displacement. 'If we consider two points, $A$ and $B$, and the displacement of "one of these points", everything that the eye observes, everything that science is able to note, is the change of the length of intervals.' The reciprocity of movement is thus a fact of observation: 'It would be possible to state this *a priori* as a condition of science; since science operates only upon measurements, and measurement generally concerns lengths and, when a length increases or decreases, there is no reason to privilege one of the end-points: all that we can say is that the distance between these end-points widens or shortens.'

However, this does not meant that, for Bergson, all movement can be reduced to what we see in space. Relative movement is 'the manifestation of an absolute internal change, which is produced somewhere in space.' The absolute character of this change reveals itself to us in our own movements as a sensation of effort and, by sympathy for analogical reason, we must admit something similar for all movement. Thus interpreted, however, the consideration of absolute movement is of interest only to our knowledge of the insides of things, i.e. to a psychology which extends into metaphysics. Science can and should retain from reality only that which is displayed in visual, homogeneous and measurable space.

Descartes was the first to formulate the relativity of movement as a postulate of science, but this postulate could not be asserted categorically by modern science because science shocks commonsense only when strictly necessary. Furthermore, we should not forget that science itself felt incapable of extending this relativity to accelerated movement. That is why Newtonian dynamism cut short the development of Cartesian mechanism. However, Mach demonstrated the inadequacy of Newton's reasoning, and Einstein finally restored the point of view of Descartes. No philosopher could fully content himself with a theory which held mobility to be a simple relation of reciprocity in the case of uniform movement, and a reality immanent in a moving body in the case of accelerated movement.

Bergson also recalls his theory of the continuity of matter. If matter is presented to him as a reality, it is not the same in the case of this or that body set up as a more or less independent entity. The fragmentation of matter is accidental, and it is probably carried out in different ways by different animal species.

But if each body, taken separately and stopped there where our habit of perception would end it, is to a large extent a conventional being, why would it not be the same in the case of the movement considered as affecting this body separately? When we see a movement occur, the only thing of which we are certain is that some modification is taking place in the universe. The nature and even the exact place of this modification escape us; we can only note certain changes of position, which are the visual and superficial aspect of this [modification].

If it is a question of the notion of time, Bergson upholds the point of view expressed in his earlier works; viz., that time is at first confused with the continuity of our internal life, and that this continuity does not imply a thing which flows and the states through which we pass, since

the thing and the state are merely snapshots of the transition taken artificially. From this internal time we pass to the time of things, since the surrounding matter seems to partake of our conscious duration. We gradually extend this duration to the totality of the material world since we do not perceive any reason to limit it to the immediate vicinity of our body. Thus the idea of a duration of the universe is born, i.e. of an impersonal conscience which would form a [sort of] hyphen between all conscious minds and the rest of nature.

Bergson understands that this duration of the universe, being one and universal, is merely a hypothesis.

If we suppose that our environment endures, nothing proves rigorously that we shall again find the same duration when we change the environment: different, I mean differently rhythmed, durations could co-exist. We once advanced a hypothesis of this kind in relation to living species. We distinguished between durations of a lower or higher tension, characteristic of the various degrees of consciousness which are spaced out along the [hierarchy of the] animal kingdom. However, we did not then see, nor do we see at present, any reason for extending to the material universe this hypotheses of a multiplicity of durations. . . . Were it necessary to settle the question once and for all, we would opt, at the present state of our knowledge, for the hypothesis of a material time, one and universal.

This is the point now stressed by Bergson much more strongly than hitherto.

Finally, Bergson maintains his statement of thirty years earlier, viz. that the passage from pure duration to measured time takes place thanks to the fact that we introduce a spatial symbol into pure duration. We measure time through movement, which has a double aspect if it is we, who carry it out: [firstly,] as a muscular sensation, it forms a part of the course of our conscious life; [secondly,] as visual perception, it describes a trajectory. As soon as we get into the habit of converting time into space, we arrive easily at the idea of an instant, which does not exist in real time. 'The instant is something that would end a duration should this duration stop, but it does not come to a stop.' Every measurement presupposes a determination of the moments of simultaneities.

This simultaneity in the instant is required (1) to note the simultaneity of a phenomenon and of a moment on the clock, (2) to tick off, over the length of our own duration, the simultaneities of these moments with those of our duration, which are created by the act of ticking off, itself. Of these two acts, the former is essential for the measurement of time. However, without the second one, there would be no measurement whatever, we would come to a number *t* representing anything at all, but we would not think of time.

It is from this point of view that Bergson will criticize and explain certain theses of the Theory of Relativity. The theory places us before a plurality of times. But here the problem is posed, in what sense are the terms taken, if the observer of one of the reference systems attributes to the other system the slowing-down of time, the contraction of lengths and the dislocation of simultaneity into succession. All these consequences result from the movement of one system in relation to another, (and from the constancy of the speed of light). As long as the systems are immobile, there is no difference between them; the time experienced and noted by one observer is identical with the time lived and noted by the other one. The differences appear at the moment when the observer in one of the systems attributes movement to the observer in the other system.

Let us call these two observers Peter and Paul. Peter, who is in system *S*, attributes the slowing of time to Paul's system, *S'*. Yet Paul does not record this slowing down in his system. Hence this time was not experienced by Paul. Nor was it lived by Peter, because Peter arrives at this slowed-down time of Paul only by means of calculation. The time attributed by Peter to Paul's system is lived neither by Peter nor by Paul; is it at least conceived by Peter as being lived, or capable of being lived, by Paul?

If we consider this closely we will see that there is nothing of the kind here. Of course, Peter labels this time with Paul's name, but he has depicted to himself a sentient Paul who lives his own life and measures it, and precisely in this way he would see that Paul had taken his [Peter's] own system as a reference system, and has then put himself in this unique time, internal to each system . . . thereby as well, Peter would moreover forsake provisionally his own reference system and consequently, his existence as a physicist, and consequently also his consciousness. Peter would see himself solely as a vision of Paul.

Thus in summary, whereas the time attributed by Peter to his own system is the time lived by himself, the time Peter attributes to Paul's system is neither the time lived by Peter nor that lived by Paul; nor [is it] a time which Peter conceives as that lived, or capable of being lived, by a living and conscious Paul. Nothing is real but time lived. If the slower time is therefore not lived by anyone, it should be considered as unreal, as fictitious time, a simple mathematical expression destined to note that the system of Peter, and not that of Paul, is taken as a reference system.

The differences of time in these two systems should be treated like the differences of the magnitude of objects seen from afar. Just as a distant object seems to us to be smaller than in reality, so also the time of another system seems to us to be slowed down, in proportion as the velocity of this system is greater, with respect to ours. 'The multiplicity of times I get in this way does not preclude the unity of real time; it would rather presuppose it, just as the diminution of size with distance, on a series of canvasses wherein I would represent James as more or less distant, would indicate that James [himself] remains the same height. . . .'

The paradoxical idea of multiple times asserts itself only in the hypothesis of a privileged system. If $S$ were this privileged system in the state of absolute rest, Peter, really immobile, would live a certain duration; Paul, really in motion, would live a slower duration, and it would be impossible to demonstrate that their times were identical. However, if we admit, according to the Theory of Relativity, that the roles of the systems are interchangeable, 'we have a least as much reason as has common sense to believe in a unique time. . . .' The slower duration is to be attributed by Peter to Paul or by Paul to Peter, depending upon whether Peter or Paul is doing the referring, [and] depending upon whether Paul or Peter is referred to. Their situations are identical; they live one and the same time, but they attribute to themselves reciprocally a time that is different from that [unique] time, and they thus express, according to the rules of perspective, that the physics of an imaginary observer in motion should be the same as that of a real observer at rest.

From this point of view Bergson gives a solution to the familiar paradox.

Suppose there is a traveller housed in a projectile which were to be launched from the earth at a velocity inferior to one twenty-thousandth part of the speed of light, and which were to encounter a star and be sent back to the earth at the same velocity. Having aged, e.g., by two years, the traveller will find on leaving the projectile, that our globe had aged by two hundred years. The traveller who is shut up in the projectile $S'$ is our friend, Paul. Peter remains near the launching device, the earth then being our system $S$. Now, Paul does not know anything about the slowing down of this time; two years which Peter attributes to it do not express the real time lived by Paul. Peter and Paul, the earth and the projectile will have lived through the same duration and will have aged in a like manner. Since the movement is reciprocal, both persons are interchangeable. If the time of Peter was one hundred years, the time of Paul would be the same. The period of two years is not the time lived, it is the time attributed to Paul by an observer placed outside of Paul's system.

Bergson then takes into consideration the relativity of simultaneity and rightly observes that it is the intuitive simultaneity that serves as the foundation for establishing simultaneity by means of signals. As long as the systems are immobile, conventional simultaneity introduced by definition is in accordance with intuitive simultaneity. However, this agreement disappears as soon as the distance between the systems varies. Therefore we have to distinguish between two kinds of simultaneity. The first kind of simultaneity is connected with the framework itself, it expresses something of the system itself; it is absolute. The other is 'laid upon the systems by an observer who is outside the system'; it is changeable, relative, and fictitious. 'It stands at the distance, variable within the scale of velocities, between the immobility which this system has for itself and the mobility it presents with respect to another.' However, there is only one convention here, 'a convention that is, moreover, necessary, if we want to preserve the integrity of the laws of physics.' This conventional and 'learned' simultaneity, can even be called real

if we agree to call any convention representative of reality, once it has been adopted for the mathematical expression of physical facts. This is well and good; but then, let us no longer speak of time; let us say that it is a question of succession and simultaneity which have nothing to do with duration, because there is no time without a before and an after, either ascertained or ascertainable by consciousness.

In his final chapter, Bergson considers the conception of space-time in four dimensions. Since every measured duration is merely a spatialized time, the idea of a four-dimensional space-time, embodying time and space, was already prepared by classical physics and even by our language. However, in classical physics, the assimilation of time to the fourth dimension is implied, and it does not go as far as in the Theory of Relativity. In the latter theory we deal with a 'space-time amalgam' because the measurement of time is connected with the system of reference, and only the sum of four squares '$\Delta x^2 + \Delta y^2 + \Delta z^2 - \Delta t^2$' is invariant.

Bergson perceives well the advantage of considering things from this point of view, because what is given as movement in a space of some number of dimensions can be represented as a form in a space having one more dimension, thanks to which all of the temporal particularities of the change can be portrayed to us by a geometrical figure. However, according to Bergson, this representation is not adequate to what is represented. It comprises more from one side, and less from another. The future *per se* is eliminated by this representation, but in another sense, it represents more, because the figure completely drawn implies an infinite number of modes of generation.

Since we have substituted the *ready made* for what we see *as it develops*, we have eliminated on the one hand, the future inherent in time, and on the other hand, we have introduced the possibility of an infinity of other processes by which the thing might just as well have been constructed.[102]

Thus 'whosoever does not here bring a double correction, would run the risk of being deceived as regards the philosophical meaning of the theory of Relativity and of constructing a mathematical representation in transcendental reality.'[103]

The invariance of the sum of the squares $\Delta x_1^2 + \Delta x_2^2 + \Delta x_3^2 + \Delta x_4^2$

relates, according to Bergson, only to fictitious time and to fictitious space. Space and time only amalgamate in the case of fantastic observers, since this amalgamation begins to occur at the moment when the observer puts the system in motion; i.e. at the time when the observer leaves his system in thought. To a real physicist, the system is always immobile and, in the immobile system, the square of the spatial distance and that of the time interval remain independent of one another and do not mix with one another. Hence, space and time are unable to mix other than by the effect of a mathematical fiction. 'The essence of the Theory of Relativity consists in putting at the same level the real vision and the virtual visions.'

In his final remark, Bergson mentions briefly the General Theory of Relativity, by observing that the Special Theory and the General Theory are not endowed with exactly the same spirit and have not a completely identical meaning. Should time in the former be a mathematical fiction, this does not mean that the same would obtain for space in the General Theory. The peculiarities of the latter can be checked by experiment, while the peculiarities of time in the Special Theory are: 'unverifiable *en droit*'.

141. We will now present our critical observations on Bergson's point of view. In his 'Preface', he points out that he wants to defend not only his conception of duration, but also the idea of a universal time, accepted by everybody. Now, we must observe that he does not proceed in a consequent manner. Taking physical time as the foundation of his notion, he closed off the path by which one might arrive at the idea of a unique time. In *Matière et Memoire*, he explicitly says that there exists no unique rhythm of duration. Now, in *Durée et Simultanéite*, he justifies himself by saying that he had thought at that time of organic beings, of the animal kingdom, endowed with various degrees of consciousness. However, let us not forget that he also attributes to the inorganic world an impersonal consciousness, and nothing proves that the 'diluted' [104] duration, which he opposes to the more condensed duration of a more intensive life, is the same throughout the entire inorganic world.

However, this observation does not have too great a value for the general appraisal of the point of view accepted by Bergson with regard to the Theory of Relativity. Bergson wants to make use of the statement of physics that the opinions of the observers agree with one another as long as the systems are immobile, and that, when movement does appear and their measurements no longer agree, then their role is interchangeable. The disagreements between the measurements, he says, ought to be treated as relating to fictitious time. Of course, this interpretation cannot be excluded, but it encounters certain difficulties.

First, it is difficult to indicate a demarcation line that would show us where fictitious time begins and where true time ends. Bergson says that we should use as a criterion that time which has been really experienced. The time which is the result of a mathematical construction is not real time, since time has not been lived by anyone. However, we shall see that this criterion is only clear in theory. Let us take the example discussed by Einstein and Bergson: if two flashes of lightning are simultaneous relative to the track, are they also simultaneous with respect to the train? Bergson is satisfied with the remark that the observer within the train would degrade himself to the role of a fiction if he wanted to consider himself as being in motion. However, this observation does not eliminate the difficulty: let us see what happens in the times lived by the observer inside the train and by him who is outside the train. The former sees in his mirror the lightning flashes one after another, and this succession belongs to the time he experienced immediately before; this is not a mathematical construction. The same lightning flashes are seen by the observer simultaneously outside the train, and this is also experienced time. Differences do exist in the experienced times and not only in the calculations, and it matters little which of the two observers wants to consider himself as being at rest.

The paradox of the two men, one of whom has left the Earth, is also erroneously represented by Bergson. He supposes that the two years of the traveller in the projectile is a fictitious time, and that the true time is that of the inhabitant of the Earth, who had aged by two hundred years. 'Had we said that the former flow of time was two hundred years, then the other flow would be of two hundred years as well. Peter and Paul, the Earth and the missile would have lived through the same duration and would have aged in like manner.'[108] This example can be studied thoroughly only from the point of view of the General Theory and, if we do this, we see that the opinions of these two persons are not contradictory; that not only has the inhabitant of the Earth lived, and aged, 200 years, while attributing 2 years to the space traveller, but also that the traveller lived and aged 2 years while assigning 200 years to the Earth. Since this example gave rise to many misunderstandings, Einstein devoted to it a special article: '*Dialog über Einwände gegan die Relativitätstheorie*'. The paradox found there is a quite satisfactory solution, different from that of Bergson.

The defender of universal time can seek another explanation, by affirming that both of these persons lived and counted their time poorly, since one of them aged too fast and the other too slowly, but that they had nonetheless passed the same absolute time. This answer deserves being considered more precisely, but first of all, we must establish that Bergson's criterion is insufficient and that it is not true that two years only existed in the calculations of the Earth dweller.

This example brings out well the difference between the Theory of Relativity and the traditional point of view. Let us suppose that the persons in question, Peter and Paul, were born simultaneously on Earth in the same place. After a certain time, wherein one of them made his cosmic voyage, they meet again; are they not entitled to say that they had lived the same period of time, even if one of them is already an old man, while the other one is still a baby? Doubtlessly yes, but this equality will have a different meaning than that of which we speak of in physics. To which of these two meanings should we give preference? Which one corresponds to reality? If we agree as far as the meaning of the words is concerned, a satisfactory answer is possible. The first duration expresses nothing but the coincidence of the end-points of the intervals; it does not say anything about what happened inside the duration between the two encounters; it disregards the difference of the rhythms of the duration, and hence it will only be an abstract, an ideal, duration. It will even be an imprecise duration, since one of these persons could disappear and be later resuscitated; the former meaning does not take measurement into account. The second meaning is more complete; it gives an account of the fact that the measurements were carried out under different conditions and that the organic rhythm and the tension of the duration were absolutely different, hence the real durations, organic and measured, were not identical. Since we decided to give the second meaning to the word 'duration', we shall say that the durations of these two persons are not equal.

Let us return to our criticism of Bergson. His observation that the mathematical representation, in which the time axis appears as one of the four dimensions of the universe, is one-sided and inadequate, is fully justified. However, on the other hand, the assertion that the invariant form of the laws of nature refers only to fictitious time is erroneous. True, this form does not well express the entire qualitative difference between time and space, but it does account very well for the metrical relations between space and time, and these relations cannot be considered as indifferent or irrelevant with respect to the nature of time. The metrical properties are just as essential with regard to time as the qualitative, topological properties.

Bergson wants to place before us the following alternative. Either we recognize the independent role of time, and then the invariable form of the laws of nature must be considered as relative to fictitious time, or else, if we identify this invariable form as real, we shall be forced to invalidate the nature of time, by degrading it to the rank of one of the dimensions of four-dimensional hyperspace. On the other hand, we show that there is a third solution; viz., the distinction between the qualitative and the metrical side of time. Qualitatively, time is absolutely different from space, but its metrical properties are connected with the metrical properties of space, and this connection is not fictitious. However, this reply is not complete, as one detail complicates it: the simultaneity of the distant events has revealed itself as a concept which is not only qualitative, but also metrical. However, one finds here no conflict with direct data since the simultaneity of distant events is not perceived directly by any consciousness (except the divine consciousness). And the superhuman consciousness, were it even able to take in the entire universe, would probably establish that the phenomena which our consciousness would consider to be simultaneous, would not behave as simultaneous. This behaviour from the dynamic point of view was recognized by ourselves as essential for temporal relations.

Finally, one should note one more inconsistency in Bergson. The last chapter of his work shows that his appraisal of the General Theory is completely different from that of the Special Theory. The reality of the non-Euclidean space in the General Theory does not seem to him to be uncertain. Obviously, he does not realize the close connection which exists between these two theories. The example by means of which Einstein demonstrates the need of introducing the formulae of non-Euclidean geometry, appeals to contraction which follows from the special theory. If this contraction is fictitious to Bergson, then the non-Euclidean character of space ought to be considered to be a mathematical fiction as well.

Beside, Bergson does not exclude other possible interpretations of the theory of relativity. In a reference [107] he mentions a book by Wildon Carr, who interprets this theory in an idealist sense, and he adds: 'We shall not proceed thus far; but we think that it is certainly in the idealistic direction that this physics ought to be oriented if one would exalt this physics to the state of philosophy.'

We are of the opinion that it is unnecessary to appeal either to the idealistic point of view or to the realistic one, which would be compelled to assign a fictitious character to the results of the Theory of Relativity. The distinction between relative properties and those that are not relative provides the best way out of the difficulty.

142. The attitude of Bergson with regard to the Theory of Relativity somewhat approaches the point of view of the German philosophers who interpret this theory in conformity with the ideas of Vaihinger's philosophy of fiction. The society of friends of this philosophy organized, in 1920, a competition for a study entitled: 'The Relation of Einstein's Theory of Relativity to Modern Philosophy from the point of view of the Philosophy of Fiction'.[108] The prize was bestowed upon the study of Alois Wenzl, who elaborated his subject very thoroughly but whose result was rather negative for Vaihinger's philosophy. In his work, Wenzl made use of vol. II of the *Annalen der Philosophie*, wherein some followers of Vaihinger's philosophy had already criticized relativistic theory from this point of view (Oskar Kraus, Lipsius, Höpfner).

Fiction is not identical to scientific hypothesis. The hypothesis is always more or less probable; we are convinced in advance of the falsehood of fiction, because it contradicts experience, and sometimes it even contains internal contradictions. In spite of this, fiction, according to Vaihinger, can be of service to us in the discovery of reality because, by introducing new fictions opposed to the former ones, we arrive at the conclusion that either all these fictions eliminate each other or else they mutually compensate one another in such a way that the final result is correct.

Höpfner,[109] having analysed Einstein's theory from the point of view of Vaihinger's philosophy, already covered six fictions of which the last three relate to time. These are: (1) the fiction of the clocks and of the observers situated in the space, (2) the definitions of simultaneity, first admitted for immobile space and later considered as valid in the changed conditions of a movable system, and (3) the fiction of the different operation of clocks immovable and immobile systems. Among the fictions which have no reference to time, we find the principle of Special Relativity and the principle of the constancy of the speed of light.

Now, Wenzl rightly observed that the supposition of this existence of observers and clocks in all of space is doubtlessly a fiction, but that this fiction cannot bring about any conflict. The latter fiction is merely the consequence of the second one, and therefore it should not appear as an independent fiction. Insofar as the second one is concerned (which refers to simultaneity), we should examine first whether it is not connected with the principles of Special Relativity and of the constancy of the speed of light, which had also been arranged by Höpfner among the fictions in an absolutely unjustified manner. These principles were suggested by experience and therefore they are no more fictitious than any law of physics.

The connection between the relativistic conception of simultaneity and these two principles were the object of a vehement criticism by Oscar Kraus, in the article '*Fiktion und Hypothese in der Einsteinschen Relativitätstheorie*'.[110] He even objected that the Theory of Relativity contained a series of contradictions. Now Wenzl, having subjected these objections to detailed analysis, established that they were not well-founded. The contradictions, if they did exist at all, could belong to one of the [following] three groups: (1) internal or logical contradictions, in the proper meaning of this word, (2) theses which contradict reality, (3) theses which contradict our intuitive principles. It is impossible to speak of a contradiction with respect to reality, because we do not know what this reality is, independently of our fictions. The internal contradiction does not exist either. Kraus found that the relativization of simultaneity made it impossible to formulate the principle of contradiction. However, there is no need to identify the meaning of the conjunction 'and', (or that of 'at one and the same time', 'together', '$\alpha\mu\alpha$', in Aristotle) which expresses the logical product of two affirmations, with the meaning of the adverb 'simultaneously', and to try to derive therefore a logic of the science of time. Wenzl's answer agrees with our opinion expressed in an article of 1920 'Philosophical Reflexions on the Theory of Relativity' (in Polish).[111]

Let us, however, suppose that there exist internal contradictions in the Theory of Relativity. Where do we find, then, the opposed fictions which compensate or eliminate the preceding ones, as Vaihinger demanded? The admitted suppositions do not disappear, on the contrary, they influence the final result; whence it follows that the Theory of Relativity does not correspond with the perspective of Vaihinger's philosophy.

There remains the third group of contradictions. The phenomenologists of Husserl's school assert that a proposition can be absurd not only when in contradicts the laws of formal logic, but also when it contradicts the 'essence' of an object, conceived as an idea. Wenzl does not question the value of our statements about ideal objects, but he correctly emphasizes that it is impossible to decide that in the facts corresponds to our ideal objects. We made the same observation in our work 'The Axiomatic Method and the Science of Nature',[112] adding that knowledge of the Theory of Relativity and of non-Euclidean geometries likewise reveals to us an unexpected field of new, ideal objects. Reichenbach noted well that lack of evidence can result from the lack of sufficient initiation, and that this evidence does not always exist, even in cases when we supposed it to.

We omit other observations of Wenzl, since they do not introduce anything new from the point of view of principles. Wenzl ends his study by remarking that we have the impression that, concerning the Theory of Relativity, it is rather a question of a verifiable hypothesis, than of a fiction.

143. There remains only for us to enquire into what the German neo-Kantians tell us with respect to the problems discussed. Kant's time, in spite of its different origin, preserved all the properties of Newtonian time. No wonder that the partisans of Kant's philosophy point to the disagreement between this philosophy and the relativistic conception of time. All the more interest was brought about by the book of Cassirer, *Zur Einsteinschen Relativitätstheorie*,[113] which aimed at persuading us that the foundations of Kantian philosophy are sufficiently extensive for the new theory to find therein a well-deserved place. Of course this theory rather exceeds Kantian philosophy, but, in Cassirer's opinion, it does not contradict that philosophy. The freedom with which the Marburg school corrected and interpreted the ideas of Kant made it easier for Cassirer to reconcile the new theory with the ideas of transcendental philosophy. *A priori* time only provides us with a general scheme for the coordination of events, which excludes neither the relativity of duration nor that of simultaneity. The unity of time does not require the existence of the same metric values in all the systems.

Cassirer states with satisfaction that the Theory of Relativity attests the correctness not of the sensualist conception of experience, but of the critical one, since it confirms the involvement of constructive elements in experience. The theory owes its existence to the attempt to reconcile the results of two kinds of experience, apparently contradictory. This attempt was successful, but only due to the modification of the fundamental principles relative to time and space. The idea of the unity of nature was saved. Moreover, we finally arrived at the conception of a truly universal law of nature, independent of all the particularities of reference systems.

According to Cassirer, the philosophical notion of truth cannot be defined as a concordance of the copy with its original; it can only consist in a connection of the results of experience in all the possible systems, according to certain rules. Knowledge is true, if the different aspects of the world correspond to each other functionally in all internal relations and in their general form.

We shall take the liberty of adding here some critical observations on Cassirer's point of view. The opinion that the Kantian doctrine of time can be reconciled with the relativistic conception, was expressed for the first time by the German physicist, Laue. According to him, to all relative times, there corresponds a temporal scheme of general order which, however, can acquire empirical reality only through its connection with a system of reference This general scheme represents neither an absolute nor a relative time, since it does not tell us anything about whether it is possible or not to deal with the plan independently of the reference system. Thus the meaning of the Kantian *a priori* is reduced to a few topological properties. We cannot counter this with anything, provided that the partisans of Kantian philosophy first explain to us the significance they attach to the term *a priori*. Indeed, we have seen already that, what the Germans call '*Geltung*', or validity for all experience, does not tell us anything about either the origin of the ideas of about its reality; or, at least, the consequences we want to draw from it, are not decisive.

The second point which interests us here, as far as Cassirer's attitude with respect to the Theory of Relativity is concerned, is the definition of truth. It seems to us that nothing obliges us to radically change our conception of truth, and that the traditional conception can be saved, provided we introduce relative properties. But perhaps Cassirer was not very far from this idea. If different aspects of reality correspond functionally to one another – and this allows us to reach a general invariant form, valid for all times and for all observers – all this can only confirm the traditional conception of truth. However, this invariant form does not comprise everything, it expresses the invariance of the metrical spatio–temporal forms and does not excuse us from the need to work with the relative properties connected with particular systems. Bergson has well emphasized that the invariance of the mathematical forms of four-dimensional hyperspace does not express the whole of reality.

Among the followers of Husserl's phenomenology we find the great relativist, Hermann Weyl. However, he is an exception since most phenomenologists, we think, assume a rather negative attitude with regard to the new theory. Weyl's point of view would have a greater value for us had it been better specified. Unfortunately, this great physicist and mathematician does not attribute too great an importance to the philosophical subtleties of various schools. In his work, *Philosophie der Mathematik und Naturwissenschaft*, he says sincerely that it is completely immaterial, as far as he is concerned, whether somebody speaks the language of Husserl or of Helmholtz, because he sees concordance in all of the philosophers who think and work seriously.[114]

144. To end this chapter, we shall focus our attention upon the attempts of those philosophers who sought to axiomatize our science of time. A definition of time, by means of the usual processes of formal logic, seems to us to be impossible. Of course, time does possess properties which it shares with space, it is extensive and has a dimension; hence we can find for it the 'nearest kind' (*genus proximum*) and the difference (*differentia specifica*), but this difference can not be expressed by any other term than that which points to the temporal character of one-dimensional length. It does contain a certain order, hence an asymmetrical and transitive relation. However, there exists an infinite number of asymmetrical and transitive relations, which have nothing in common with the temporal order. The possibility of distinguishing between the past, the present and the future in this order of temporal succession is unique in its kind, and it is impossible to deduce this difference from the sole idea of an order. The term 'succession', if it is not identical with the temporal succession, indicates only an asymmetrical and transitive relation.

Attempts were made at defining at least some of the temporal notions, such as, e.g., those of succession and simultaneity. Now, we must accept that at least one of these terms is a primitive term, in order to express the meaning of another term by means of this term and by way of primitive terms of logic.[115] The primitive terms of logic do not suffice by themselves.

Definitions by axioms are often used in modern theories. We formulate a series or propositions wherein we find the primitive terms of the theory constructed, connected to each other by means of logical terms. Thus, the primitive terms are defined 'implicitly' by the role they play in these axioms or primitive propositions. This method of defining implicitly has often been used by the Italian mathematicians. Peano defined the notion of the integer in this way by means of five axioms and three primitive terms. Hilbert applied this method to geometry. The Theory of Relativity provided philosophers with the impetus to establish the axioms of a science of time, connected with the theory of space. We enumerate the primitive terms of the theory, thereafter, we formulate its axioms or primitive propositions, and then we define each new term of the theory using only these primitive terms, and we deduce new propositions on the basis of the primitive propositions. When constructing new notions and deducing new propositions, we make use solely of certain rules of formal logic.

In his book, *Abriss der Logistik*[166] Carnap gives us several examples of the application of the axiomatic method to the topology of space-time, according to the Theory of Relativity. The same topology is *axiomatized* once by means of two primitive terms, and again using a single primitive term. The primitive terms of the former example are (1) the symbol of coincidence '*K*' encountered in the expression 'x*K*y', which reads as follows: point x of the universe coincides with point y of the universe, and (2) the symbol *Z* which indicates, in the expression 'x*Z*y',that two points – x,y – of the universe belong to one world-line, and that x precedes y in time. In the second example, these two terms appear as defined by means of the symbol 'ylin', which is the unique primitive term. It designates a class of relations which originate from the limitation of the relation *Z* to the singular world-lines of the universe. Another variant of this second example is based on the primitive term '*W*' which denotes a relation of action which is transitive, nonreflexive and dense, and possesses neither a beginning nor an end.

145. But the most interesting of all the attempts at *axiomatizing* the science of time is that of Whitehead, who adopts as the basis of his theory, not the points of the universe, but the natural elements which are the events, and constructs the geometry of the events. The events are the members of a homogeneous relation of extension, understood in the widest sense of the word. The content of this notion of extension is determined by the following axioms: (1) if the event *a* extends over the event *b* (or if *b* forms a part of the event *a*), *a* is different from *b*, (2) each event overlaps other events and forms a part of other events, (3) if parts of the event *b* also form parts of the event *a*, and *a* and *b* are different, than *a* always overlaps *b*, (4) the relation of the extension is transitive, (5) if the event *a* overlaps the event *c*, there always exists an event *b* such that the event *a* overlaps *b*, and *b* overlaps *c*; (6) if *a* and *b* are any two events, there always exists an event *c* which overlaps *a* and *b*.

The construction of time notions and space notions is based upon this relation. However, we have to use *herewith a special method, called a method of extensive abstraction*. The essence of this method consists in that we form abstractive classes and define scientific notions by means of these classes. A set of events forms an abstractive class:L (1) if any two events of this set are connected by the relation of extension in such a way that one of them contains the other; (2) if there exists no event that would be a common part of all the events or of that set or, in other terms, if all of the events were to overlap it. We can see from the definition that the elements of an abstractive class form a series, that the extension of elements of this series decreases without cease in such a way that, thanks to this series, we can approach an ideal state of nature, which is not an event, but which is determined univocally by the internal character of this series. Two abstractive classes are equal if each element of one class overlaps an element of the other one, and vice versa. A class of these classes is called an abstractive element.

Further on, Whitehead gives two applications of his method, one for finite events, the other for durations or indefinite events. The point is that Whitehead understands by duration the totality of nature, indefinite with regard to space, but definite with regard to time, thus a temporal layer of all simultaneous nature. Through the application of the abstractive method to duration we arrive at the concept of a moment (or instant), or 'of nature at a moment'. The definitions of the spatial forms of momentary space can be deduced from the relations between these moments in such a way that the spatial order of momentary space derives from the temporal order.

By using the abstractive method in the case of limited events, we arrive at the concept of the point-event and, by means of the latter concept, we construct the four-dimensional geometry of events. In order to define metric relations, Whitehead introduces six axioms of congruence; two of them relate to time; viz., the axiom of kinetic symmetry and the axiom expressing the transitiveness of the temporal congruence. By means of these axioms, Whitehead arrives at very general transformation formula which contain a constant, possessing the dimension of velocity. This constant can be endowed either with an infinite value, or with a finite, positive, or negative one. Whitehead analyses these three cases one after another; the first one, wherein this constant assumes an infinite value, corresponds to Newtonian mechanics. The second one, in which this constant assumes a finite and positive value, is that of relativistic physics; the third has no physical meaning. Thus we arrive at the formulae of relativistic physics by means of deduction, as one of the three logically possible cases. The speed of light does not play any role in this deduction.

Whitehead's geometry of events is also interesting from another point of view. It shows the influence of Bergson's philosophy upon the conceptions of mathematicians. Whitehead has shown that the geometry of space-time need not necessarily take geometrical points as its foundation, but that it can also be constructed using the notion of an event. It is beyond doubt that, in his philosophy of continuity, Bergson caught hold of something that brought him close to the modern scientific conceptions of the universe.

Now English literature possesses also another axiomatization of the science of time, created by Alfred Robb and based on the supposition (rather artificial and arbitrary) that only those phenomena which occur in the same place can be simultaneous. It is usually admitted that the phenomena *A* and *B* are simultaneous, if they take place in such a way that neither *A* precedes *B*, nor *B* precedes *A*. Robb finds that this reasoning is true only if the phenomena occur in the same place.

## 4. THE LAST GROUP OF TIME *APORIAS*: CONTINUITY, INFINITY, IRREVERSIBILITY

### *Continuity*

146. Continuity had almost always been attributed to time, as well as to motion and to space; thus, the problem arises of understanding what this continuity means. To begin with, in order to distinguish continuity in space, we indicate it is a unidimensional continuum. The definition that such a continuum is an ordered set, without jumps and gaps, is merely the product of modern mathematical analysis. This definition has given rise to much protest, as it defines a continuous magnitude by means of non-continuous elements, such as numbers, points, and instants. For this reason we prefer to take, from the outset, as our starting-point, the much simpler conception found in Aristotle; viz., that the continuity of time consists in that time is infinitely divisible and that each part of it is time. Instants are not parts of time, they limit the intervals and connect them. By limiting the parts of time, they 'measure' time says Aristotle, but, since they connect the divided parts, they [also] render time continuous. Therefore, time contains not only parts, intervals, but also the moments or instants, as point-limits of these parts. Since the division into parts can be, according to Aristotle, infinitely continued, the shortest part of time does not exist. Lastly, we recall that, according to Aristotle, this division exists only *in potentia* and not *in actu*.

This opinion of Aristotle has almost become the common property of all of mankind; it was probably the best expression of our intuitions. Only, from the psychological point of view, we sometimes observed that the present was not identical with the instant, conceived as a limit-point and that from the psychological point of view the instant appeared as an element whose parts could not be distinguished by consciousness. We find in Locke, for example, a correction of Aristotle. By means of this correction, the instant acquired a more real meaning, though at the same time a less precise one, since the incapacity to distinguish among the parts is probably not the same for all men and, obviously, it also varies within the same individual.

The first attacks against continuity and infinity were launched by Zeno of Elea, in his famous arguments against the existence of motion. Aristotle objected to these arguments, (1) that they do not distinguish between the infinity of an endless extension, (*apeiron tois eskatois*), and the infinity resulting from an endless division (*apeiron kata diairesin*); (2) that they do not distinguish between actual infinity and potential infinity, and (3) that Zeno, by infinitely dividing the space covered, does not do the same with time. This last observation is completely unjustified, since the entire reasoning of Zeno, in the argument of Achilles and the tortoise, and in that of dichotomy, presupposes as much the infinite division of space, as that of time.

147. The distinction between actual and potential infinity played a very great role in history, but even this distinction had often been considered as insufficient, since it was doubted whether the question was of *potentia* in nature or of *potentia* only with respect to our mind. That is why modern criticism has not always been happy with Aristotle's answer. One of the most interesting articles in modern times on Zeno's arguments, viz. that of G. Noel in the *Revue de Métaphysique et de Morale*, of 1893, gave another solution quite akin to that of Bergson. Let us see how he presented this problem.

Two suppositions can be made: either time, motion and space are continuous and infinitely divisible, or else they are composed of a finite number of indivisible elements. Zeno's arguments aim to show that both these suppositions are false and, consequently, that movement is impossible. The absurdity of infinite division is shown by the arguments of dichotomy and of Achilles, the absurdity of the second supposition becomes obvious by the arguments of the stadium and of the arrow. The usual refutations of Zeno's arguments do not seem to be sufficient, for they do not consider Zeno's suppositions. We note that the infinite series of the two former arguments represents a finite sum which can be calculated by a mathematician. But a mathematician does not add the elements of the series, 'he uses a reasoning which is not within the reach of the changeable.' Moreover, the assertion that the sum is finite in spite of the infinity of the elements, is not true, since it is the limit of this sum, and not the sum itself, that has a finite value.

Similarly, the arguments of the arrow and of the stadium are not as easily refuted as is usually supposed, according to Noel's opinion. The argument of the stadium 'is generally assumed to be a crass sophism. Indeed, it is that, if we suppose that the double continuity of space and time is accepted. On the contrary, it is irrefutable if we admit the opposite hypothesis.' Similarly, the argument of the arrow is irreproachable, according to Noel, if we allow that time is discontinuous, since, with this supposition, instants are not zero values of duration, as in the usual hypothesis. The argument should be expressed as follows: 'The mobile, whether it is at rest or in movement, is always in a space equal to itself; moreover it occupies this space at least in the course of an instant; viz., a finite time. It is in a space equal to itself during a finite time, hence, by definition, it is at rest.'

Unfortunately, the partisans of the discontinuity thesis do not always admit that the instant is a finite duration. Evellin tested this supposition. And even if the indivisible instant has a finite duration, we do not see why the mobile should not occupy different positions at different instants. The arrow will not fly better in discontinuous duration than in continuous time.

However, Noel believes that the arguments of the arrow and of the stadium are irrefutable, and that the error is found rather in the two former arguments.

It seems that a quantity can be given only in two ways, either the entire quantity all at once, or by successive fractions. For movement as well as for time, the former supposition is obviously untenable, and Zeno shows that we could not accept the second one [either]. But is it really true that a third one could not be conceived? A careful examination of the two opposing hypotheses, beyond their apparent contradiction, reveals to us their fundamental identity. Indeed, the parts which we unite in order to form a whole are, when taken in themselves, quantities complete for the same reason as this whole. Thus, the two suppositions which we distinguished contain a common assertion. When establishing them as the only conceivable ones, we assert that quantity is necessarily given to us ready made, always *in actu*, never *in potentia*. We forbid ourselves to apply to magnitudes the category of becoming. We remove in advance the notion of a quantity in the state of becoming.

The mathematical operations concern only clearly defined quantities, i.e. perfect [quantities]. Understanding composes and decomposes them in turns, but it forms them only from pre- existing elements and resolves them always only into parts which they already contained. Understanding is powerless to grasp continuity as such.

Noel, like Bergson, may be right when saying that mathematical operations do not given an account of becoming, but nevertheless Zeno's arguments contain false reasonings and are by no means irrefutable. Let us suppose that the velocity of Achilles and that of the tortoise are in a relation of *n:1*, let us further take their distance as a unit, and, let us likewise take as a unit of time that which Achilles requires to cover this distance. Reasoning proves that, if we divide a finite distance and the finite time according to the law:

$$1 + \frac{1}{n} + \frac{1}{n^2} + \frac{1}{n^3} + \ldots\ldots \text{ in inf.}$$

and if Achilles and the tortoise are forced to run according to this law, a finite number of their paces or a finite number of time fractions will never exhaust either this distance or the time indispensable for covering this distance. However, this is not the proof that motion is impossible or that Achilles might never overtake the tortoise in any other condition. The argument of dichotomy contains the same error. That of the arrow is of another nature, but even in this case, Noel's observation is not completely correct.

According to Noel, the argument of the arrow is irreproachable provided we accept Zeno's supposition. Now according to Noel, in the arguments of the dichotomy and of Achilles, Zeno wants to prove that movement is impossible if we assume the division to infinity, and, in the arguments of the arrow and the stadium, that movement is impossible if we suppose the contrary, i.e. that time and space are composed of a finite number of elements. We do not see whether the latter supposition really gives a greater force to the argument of the arrow than the contrary supposition, however. It seems to us that, whatever the number of moments or positions (finite or infinite), reasoning will point to the difficulty in understanding movement. The point is that, by assigning a position to every instant, we obtain a series of positions and not what, according to our intuition, is movement. Noel and Bergson are right in that mathematical analysis does not explain how movement is accomplished, but Zeno is wrong to affirm that it is only an illusion.

148. Since the appearance of quantum theory, Zeno's arguments have lost their interest, because we see that movement is possible, even if it is

not continuous. In any case, the philosophical discussion dealing with these arguments does acquaint us with all the possible opinions relating to the subjects of the continuity of time and of change. They can be divided into four groups:

(1) the opinion of the finitists, to whom each duration is composed of a finite number of indivisible elements,
(2) the opinion of the infinitists who accept, in continuity, an actual infinity,
(3) that of the infinitists, with the infinity of the parts *in potentia*,
(4) and the point of view of the intuitionists, such as Bergson, who see in concrete duration an invisible continuity.

The opinion which is most widespread among philosophers is, without doubt, that of Aristotle, which we find under (3). The second opinion is that of mathematical analysis, the point of view of Cantor, who created the theory of sets. However, modern physical science seems to plead in favor of the finitist point of view, though not completely. We shall therefore examine the finitist doctrines more closely.

The finitists do not agree on the nature of their indivisible time elements. We can distinguish three forms of finitism (at least theoretically): (1) According to some, the indivisible elements are the zeros of duration, which create duration only by their succession. Obviously, these temporal points cannot be continuous, and therefore there are 'hiatuses' between the two points, because they are finite in number. This theory would be the most enigmatic one of all the theories. (2) The second form supposes that the indivisible elements of time possess a certain duration, but that these durations are separated from one another by holes in the time, by hiatuses similar to those of the preceding theory. We are not sure whether this theory really had a representative in philosophy, but the theory is possible. (3) The third form also attributes to time atoms a certain duration, but it does not accept the hiatuses, since the shortest durations are contiguous. In our opinion, this theory does not deserve the name of atomism; it can very easily agree with the theory of continuity. Still we should distinguish between two kinds of this third form. According to the first modification, the shortest durations are indivisible, not because of our inability to divide them, but because there exists no shorter process in nature. This opinion has acquired in recent time the character of a scientific hypothesis. Its origin is as follows.

The density of energy possesses a maximum, and therefore the frequency of oscillations $\nu$ which create electromagnetic waves, has its upper limit. These oscillations represent the shortest processes occurring in nature. But theory affirms something more: it says that every other duration [that is,] every slower oscillation, is achieved in a time which is an integral multiple of the shortest duration called a 'chronon'. It contains barely a few quadrillionth parts of one second: $\vartheta = 4.3 \times 10^{-24}$ sec or $4.5 \times 10^{-24}$ sec.[117] The durations of this frequency appear in the 'hardest' components of cosmic radiation, and it is claimed that the hypothesis of the chronons was verified for the shortest waves. Thus chronon would be the least interval or the smallest unit of the rhythm of nature.

But the hypothesis of the chronons is not yet quite certain. The weak side of this idea consists in that it had been created independently of quantum theory and of the theory of relativity. We do not see clearly the relation between these theories, and even the chronon hypothesis agrees neither with the theory of relativity, nor with the quantum theory. In any case, this hypothesis does not contradict the theory of continuity for in our opinion, the continuity of time would be abolished if a 'hiatus' were established in the flow of time. Now, such a hiatus, a 'hole' in time as it were, does not exist. All that the theory asserts is that each real duration is an integer multiple of that shortest duration, which is not further divisible. This theory concerns only vibrations, i.e. periodical processes, and nothing has been established for other processes, unless it be supposed that each process of nature is a rhythm or a fragment of a broken rhythm.

The chronon theory does not abolish the difference between a time interval and the moment, conceived as a limit-point of this interval. Atoms have a duration, and, during the duration of a chronon, another chronon, in another series, can begin to exist and thus even the determination of a temporal point within the limits of the duration of a chronon is not absurd. Finally, we should note that the chronon theory does not confuse the physical atom of time with the psychological atom, as has so often been done. The chronon has nothing to do with the elements of time wherein our consciousness is unable to distinguish the parts.

The atomism of time can, moreover assume different forms, depending upon the point of view we accept concerning the question of the realism of time. For moderate realism, this problem is reduced to that of the atomism of events. Now, it is beyond doubt that certain phenomena are not continuous. The impressions of some senses produce discontinuous series; e.g., sounds, tastes, smells, etc., as Aristotle already observed. In order to abolish the continuity of time, understood in the sense of moderate realism, it would be necessary to show that all events, all movements, without any exception, take place by fits and starts, producing 'hiatuses' which are not filled by any event. Modern physical science really does expose us to the danger of such an adventurous theory, but prior to tackling this problem, we shall first take a look at the mathematical theory of continuity, created by Cantor.[118]

149. This theory constructs the continuum by means of discontinuous elements. It disregards the continuum of space or of time and speaks, first and foremost, of the continuum produced by numbers. The entirety is just as abstract as the elements. Instead of the intuitive term 'hiatus', we introduce the well-defined terms of the 'leap' or of the 'gap'. Let $U$ be an ordered set. We call the cut of the set $U$, each decomposition of $U$ into two non-empty classes $A$ and $B$, such that any element of class $A$ precedes, in $U$, any element of class $B$. If, in a cut, class $A$ possesses a last element, and class $B$ possesses a first element, we say that the cut under consideration determines a leap. For instance, in the ordered set of all integers, each cut determines a leap.

Let us now take the set ordered with respect to the magnitude of all rational numbers, i.e. of integers and fractions. In this set no cut yields leaps. An ordered set of which no cut will produce leaps is called a dense set. Hence our set is dense. Yet it is not yet a continuous set, as it possesses gaps. What is a gap? If, in a cut, class $A$ does not possess a last element, and class $B$ does not possess a first element, we say that we have a gap. Now the set of all rational numbers possesses such gaps. An ordered set without leaps and without gaps is called continuous. If we complete the set of all rational numbers by filling up the gaps, we obtain a continuous set. The theory of irrational numbers, given by Dedekind, indicates the method of filling up the gaps of the set of rational numbers, by using irrational numbers. Almost simultaneously with Dedekind, Cantor created his theory of irrational numbers, equivalent to that of Dedekind. Mathematicians are very proud of their work in defining the continuum, and of their theory of irrational numbers, and it is from this moment that there was discussion of the arithmetization of mathematics.

This theory gave rise to many protests on the part of intuitionists. Is it possible to construct the continuum by means of discontinuous elements? Let us not forget, however, that the whole can possess properties which are not possessed by the elements; hence the above objection is not dangerous. Moreover, it is important to note that the theory of sets discovered various kinds of infinity. An infinite set, the elements of which can be numbered, is called denumerable or, more accurately, an infinite set is called denumerable if there exists a mutual correspondence between its elements, on the one hand, and the natural numbers on the other; in which, [that is,] to any element of the set under consideration, there corresponds a well-determined natural number, and every natural number corresponds to a unique element of the set under consideration. Such a correspondence is called a one-to-one [biunivocal] relation (univocal and reciprocal). We also say that the denumerable sets have the same power as the set of all natural numbers.

Now, it was established that, beside the denumerable, infinite sets there are non-denumerable sets. The set of all rational numbers is denumerable, but the set of all real numbers is no longer denumerable. The real numbers, and among them the transcendental numbers, are so numerous that, in this whole, the rational numbers are only exceptional and almost disappear. Now, the sets which are of the same magnitude as the set of all real numbers, are said to be of the magnitude of the continuum.

However, we can make the following objection to the mathematicians: for them, the continuum is always a set, but the intuitive continuum has nothing in common with a set. What entitles them to consider the continuum of a line as a set? The mathematicians may reply: (1) that what they call a line is not only a set, but an ordered set; (2) that this line is not identical with a physical line.

150. At any rate, the mathematician is obliged to admit that difficulties may arise at the moment when he begins to apply his idea of the continuum to the real world. Up until now, this application has created difficulty, [and] no contradiction However, let us not forget that the 'up until now' means up until the appearance of quantum physics. With the emergence of the latter, difficulties do arise. We shall speak here of the application of the notion of the mathematical continuum to temporal duration. This application did not at all require that each moment, 'understood as a mathematical point,' might exist apart, independently of the others. All that we could ask is that to each temporal point there should correspond well-determined physical state. Now, it is obvious that this supposition could never be verified by experiment, due to the lack of precision of our measuring instruments. Yet we always believed that we could draw nearer to the limit-point, as our instruments became more and more accurate. Quantum theory dispelled these illusions; we found ourselves facing impassable boundaries which render absurd all our faith in the possibility of checking our supposition. According to Heisenberg's principle, it is impossible to verify whether, to each determined temporal

point, there corresponds a precise state of physical reality. The consequence is dangerous for all of the conceptions of continuity which have existed up until now. As we narrow down the interval to arrive at an instantaneous section of the temporal flow, reality evaporates, as it were, before our eyes, because the value of energy becomes completely indeterminate. No atomism of time foresaw this consequence, nor did the theory of continuity.

This state of affairs entails dangerous consequences also for the establishment of continuity in psychical life. Heisenberg's principle applies, of course, only to the physical world. But there are reasons which force us to believe that the determination of psychical states disappears even before the disappearance of the determination of the states of our physical body.

Bergson remarked very correctly that psychical duration is highly condensed; the sensation of red light, experienced by us during one second, corresponds, in itself, to a succession of phenomena which, if they occurred in the course of our duration, would occupy more than 250 centuries of our history. It is extremely improbable for a quality of some kind, sensual or emotional, to exist in us if we descend to the duration of one chronon, i.e. of a few quadrillionths of a second. It is probably that everything disappears already below a thousandth of a second, while experimental psychology expresses the duration of the shortest physical phenomena in multiples of thousandths of a second.

Some philosophers have affirmed that intellectual and volitional acts have no duration, hence that they are momentary. However, no act exists without content, it seems to be co-extensive with the duration of this content; even if it were shorter than the content which explains it, it would be improbable for the act not to have a duration. Let us, however, admit that everything we are asserting with regard to this matter is merely a probability. The continuity of our psychical life, combined with the unity of consciousness, is always enigmatic.

151. If the mathematical notion of continuity cannot be verified for the physical world because of the discoveries of quantum physics, and if its application to the psychical world has been impossible for a very long time because the duration of psychical life presents itself as an enormous condensation of physical duration, none of this proves that the continuity of time has lost all of its intuitive meaning, especially that of Bergson. On the contrary, it is a confirmation of his theory, at least as we understand it. Moments, when understood as mathematical points, do not indicate

any determinate and verifiable reality; it is necessary that the world endure, [and] that this duration have extension, for otherwise this world could assume determinate form. Is this the atomism of duration? No, it is not.

Let us recall, once more, all the possible points of view with respect to the continuity of time: Aristotle's theory; the theory of modern mathematics, based on the theory of sets; the different forms of atomism (we have distinguished three such forms); and the intuitive continuum of Bergson or at least similar to his idea. We should also add to this the new mathematical conception of the continuum, created by the intuitionists, Brouwer and Weyl, who see the essence of the continuum 'not in the relation of the elements to the set, but in that of the part to the whole'.[119] This conception seems to us to approach that of Aristotle.

We have seen that the continuity of modern mathematics, applied to time, is unverifiable because of Heisenberg's principle. When we specify the moment of appearance of a phenomenon, by drawing nearer to the limit $\Delta t = 0$, the imprecision of the energy $\Delta E$ becomes infinitely great. The momentary section of the temporal flow of the world gives us nothing. Aristotle's continuity does not remove this difficulty, because it differs from Cantor's continuity in that it admits the infinite divisibility of the continuum only *in potentia*. Now, it is of little value to us whether or not the number of realizable cuts is finite; we want to know whether a determinate physical state corresponds to each section. Since the constant '*h*' shows that this supposition is untenable, Aristotle's conception of the continuum is just as threatened as that of Cantor.

Let us now take the atomistic theories. The theory of the chronons is highly problematical; it is based on the suppositions of classical physics and, consequently, it contradicts quantum theory. But even if it prevailed over the latter, it would not merit its name, for it affirms only that each duration is an integer multiple of the shortest duration. Truly atomistic is the theory that supposes that continuous temporal series do not exist, that the events of the microcosm are separated from each other by the time hiatuses. It seems that the modern state of physical science does not justify such a supposition. Only a false interpretation of Heisenberg's principle could suggest a comparable idea. The physics of discontinuity does not assert (or, at least, has no right to assert) that, between two values of the energy of an atom, there exists a complete void wherein nothing happens. Along with the corpuscular theory, there also exists the wave theory, in which the continuity of space is connected with the continuity of time. Is it possible to suppose that two states of a field are separated from each other by nothingness, in which nothing occurs and nothing exists? The void, the nothingness, the indeterminacy, opens up before us, the moment we want to check the momentary state with an error equal to zero. Now this bears out the theory of continuity according to which the continuum of changes forms an indivisible whole. The sudden leaps of energy do not prove that they are separated by a hiatus of time. We can evaluate with approximation the time in which the atom stays in a stationary state; is it possible to suppose that, during that time, nothing happens in the interior of the atom? All we can say is that, the closer energy is determined, the less determined is the time, i.e. the moment of change.

152. Thus, modern physics seems to threaten the application of the mathematical continuum to the real world, but not that of the Bergsonian continuum. It is probable that some forms of atomism, which do not directly contradict this continuum, can also be preserved with some modifications. However, we shall go still farther. Modern physics seems to us to confirm Bergson's theory in another respect. According to Bergson, there exists no absolute distinction between a substance and its attributes, a thing and its states. There exist continuous changes without an absolute invariant substrate. Now, since in quantum physics invariance refers only to the product of energy and time (the constant '*h*' is the product of fractions of energy units and of time units, $6.55 \times 10^{-27}$ erg sec), this theory unexpectedly finds its confirmation in the results of modern physics. Heisenberg's principle is only the consequence of this state of affairs. Just as the product of energy and the time coordinate, so also the impulse (the product of mass and velocity), and the coordinates which determine local position, form associated couples inseparable from the micro-structure of the universe. The best expression of this is the parallelism of corpuscular and wave physics. Moreover, we recall that every mass can be transformed into energy: even electrical charges are considered in some physical theories as things which can disappear during transformations wherein all the matter will be dissolved into radiant energy.

Distinct substances exist only approximately in the macrostructure of the universe. Of course, some fragmentation already occurs within radiant energy, in the form of photons, but these photons are probably not absolutely distinct individuals, if we consider that wave theory preserves its '*droit de cité*' in physics beside corpuscular theory. However, we admit that everything we have said in this chapter might be only no more than a great question-mark, since the results of quantum physics are as yet too recent to permit us to consider as definitive any answer to all of these problems.

*Infinity*

153. The second problem to which we devote this chapter is the problem of the infinity of time. The criticism of Kant's antinomies, in the historical section of our work, prepared the reader for the manner in which we treat this [problem]. Its solution through progress in physical science is not out of the question. The principle of the degradation of energy, unknown at Kant's time, throws new light upon the question of the beginning and the end of the universe. However, we shall only touch lightly on this problem here. We have already devoted a separate study to it in our work on the problem of eternal returns.[120] In it we opposed the point of view of cyclism to that of asymptotism. According to the former, time would appear as a closed line; according to the latter, as an open one. The open form can assume three aspects: either it will be limited '*a parte ante*' and '*a parte post*', or it will be limited only at one of its ends, or else it will have neither beginning nor end.

The first and second cases oblige us to admit a supranatural origin of the universe, and this would mean the resignation of scientific explanation. The scientific mind would seek refuge in this contingency only if all of the natural explanations had failed. Thus, if we prefer scientific explanations, we shall be inclined to make our choice between cyclism and the asymptotism which accepts neither the beginning nor the end of the universe. The term "asymptotism" is borrowed from the tangent which approaches its curve indefinitely, without touching it however. This analogy should express the fact that the universe tends toward thermal equilibrium, without ever reaching this state in a finite time. This term can even be applied to the case which admits the absolute beginning of the world, but does not suppose its end in a finite time. The term cyclism, as given to the closed form of time, is self-explanatory.

All discussion on the possibility of eternal returns went out of date with the advent of Lemaître's cosmology. The so-called indeterminism of quantum physics likewise presents the problem in a new light. The only argument advanced in favor of the theory of eternal returns was that of the existence of phenomena which presented an exception to the law of energy degradation. Brownian movement demonstrated that there were reversible processes in cases where irreversibility should have existed according to the second law of thermodynamics, the law of entropy or, to use another expression, the law of energy degradation. This exception was explained by the kinetic theory of matter, according to which physical processes are in reality mechanical processes, and the latter are always reversible. Hence every irreversibility is only apparent, and the law of entropy is only a statistical law with an approximate validity, arising from the fact that the reversibility of some processes is extremely rare owing to their complexity. The German physicist, Boltzmann, who had contributed most of the development of the kinetic theory of matter and who was first to give a statistical interpretation of the law of entropy, also foresaw the existence of possible exceptions thereto prior to their discovery. The explanation of the Brownian movement, given almost simultaneously by the Polish physicist Smoluchowski and by Einstein (1905), confirmed Boltzmann's predictions. Smoluchowski devoted virtually all his life to the study of all the exceptions to the second law of thermodynamics, generally called 'thermodynamic fluctuations' (*Schwankungserscheinungen*).

154. Thermodynamic fluctuations were established in the microcosm only: is is possible to suppose that the entire universe is subject to similar fluctuations? Abel Rey expressed this opinion without hesitation in his work, *Le retour éternel et la philosophie de la physique* (1927), basing his idea on the results of the kinetic theory of matter and on the exceptions to the law of entropy. In the same year, our above cited work appeared, devoted to the same subject; we studied there the question to what degree can a mechanical model of phenomena be looked upon as an adequate representation of physical reality? We focused our attention upon the theorem of the 'phase', formulated for the first time by Poincaré (1890) in the work 'On the Problem of Three Bodies and on the Equations of Dynamics',[121] wherein the author proved that a purely mechanical system, however complex it might be, repeats itself in general an infinite number of times given a sufficiently long period, to pass through a configuration which is as close as we like to its initial state. This theorem occasioned a controversy between Boltzmann and Zermelo who declared, in an article[122] published in 1896, that the kinetic theory of matter should be considered false if one of its consequences, the theorem of Poincaré, contradicted the second law of thermodynamics, since Zermelo realized the possibility of applying this theorem to cosmological speculation. It was then that Boltzmann appealed to his statistical interpretation of entropy. Zermelo[123] likewise simplified Poincaré's

proof, yet the most exact form of this theorem was given by Caratheodory[124] (1919) and from that time on bears the name of the 'theorem of Poincaré-Zermelo-Caratheodory'. This theorem allows us to predict the return of every mechanical state which fulfils the conditions of the kinetic theory of matter or, to put it more precisely, those of the kinetic gas theory, since that theory had been worked out completely only for gases. The time of return plays no role there; for an uncomplicated state it is shorter, for a complex state it is longer; but it does exist, and its existence can be affirmed with a probability almost equal to certainty. If the statistical interpretation of entropy led us to say that the return of some mechanical states was of an extremely low probability, then this would signify that the time of return was extremely long. But in Poincaré's theorem, the length of the time of return is immaterial, it is thus that we can explain the differences in the statement[s].

In order to apply this theorem to the world, it is necessary (1) that the world be finite, (2) that there exist neither ether, nor radiant energy, (3) that the universe be considered as a set of gas molecules, since the kinetic theory of matter was satisfactorily elaborated for the gaseous state only. The theory of relativity admits that the universe is finite, but this theory was constructed upon principles which do not agree with the laws of classical mechanics.

The application of Poincaré's theorem will only yield results very distant from reality, for this theorem disregards the existence of all the phenomena for which the cosmic ether of the classical theory was the basis.

Finally, as regards point (3), we must observe that attempts to explain all phenomena of the solid and the liquid state by means of the kinetic theory of matter have failed. Quantum theory shows that even the theoretical suppositions of Poincaré's theorem are false.[125]

Thus the considerations based on this theorem are absolutely insufficient.

155. Already Smoluchowski realized that there were phenomena which seemed to be truly irreversible, viz. the phenomena of radiation, those of radioactive transformations, and the appearance of retarded potentials. The progress of the sciences led to the discovery of new complications. Modern theories suppose that all matter can disappear through its transformation into radiant energy. Should we hope that matter would know how to rebuild itself from this energy? Some German physicists suppose that it could. Nernst expressed this possibility in his

work of 1921, 'Das Weltgebaüde in Lichte der neuren Forschung' ['The Structure of the World in the Light of Recent Investigations']. His theory has been frequently criticized,[126] but we are not in a position to say whether these criticism have been decisive. We might sooner suppose that they were not, as this idea reappeared several times in the last few years,[127] and attempts were even made at explaining the origin of cosmic rays by this resurrection of matter. 'Cosmic radiations which we receive would thus be only bursts of light from the formidable forge.'[128]

The new stage of quantum theory and the idea of an expanding universe have added new difficulties. Quantum theory shows us the futility of all of the models of physical processes, and the theory of the expanding universe places before us the possibility of a new rhythm of the universe, that of contraction and expansion. If such a rhythm really does exist, it is probably not without relation to the process of energy degradation, or to that of the degradation and reconstitution of matter. Given the present state of science we do not believe in the possibility of the return of identical worlds. The causality principle, which had played a great role in the reasoning which justified this idea, is threatened by quantum physics. The idea of the creation of absolutely new states becomes probable. If returns do exist, they would only be returns to similar states. If the asymptote is not an image adequate to the development of the universe, then such an image would be sooner a spiral than a closed circle.

### *Irreversibility*

156. When speaking of time we say not only that it has a dimension, but also that it is irreversible. We have a direct intuition of this irreversibility. Physicists have tried to define it objectively by means of the second law of thermodynamics. This idea was expressed for the first time by Ernst Mach. Edgar Zilsel tried to accomplish this in a more precise way in his article entitled: 'Über die Asymmetrie der Kausalität und die Einsinningkeit der Zeit'.[129]. According to him, the irreversibility of time reduces to three facts, demonstrated by mental experiments.

1. If we compare the different qualitative states of a system, we find that they can be determined by means of a scalar, in such a way that each state would be higher or lower with respect to the other (whether we think in terms of different energy levels or of different degrees of energy dissipation).
2. If we compare different systems, we find that the higher states exist simultaneously at one moment, and the lower states at another.

3. Finally, if a change takes place in the same system, the states from the same level can never exist at different times, the same order reigns everywhere and always.

Alone, these three 'topological facts of experience' determine objectively, according to Zilsel, the single direction of temporal succession. This definition presupposes, in the second point, the knowledge of simultaneity for distant points; now, this simultaneity can be determined by means of optical signals, according to the precepts of the theory of relativity. Zilsel is convinced that his definition is better than that of Reichenbach, who appealed to the notion of causality in order to define the order of succession in an objective manner. He also replies to the possible objections.

The second principle of thermodynamics is only a statistical law. The relevant exceptions had been established in the microcosm. For these processes, the single direction of time is therefore not defined. However, our definition, says Zilsel, can be applied to the the microcosm by way of interpolation if we admit the transitivity of the relation 'after' and if we introduce the relation 'in between' with respect to time. However, exceptions are not impossible, for the macrocosm, either, if we want to apply the theorem of Poincaré-Zermelo-Caratheodory to the universe, conceived according to the model of the kinetic theory of gases. Loschmidt had also observed that it is possible to construct mechanical models which will function with a decrease of entropy. Zilsel answers: if entropy decreases everywhere, the single direction of [temporal] order will be restored; if not, then the before and after will lose their sense. This result should also be considered as satisfactory from the philosophical point of view, for the direction of time would be defined only for one period of the universe.

Zilsel draws our attention to other subtleties. If we observe the propagation of waves in the form of concentric spheres around a vibrating point, we find a unique direction there as well. We never observe concentric centripetal waves. But the impossibility of the reverse direction is only due to a technical difficulty, while in theory and, even in reality, the reverse direction is possible. Such is the opinion of Zilsel as regards material waves. However, electromagnetic waves, emanating from a material point, present more difficulty. Their unique direction seems to originate from a real cause which is tied to a difference between matter and field, and which has not yet been explained by modern science.

157. To these considerations of Zilsel we will now add our own observations. It has been maintained that the irreversibility of time, as long as we experience it, cannot be reduced to any physical process, because it contains something specific and undefinable, linked to the memory of the past and to the expectation of the future. But this observation, however just it might be, does not free us from looking for an objective criterion outside of our consciousness, which would facilitate our orientation regarding the direction of the order of change.

Yet, in this case we are prescribed with a diagnostic definition and not with a definition that completely exhausts the sense of the term defined. In order to give such a definition, we sometimes used the causality principle, sometimes the second law of thermodynamics. However, by means of these principles, we only fixed the sense of before and after, and we did not account for the irreversibility which is connected with the division of time into past, present and future. It has sometimes been affirmed that this latter division was immaterial for nature, and that for the latter, [i.e. nature] only the dichotomous division into the before and the after was of value.

We do not think that this is the case, however. The present can sometimes be immaterial for the calculations of the physicist, since he places the zero of time in his co-ordinate system wherever he wishes. The result of the calculation is not thereby changed. However, as far as the life of nature and of the universe is concerned, the present has, it seems to us, the same value as it does for our consciousness. And that is where the place of the real irreversibility of nature is to be found. In this respect we declare that Bergson and the Scholastics are right. To speak rigorously, each movement, at the moment at which it takes place, is irreversible, 'because each one [movement] has its own individuality, of an essentially momentary nature, and the movement we call reversed always constitutes in reality a new movement, which can take place only after the total and irretrievable disappearance of the movement whose reversion it is supposed to bring about.'[130]

Thus, we tolerate the definition of irreversibility by means of the law of entropy or of the principle of causality, solely as a diagnostic precept which does not exhaust the essence of irreversibility.[131] After this explanation, the problem arises of knowing to which of these two principles we should give preference. Should we prefer the law which governs all the dynamical relations, or the statistical law of entropy? Following Reichenbach's example, we chose the former in the preceding chapter, but this choice has a weak side. The causality principle is questioned in modern science, its idea is the subject of unending controversies; on the other hand, the law of entropy, though a statistical one, is much more readily applied. Its weak side consists in that energy degradation, which it explains, is not the only irreversible process of nature. The degradation of matter, the incessant emission of radiant energy, the transformation of mass into energy, this is a much more important, irreversible process. For the degradation of energy, exceptions have been found, but there is no trace of an exception for the degradation of matter. We have only vague, perhaps unverifiable presentiments. It is not impossible that these two irreversibilities be intimately connected with one another. Perhaps the discovery of such a connection will cast more light upon the relationship which exists between the statistical laws and the dynamical laws based on the principle of causality, such as the law of least action, or the law of least resistance, used also to define the meaning of becoming.

158. However, why is there becoming and change in the universe? In his mechanics, Mach put forward a casual observation which, unfortunately, was not well motivated by him. If everything were determined in the world, nothing would happen, and the universe would be immobile. However, not everything has been determined as yet, and it is for this reason that something does happen. We, ourselves, are one of the elements of nature, partly determined, and partly indeterminate.

At first glance, Mach's idea seems to be wholly trivial. He gives us the comparison of nature with a number '$n$' of magnitudes, which correspond to a number '$n'$' of equations, where $n'<n$. Now, if $n = n'$; each of the magnitudes receives a constant value, and a continuous series of values is excluded, hence change is impossible. Thus Mach's idea is reduced to a simple tautology; viz., if each element of the universe can be represented by means of a constant magnitude, any change is impossible. However, the rest of the passage discussed in the mechanics shows that Mach wanted to say something more. He assimilated this mathematical indeterminism to our sense of freedom, and we get the impression that he wished to explain by this mathematical indeterminism not only the fact of change, but also the fact that the choice of the direction of change depends sometimes upon us.

Now, this second idea is not well motivated, and yet it is this idea that renders this passage very up-to-date. In an article by Schrödinger, who is one of the creators of quantum physics, we read: [132] 'If something happens in a system, this consists, from the point of view of quantum mechanics, in the lack of precise determination of the energy of the system. In a system with completely defined energy, there are no time-dependent values, the system does not change at all with regard to its observable properties, it behaves like a system in a thermodynamic equilibrium.' This passage reminds us of Mach's ideas, but we do not understand it well. Schrödinger probably means that the lack of the precise determination of energy is responsible only for the fact of change, but not for its direction. However, both the fact of change, and its direction, seem to be determined by the fundamental laws of physics, such as the law of least action or the law of entropy. Consequently, what is the relation between the lack of precise determination of energy and these laws? Although we have no answer to this question, we should note here the idea of modern physics, viz. that the temporal character of the universe originates in the lack of the precise determination of energy.

## Conclusion

159. We devote last chapter to the question, in what sense can we say that time represents something irrational; thereafter, we shall consider the role of time in the scientific conception of the universe.

According to Meyerson, time is irrational because it is irreversible. From the point of view of the amount of energy contained, the universe is always identical with itself, owing to the principle of the conservation of energy. This quantity is always the same, but if we take into account its dislocation, its difference of levels, we find a breach of the law of identity. Energy is dissipated; it passes from higher levels to lower ones, in such a way that the universe, considered from the qualitative point of view, no longer fulfills the principle of identity. Fundamentally, the principle of irreversibility is a principle of irrationality.

According to others, time also does not conform to the principle of contradiction, which does not permit an object to possess, at one and the same time, two contradictory properties. In one of the preceding chapters, we have already pointed out that the expression 'at one and the same time' is sometimes interpreted in the sense of 'simultaneously'. Hence we say 'that this principle is only valid for the present.' Yet, since the principle of contradiction must be independent of time, some philosophers interpreted this independence in such a way that, in their opinion, [the possession of] contradictory properties [is] impossible not only in simultaneity, but also in succession. Such is, for example, the opinion of Jakubisiak, expressed in his work, *Essai sur les limites de l'espace et du temps*.[133]

This author has thus arrived at the negation of time, basing his conclusion on the laws of logic which exclude all becoming.

If we add to all this the antinomies of Kant, relative to the problem of finite or infinite time, and all the difficulties connected with the continuity of time and, finally, Zeno's paradoxes of time, we obtain the complete picture of all the arguments on which is based the opinion that time and becoming represent an irrational element for human knowledge. From this result, one of the following two consequences is drawn: either an objection to the reality of time based on the thesis that every being should be rational, or this irrationality was accepted as a given result and [the defender of this thesis] proclaimed the limits of human knowledge which can only be applied to the stable and discontinuous elements of reality.

In order to avoid repeating what has already been said when discussing the problem of the continuity of time and the antinomies of Kant, we shall limit ourselves to the following remarks: In the first place, we must agree on the meaning of the terms 'rational' and 'irrational'. The most proper meaning of the latter term would be: that which does not conform to the laws of logic. Now, even if it were true that these laws do not suffice, the idealistic consequences do not seem to us to be justified. We have to take a being such as it is, and look always for the forms of thought, which permit us to understand it. What do we gain by answers such that the irreversibility of time is irrational, or that the continuity or infinity of time, or becoming and changes, are irrational?

The strongest of all these objections would be that each of these notions implies a contradiction. However, we failed to establish this either for irreversibility, or for continuity, or for infinity. We have still to give an answer to the question raised by Jakubisiak, viz., that the law of contradiction excludes contradictory properties not only for the same instant, but for different instants of time. This idea seems absurd to us. We are confronted with contradictory judgements when one of them negates what the other affirms in the same respect ('*kata to auto*,' according to Aristotle), and in the same proportion. Now let us take a proposition which refers to a concrete fact, to an event: 'the object *x* does not appear at the moment *t* with a feature *y*'. The indication of time therefore enters into the content of contradictory properties or objects. Without it, the objects are not contradictory. But why is this indication often omitted?

We shall answer. We often speak of so-called ideal objects which have nothing to do with existence in time. However, if we speak of concrete facts, the proposition that does not indicate the time of this fact is elliptical. If the proposition is not always completed, it is because we suppose it to be self-explanatory.

Thus, we do not see a contradiction anywhere. True, our thought is best adapted to stable and discontinuous elements, and this was very well observed by Bergson. Yet Bergson's solution is not satisfactory. The intuition which he substitutes for science and for scientific philosophy is too vague and too equivocal a concept to be able to satisfy the intellectual needs of mankind. Whatever is suggested by intuition should and can be expressed by means of a language subject to the laws of logic. Moreover, Bergson does not want to undervalue science. 'The definitive effort of the distinct intuition would be impossible for anybody who has not brought together and confronted a very large number of analyses. The masters of modern philosophy were men who had assimilated the entire material of contemporary science. And the partial eclipse of metaphysics during the last half-century has obviously no other cause that the extraordinary difficulty which philosophy experiences today when establishing contact with a science become much too dispersed. . . . Metaphysics has nothing in common with the generalization of experience, but nevertheless it could define for itself the integrality of experience.'[134]

All this is very true. The intuition which originates in science and which tries to express itself in the language of science, cannot be dangerous to science, since it will not be able to do anything other than science, nor anything that differs from it [science]. But here we have another passage of Bergson to which we cannot subscribe: 'It (viz., metaphysics) is properly itself only when it oversteps the concept, or at least when it throws off the yoke of rigid and ready-made concepts in order to create concepts which are quite different from those we usually handle, I mean the supple, almost fluid representations, always ready to be modelled on the fleeting forms of intuition.'[135]

What does this supple, mobile, fluid representation mean? From the fact that the object of our knowledge is mobile and fluid it does not follow that our representation should possess the same quality. Using vague, imprecise and equivocal terms, it is impossible to express any thought. Hence any appeal to terms of this kind ought to be absolutely rejected. If Bergson meant that analysis can vitiate the image of reality because it crumbles and immobilizes it, then his objection is correct, but only on condition that we substitute the products of analysis for reality. However, there is no need to do this. Whitehead's geometry of events and the work of Nicod on 'Geometry in the Sensible World' indicated that science can elaborate a method which will not be exposed to Bergson's objections.

160. If the question is that of the role of time in the scientific conception of the universe, Bergson pointed out very well the existence of two tendencies which manifested themselves in this respect in the history of the sciences and of philosophy. One of them has come into existence in the mathematical sciences, the other in the biological sciences. One of them eliminates time from the field of its occupations, it considers the universe *sub specie aeternitatis*; the other does the opposite, and considers the universe *sub specie durationis*.

Bergson has called the former 'platonism'; it lives in [the] ideal of universal mathematics. The other ought to be called 'bergsonism'. But this solution is likewise one-sided. However, if the evolution connected with indetermination, and even with the impossibility of predicting in a univocal manner, does exist, [and does so] even in the inorganic world, then it is no less true that Descartes' ideal of a universal mathematics, in the widest sense of the word, cannot be abandoned. The point is that 'the geometrical mind' is also capable of evolutions and creations without end.

Should we affirm that there exists no object, no component of reality outside of time? Prior to doing this, we must understand the different meanings of this assertion. For Aristotle and the Scholastics, non-temporal being is a permanent Being, always identical with itself, which co-exists with the universe and is thus its contemporary. However, the duration of the universe does not determine it, being extrinsic to it. For Kant, the thing-in-itself 'is neither identical at different times nor different at different times.' [136]

Perhaps Kant would object to the affirmation that God is contemporary with the universe. For Bergson, God endures just as we do, but with a duration which is much more concentrated and more intense ('living eternity').

If we accept the definition that each temporal object exists in a relation of succession or simultaneity with every other, then it is necessary, we believe, for God – if he exists – to be temporal since he co-exists with the universe and is contemporary with it. Yet this answer does not yet tell us all that we would like to know.

# NOTES

[1] Of some help to the author were the articles concerning the reports, which were published in *Le Temps* of September 18–22.
[2] This happened, indeed, in June 1935.
[3] $v(p)$ signifies: the value of the proposition $p$.
[4] Mme. Hosiasson was astonished when we saw in the iteration of probability a difficulty for the two-valued logic. We agree that there is no difficulty if the iteration appears only a finite number of time, however, if it goes up to infinity, we shall never obtain a proposition which will be able to belong to the two valued logic.
[5] 'Intuitionisme en formalisme', Amsterdam, (1912), in English, *Bull. Amer. Math. Soc. 20,* (1913).
[6] As early as in the Dutch-language work of 1907, Brouwer pointed to the necessity of distinguishing between the mathematical language of the formalists and the mathematics proper as the intuitive theory of the content of laws written with the aid of the formulae of that language ('Over de grondslagen der wiskunde'), Amsterdam 1907.
[7] *Bull. Amer. Math. Soc., 20,* No. 2, p. 85, paper by Brouwer: 'Intuitionism and formalism'.
[8] For Brouwer, an elementary series is a well-ordered series (i.e. a series every part of which has a first element), with the natural number of ω.
[9] Brouwer did come out against the 'mindless' application of the principle of the excluded middle in similar cases as early as a paper written in Dutch in 1908, 'De onbetrouwbaarheit der logische principes'.
[10] People do not understand each other because they do not speak the same language and because languages exist which cannot be learnt.
[11] According to this distinction we should, e.g., when talking about points, distinguish correctly according to the author's rule, between the set of points to be constructed and the more general notion of the point species.
[12] The point is that this postulate could be obtained from the law of excluded middle by substituting '∼a' for 'a'.
[13] In view of the fact that Heyting's axioms and matrix do not designate the same system, one can preserve the matrix and change the axioms instead of changing the matrix and preserving the axioms. This was the course taken by Łukasiewicz who thus obtained a new three-valued logic which, however, fails to carry out – or carries out only in part – the mathematical reform required by Brouwer.
[14] Grundlage der allgemeinen Relativitätstheorie. 2. Über die Gründe, welche eine Erweiterung des Relativitätspostulates nahelegen.
[15] *Raum und Zeit*, dritte Auflage, 1920, p. 41, Note.
[16] 'Let us hang a vessel using a very long string which we shall keep twisting until it becomes very rigid due to twisting; next let us fill the vessel with water and hold the filled vessel for some time. If then the vessel is set revolving in the opposite direction owing to a sudden force which will be maintained for a longer time owing to the unwinding of the string, then the surface of water will be at first plane, just as before the vessel had been set in motion;

later, when the force will gradually act on water, the vessel will bring about an evident rotation of the water. It will recede more and more from the centre and rise on the walls of the vessel, assuming a hollowed form.

'At first, when the relative movement of the liquid in the vessel was at its maximum, it did not cause any tendency to recede from the axis. Water did not attempt to draw near to the periphery climbing up the walls, and it remained plane, and the true rotating movement did not yet begin. However later, when the relative motion of the water decreased, the rising of water on the walls of the vessel pointed to its tendency to recede from the axis, and this tendency pointed to the ever increasing true rotating movement of water, which finally reached its highest value when water itself was at rest with respect to the vessel'.

In addition to this example Newton quotes also another one which enables one to differentiate the absolute rotary movements from the relative ones. 'If we connect two spheres whose mutual distance is known, using a string and if we rotate them round their common centre of gravity, then the tension of the string will elicit the tendency of the spheres to recede from the axis of rotation, from which it will be possible to calculate the magnitude of the rotary movement. By applying new forces of equal value at both sides and by measuring the so generated variations in the tension of the spring, it would be possible, according to Newton, to find not only 'the value but also the direction of such a rotary movement in any infinitely great empty space, even if it did not contain anything outside capable of being recognized as something with which the spheres could be compared.'

This second example of Newton's, just as the first one, does not prove anything. Newton assumes here beforehand that the centrifugal forces which stretch the string could appear also if an empty space existed around. A certain modification of this second example is the example adduced by Kant in his *Metaphysische Anfangsgründe der Naturwissenschaft* (paragraph: *allgemeine Anmerkung zur Phänomenologie*) for the justification of the difference proposed by Kant between the reality of the rotary and progressive motions.

[17] Even before Mach the contemporaries of Newton – Huygens and Leibniz – considered Newton's argument to be erroneous. Cf. correspondence of Huygens with Leibniz in Leibniz's *Hauptschriften zur Grundlegung der Philosophie*. B. 1. *Philosophische Bibliothek,* vol. 107.

[18] 'I answer that the motion is independent of observation, but that is not independent of observability' Fifth letter to Clarke, *Phil. Bibl.*, vol. 1, 107.

[19] Josef Winternitz, 'Kausalität, Relativität und Stetigkeit', *Kantstudien,* 1920; Schlick, Naturphilosophische Beobachtungen über das Kausalprinzip', 1920.

[20] *Annalen der Philosophie, 2*, p. 382, note.

[21] 'On the Notion of Cause', printed in the *Proc. of the Aristotelian Society*, in 1912; The same in *Mysticism and Logic*, London, George Allen and Unwin Ltd. p. 195, first published 1910, 5th edition 1932.

[22] Cf. Silberstein, *Wykłady Zakopiańskie* (*The Zakopane Lectures*). 'Remarks on the isolated, undisturbed systems' quotes likewise Gawecki: *Przyczynowość i Funkcjonalizm* (*Causality and Functionalism*). Supplement.

[23] The detailed form of the laws of nature changes in every single system depending on the values assumed by the so-called components of the gravitational potential. They can be determined from the disposition of masses in the given system. Now in the area of the small theory of relativity these components have a constant value, and consequently even the detailed form is invariant in that case. Cf. Neumann, Ernst Richard: *Vorlesungen zur Einführung in die Relativitätstheorie.* Jena, 1922. p. 147. Notes on the terms 'invariant' and 'covariant'.

[24] C.D. Broad, *Mind* 1923, No. 126.
[25] Cf. the beautiful work by Gawecki 'Przyczynowość i Funkcjonalizm we Fizyce' (Causality and Functionalism in Physics), in the first yearbook of *Kwartalnik Filozoficzny*.
[26] *Kantstudien*, vol. XXIV, 1921. Laue: 'Die Lorentz-Kontraktion'.
[27] According to the opinion of the Polish philosopher, T. Kotarbiński.
[28] *Kants Briefe*, herausgegeben von F. Ohmann, Leipzig 1911, 50; lettre à Marcus Herz, 21.II.1772.
[29] Wartenberg, M., Le problème du temps, Cracow 1916.
[30] *Revue de philosophie,* 1911.
[31] *Estin ho chronos kineseos tinos arithmos e kai katholou diastema tes tou pantos physeōs.*
[32] Joseph Fabre, *La pensée antique*, p. 198.
[33] Emile Bréhier, *Histoire de la philosophie*, I, 66.
[34] Aristotle, *Physics*, transl. by P.H. Wicksteed, vol. II, p. 187.
[35] Plato, *Timaeus,* transl. by R. G. Bury, p. 73.
[36] *Loc. cit.,* pp. 75, 77.
[37] *Loc. cit.*, pp. 77, 79.
[38] *Viz.*, the revolution of the five planets, the sun, the moon and the celestial sphere.
[39] *Loc. cit.,* p. 83.
[40] Aristotle, *Physics*, transl. by P.H. Wicksteed, vol. I, 379.
[41] *Loc. cit.*, p. 385.
[42] *Loc. cit.*, p. 387. *arithmos kineseōs kata to proteron kai husteron.*
[43] *Loc. cit.*, p. 389.
[44] *Loc. cit.*, p. 389.
[45] *Loc. cit.*, p. 391.
[46] *Loc. cit.*, p. 391.
[47] This passage is quoted according to Duhem.
[48] Aristotle, *loc. cit.*, p. 399.
[49] Aristotle, *loc. cit.*, p. 401.
[50] Aristotle, *loc. cit.*, p. 403.
[51] Aristotle, *loc. cit.*, p. 405.
[52] Aristotle, *loc. cit.*, p. 407.
[53] Aristotle, *loc. cit.*, p. 413.
[54] Here we are following the opinion of Nys.
[55] English translation by S. MacKenna, 3rd ed., Faber & Faber, 1962.
[56] Cf. the summary of this chapter, given by Brehier.
[57] St. Augustine discusses time also in his main work *De Civitate Dei* (book XI, chapt. 5–6). He states there that time was created together with the world and that it cannot exist without something movable and capable of changes.
[58] Desire Nys, *La Notion du Temps.*
[59] Desire Nys, *La Notion du Temps*, 3e edition, 240.
[60] Third reply.
[61] Fourth reply.
[62] Tempus et aeternitatem non esse idem. Et hujus diversitatis rationem quidam assignaverunt ex hoc, quod aeternitas caret principio et fine, tempus autem habet principium et finem. Sed haec est differentia per accidens et non per se. Quia dato quod tempus semper fuerit, et semper futurum sit, secundum positionem eorum, qui motum coeli ponunt sempiternum, adhuc remanebit differentia inter aeternitatem et tempus, ex hoc quod aeternitas est tota simul, quod tempori non convenit. *Summa Theologiae*, Pars prima quaestio X, art. 4.

[63] A. Marty, *Raum und Zeit*, 1916, p. 241.
[64] *An Enquiry Concerning the Principles of Natural Knowledge* (1919) and *The Concept of Nature* (1920).
[65] Cf. the explanation of the 'Transcendental Aesthetic' (p. 7), as analysed by us below.
[66] F. Ohman, *Kant's Briefe*, Leipzig 1911, p. 50. Letter to Marcus Herz.
[67] Zawirski, *Le rapport de la logique à la mathématique,* Cracow 1927.
[68] *Bulletin of the American Mathematical Society,* 1913.
[69] 'This intuition of two-oneness, the basal intuition of mathematics, creates not only the numbers one and two, but also all finite ordinal numbers, inasmuch as one of the elements of the two-oneness may be thought of as a new two-oneness, which process may be repeated indefinitely'. *Loc. cit*, p. 85–86.
[70] 'Wesen und Bedeutung der Synthesis in Kants Philosophie.' *Zeitschrift für Philosophie und philosophische Kritik,* Vol. 94, 1888.
[71] This passage, which we had made use of so often in our work, looks as follows in its full extent: 'That the changes represent something real, this I gainsay just as little as that bodies are something real, although I thereby only mean that something real does not correspond to the phenomenon. I cannot even say that the internal phenomenon undergoes a change, since by what means would I observe this change, if it would not appear to my inner sense? If one would say that it followed from this that: everything in the world were objective and unchangeable in itself, then I would answer: they are neither changeable nor unchangeable, just as Baumgarten *Metaph.*, 18. says, that the thing which is absolutely impossible is neither hypothetically possible nor impossible, since it cannot be observed at all, under any circumstance, and again: the things of the world are objective or in themselves neither in a monotonous state at different times nor in different states, because in this sense they are not at all represented in time. Letter to Marcus Herz. 21 February 1772.
[72] *Critique de la raison pure,* 2nd edition, p. 152.
[73] Schopenhauer, *Parerga and Paralipomena,* I.
[74] Victor Cousin, *Phil. de Kant*, 3me édit. 1857. Quotation extracted from Vaihinger's *Commentary,* Vol. II, p. 485.
[75] Cf. also *Prolegomena* 13, III 32.
[76] *Prolegomena* 13, Remark III, *Critique of Pure Reason*, p.3 50.
[77] *Critique,* p. 670.
[78] Sentroul, *Kant et Aristotle,* p. 117.
[79] *Les données immédiates,* p. 18, eleventh edition.
[80] *Matière et Mémoire,* p. 278, Seventh edition.
[81] *Die Naturwissenschaften,* 1926, p. 989.
[82] Bergson, *Durée et Simultanéité,* 1922; Préface, pp. VI-VII.
[83] Bergson, *Durée et Simultanéité,* p. 58.
[84] Rageot, *Le Problème expérimental du temps, Revue Philosophique,* 1908, p. 30.
[85] Dumas, *Traité de Psychologie,* II 54, first edition.
[86] Wundt, *Physiologische Psychologie,* III, 34, ed. 1903.
[87] Dumas, *Traité de Psychologie,* II, 54, 1st edition.
[88] Cf. especially a passage in the article 'L'ame et le corps', where we find the comparison between life and a stated sentence.
[89] Dumas, *Traité de Psychologie,* II, 64. Abbreviated as *Traité.*
[90] Reichenbach, *Philosophie der Raum-Zeit-Lehre,* 138.
[91] *Loc. cit.*, 140.

[92] Reichenbach, *loc. cit.*, p. 150.
[93] Reichenbach, *loc. cit.*, p. 151.
[94] We are here leaving aside the differences between the points of view of Leibniz and Reichenbach on the one hand, and that of Kant, on the other. While Leibniz and Reichenbach wanted to define the temporal order by means of the causal order, according to Kant, the causal relationship only renders us the service of objectifying the subjective succession. Our point of view is rather nearer to that of Kant; omitting, of course, his idealism and *apriorism*.
[95] The work by Hugo Bergmann 'Kampf um das Kausalprinzip' has been published in full English translation in *Boston Studies*, Volume 13.
[96] *Die Naturwissenschaften*, 1927, pp. 280–286.
[97] The Italian scientist Vailati devoted to the history of this principle a special article 'A proposito d'un passo di Teeteto e di una dimostrazione di Euclide'. *Rivista di Filosofia e scienze affine*. 1904.
[98] Einstein says that he introduces the constancy of the velocity of light solely as convention, but, in the reference to Paragraph 8, he admits that it is impossible in the following not to consider this convention as a hypothesis. At the end of this paragraph he emphasizes that he has also introduced the physical hypothesis that clocks run everywhere at 'the same speed.' Moreover, Einstein is not a logician. He felt instinctively that he was right, and it is the task of the logicians to examine the logical foundations of his theory.
[99] *Zur Elektrodynamik bewegter Körper*, paragraph 1.
[100] *Loc. cit.*, paragraph 2.
[101] Cf. especially the end of paragraph 8 in the popular book by Einstein.
[102] Bergson, *Loc. cit*, p. 208.
[103] Loc. cit., p. 222.
[104] *Matière et Mémoire*, p. 231, 7th ed., 1911.
[105] Bergson, *Durée et Simultanéité*, 1922, p. 103.
[106] *Die Naturwissenschaften*, 1918.
[107] Bergson, *Durée et Simultanéité*, p. 110.
[108] Wenzl Aloys, *Das Verhältnis der Einsteinschen Relativitätslehre zur Philosophie der Gegenwart*, 1924.
[109] *Annalen der Philosophie*, II, 1921, No. 3.
[110] *Annalen der Philosophie*, II, 1921.
[111] *Przegląd Filozoficzny*, XXIII, 352.
[112] *Kwartalnik Filozoficzny*, II, 44, Cracow 1924.
[113] Ernst Cassirer, *Zur Einsteinschen Relativitätstheorie*, Berlin 1921.
[114] Weyl, *Loc. cit.*, p. 161.
[115] We find in the Polish philosophical literature an interesting article on this subject, viz. Ajdukiewicz, 'Le Temps Relatif et Absolu', *Przeglad Filozoficzny*, XXIII, 1920. Cf. also *Ruch Filozoficzny*, X, 1926/7, Mehlberg's lecture, p. 192.
[116] R. Carnap, *Abriss der Logistik*, 1920, pp. 80–85.
[117] Pokrowski, 'Zur Frage nach der Struktur der Zeit', *Zeitschrift für Physik*, 1928, vol. 51.
[118] Cf. the highly interesting observations of Cantor on the various manners of understanding the continuum. Cantor, *Abhandlungen*, Berlin, 1932, p. 190.
[119] Weyl, *Philosophie der Mathematik u. Naturwissenschaft*, 43.
[120] Zawirski, *The Eternal Returns of the Worlds. Historical and Critical Research on the Doctrine of Eternal Returns* (in Polish), Cracow, 1927.

[121] *Acta mathemathica*, XIII, pp. 67–72.
[122] Cf. the title above.
[123] *Annalen der Physik u. Chemie LVII,* 1896, pp. 485–494.
[124] *Sitzungsberichte der Preuss. Akademie der Wissenschaften,* II, 1919, p. 580.
[125] The theorem is based on the 'ergodic' hypothesis which is false. Attempts were made to substitute for it a 'quasi-ergodic' hypothesis.
[126] Nölke Fr., *Entwickelung im Weltall, Kosmogonische Probleme und Hypothesen,* Hamburg, 1926.
[127] Haas Art., 'Über Frequenzerhöhungen von Lichtquanten durch Zusammenstösse mit raschbewegten Materieteilchen', 1926. *Sitzungsberichte der Akademie der Wissenschaften in Wien,* mathem.-naturwiss. Klasse.
[128] E. Persico, 'Les Rayons Cosmiques', Scientia, 1931, XLIX.
[129] *Die Naturwissenschaften,* 1927, p. 280.
[130] Nys Désiré, *La Notion du Temps,* 1925, p. 110.
[131] We think that Reichenbach is of the same opinion. Cf. *loc. cit.*, p. 164.
[132] Schrödinger, Spezielle Relativitätstheorie und Quantenmechanik', *Sitzungsberichte der preussischen Akad. der Wiss.*, 1931, pp. 238–247.
[133] A. Jakubisiak, *Essai sur les Limites de l'Espace et du Temps*, 1927, 34.
[134] Bergson, Introduction à la Métaphysique, *R. de Mét. et de Mor.*, 1903, p. 36.
[135] Bergson, *loc. cit.*, p. 9.
[136] Letter to Marcus Herz of 1772.

# LIST OF ZAWIRSKI'S MAIN PUBLICATIONS


A. *Methodological Problems of Science*

[1] *Przyczynowość a stosunek funkcjonalny. Studium z zakresu teorii poznania* (*Causality and Functional Relation. An Epistemological Study*), Drukarnia Polska, Warsaw, 1912, p. 66.

[2] *Relatywizm filozoficzny a fizykalna teoria względności* (*Philosophical Relativism and the Theory of Relativity*), Slowo Polskie, Lwów, 1921, p. 79.

[3] '*Metoda aksjomatyczna a przyrodoznawstwo*' ('*The Axiomatic Method and Science*'), *Kwartalnik Filozoficzny,* Kraków, 1923, p. 128.

[4] '*Związek zasady przyczynowości z zasadą względności. Odczyt habilitacyjny*' ('*The Connection Between the Principle of Causality and the Principle of Relativity, Habilitation Thesis*'), *Kwartalnik Filozoficzny,* Lwów, 1924, p. 23.

[5] *Science et Philosophie,* Warsaw, 1937, Instytut Mianowski, p. 16.

[6] '*W sprawie syntezy naukowej*' ('*The Problem of Scientific Synthesis*'), *Droga,* Warsaw, 1937, pp. 974–996.

B. *Logic*

[7] '*O modalności sądów*' ('*On the Modality of Judgements*'), *Polskie Tow. Filozoficzne,* Lwów, 1914, p. 108.

[8] *Stosunek logiki do matematyki w świetle badań współczesnych* (*The Relation between Logic and Mathematics from the Point of View of Contemporary Investigations*), Uniwersytet Jagielloński, Kraków, 1927, p. 38.

[9] *Logika trójwartościowa Jana Lukasiewicza. O logice L.E.J. Brouwera. Próby stosowania logiki wielowartościowej do współczesnego przyrodoznawstwa.* (*The three-Valued Logic of Jan Lukasiewicz. The Logic of L. E.J. Brouwer. Some Efforts to Apply Many-Valued Logic to Science*), Poznán, 1932, Tow. Przyj. Nauk, 232.

[10] *Les logiques nouvelles et le champ de leur application.* Paris, A. Colin, pp. 419–503. Revue de Metaphysique et de Morale, 1932.

[11] *Znaczenie logiki wielowartościowej dla poznania i związek jej z rachunkiem prawdopodobieństwa.* (*The Significance of Many-valued Logic for Epistemology and its Connection with the Probability Calculus*), Warsaw, 1934, I. Pyz i S-ka, p. 8.

[12] *Stosunek logiki wielowartościowej do rachunku prawdopodobieństwa.* (*The Relation between Many-Valued Logic and Probability Calculus*), Poznańskie Towarzystwo Przyjaciól Nauk, Poznán, 1934.

[13] '*Über das Verhältnis der mehrwertigen Logik zur Wahrscheinlichkeitsrechnung*', Leopoli, 1935, *Studia Philosophica. Commentarii Societatis Philosophicae Polonorum v. 1.*

[14] '*Uber die Anwendung der mehrwertigen Logik in der empirischen Wissenschaft.*' *Erkenntnis,* Bd. 6, Hft 5/6, Leipzig, 1937, Meiner.

[15] 'Importance des recherches logiques et sémantiques pour les theories de la physique contemporaine'. Paris 1937, Extr. *Travaux du IX-e Congrès International de Philosophie.*
[16] *Logika teoretyczna,* (*Theoretical Logic*), Kraków, 1938, Text-book.
[17] 'Donioslość badań logicznych i semantycznych dla teoryj fizyki współczesnej' ('The Importance of Logical and Semantical Investigations for the Theories of Contemporary Physics'), Warsaw, *Przegląd filozoficzny*, 1938.
[18] 'Geneza i rozwój logiki intuicjonistycznej' ('The Genesis and Development of Intuitionistic Logic'), *Kwartalnik Filozoficzny*, T. 16 Kraków, 1946, pp. 165–222.

C. *Ontological Problems*

[19] 'Wieczne powroty światów. Badania historyczno-krytyczne nad doktryną 'wiecznego powrotu'. ('The Eternal Return of the Worlds. Historico-critical Investigations of the Idea of 'Eternal Return', Kraków, Uniwersytet Jagielloński, *Kwartalnik Filozoficzny*, 1927, p. 104.
[20] (a) Teoria Kwantów a Zasada Przyczynowości (The Quantum Theory and the Principle of Causality), *Przegląd Filozoficzny* 33, 1930.
(b) 'W sprawie indeterminizmu fizyki kwantowej' ('Indeterminism in Quantum Physics'), Lwów, 1931, *Książnica Atlas,* pp. 1–28.
[21] 'L'évolution de la notion du temps', Milano, *Scientia*, Avril 1934, Extrait de *Scientia*.
[22] *Rozwój pojęcia czasu* (*The Evolution of the Notion of Time*), Kraków, Drukarnia Uniwersytetu Jagiellońskiego, 1935.
[23] *L'Evolution de la notion du temps.* Ouvrage Honoré Du Prix Eugenio Rignano, Cracovie 1936, Librairie Gebethner et Wolff. p. 357.

D. *Psychology*

[24] *Ilość praw kojarzenia przedstawień, Szkic historyczno-krytyczny* (*The Number of the Laws of Association. A Historico-critical essay*), Rzeszów 1909–1910, J.A. Pelar, p. 88.
[25] Hoefler Alvis, *Zasady psychologji* (*The Principles of Psychology*), Translated by Zygmunt Zawirski, Lwów, 1922.
[26] Walther Leon, *Psychologia pracy* (*The Psychology of Working*), Translated by Zygmunt Zawirski, Poznań, 1932.

E. *Varia*

[27] 'Hermana Weyla, Filozofia matematyki i przyrodoznawstwa'. ('*Herman Weyl's Philosophy of Mathematics and Natural Science*'), Kraków 1930, PAU, '*Kwartalnik Filozoficzny*'.
[28] '*Filozofia nauk Meyersona*' ('Meyerson's Philosophy of Sciences'), *Droga*, Nr. 6, Warsaw.
[29] 'Joachim Metallman: Determinizm nauk przyrodniczych' ('Joachim Metallman's, Determinism in Science'), Review, *Przegląd Filozoficzny* 1935.
[30] 'Les tendances actuelles de la philosophie polonaise'. *Revue de Synthese 10*, Paris 1935.

[31] 'Władysław Mieczysław Kozłowski (1858–1935). Wspomnienie pośmiertlne' [Obituary], *Przegląd Filozoficzny*, 38, Warsaw 1935.
[32] 'Ósmy Międzynarodowy Kongres Filozoficzny' ('The Eighth International Philosophical Congress'), *Kwartalnik Filozoficzny*, Kraków 1935.
[33] *Wrazenia z I Międzynarodowego Kongresu Filozofii Naukowej 15–23 wrzesień, 1935* (*Some Impressions from the First International Congress of Scientific Philosophy, held September 15–23, 1935*), *w oparciu o artykuły sprawozdawcze z Kongresu, w Le Temps*, September 18–22, 1935.
[34] *'Kazimierz Twardowski (1866–1938)' [Obituary]*, Biblioteczka Czasu, Czerwiec, Kraków, 1938.
[35] *O współczesnych kierunkach filozofii* (*Contemporary Trends in Philosophy*) Wiedza Zawōd, Kultura, Kraków, 1947.

# NAME INDEX

# Boston Studies in the Philosophy of Science

138. W.A. Wallace: *Galileo's Logical Treatises.* A Translation, with Notes and Commentary, of His Appropriated Latin Questions on Aristotle's *Posterior Analytics.* 1992 ISBN 0-7923-1578-2
Set (137 + 138) ISBN 0-7923-1579-0
139. M.J. Nye, J.L. Richards and R.H. Stuewer (eds.): *The Invention of Physical Science.* Intersections of Mathematics, Theology and Natural Philosophy since the Seventeenth Century. Essays in Honor of Erwin N. Hiebert. 1992
ISBN 0-7923-1753-X
140. G. Corsi, M.L. dalla Chiara and G.C. Ghirardi (eds.): *Bridging the Gap: Philosophy, Mathematics and Physics.* Lectures on the Foundations of Science. 1992 ISBN 0-7923-1761-0
141. C.-H. Lin and D. Fu (eds.): *Philosophy and Conceptual History of Science in Taiwan.* 1992 ISBN 0-7923-1766-1
142. S. Sarkar (ed.): *The Founders of Evolutionary Genetics.* A Centenary Reappraisal. 1992 ISBN 0-7923-1777-7
143. J. Blackmore (ed.): *Ernst Mach – A Deeper Look.* Documents and New Perspectives. 1992 ISBN 0-7923-1853-6
144. P. Kroes and M. Bakker (eds.): *Technological Development and Science in the Industrial Age.* New Perspectives on the Science–Technology Relationship. 1992 ISBN 0-7923-1898-6
145. S. Amsterdamski: *Between History and Method.* Disputes about the Rationality of Science. 1992 ISBN 0-7923-1941-9
146. E. Ullmann-Margalit (ed.): *The Scientific Enterprise.* The Bar-Hillel Colloquium: Studies in History, Philosophy, and Sociology of Science, Volume 4. 1992 ISBN 0-7923-1992-3
147. L. Embree (ed.): *Metaarchaeology.* Reflections by Archaeologists and Philosophers. 1992 ISBN 0-7923-2023-9
148. S. French and H. Kamminga (eds.): *Correspondence, Invariance and Heuristics.* Essays in Honour of Heinz Post. 1993 ISBN 0-7923-2085-9
149. M. Bunzl: *The Context of Explanation.* 1993 ISBN 0-7923-2153-7
150. I.B. Cohen (ed.): *The Natural Sciences and the Social Sciences.* Some Critical and Historical Perspectives. 1994 ISBN 0-7923-2223-1
151. K. Gavroglu, Y. Christianidis and E. Nicolaidis (eds.): *Trends in the Historiography of Science.* 1994 ISBN 0-7923-2255-X
152. S. Poggi and M. Bossi (eds.): *Romanticism in Science.* Science in Europe, 1790–1840. 1994 ISBN 0-7923-2336-X
153. J. Faye and H.J. Folse (eds.): *Niels Bohr and Contemporary Philosophy.* 1994
ISBN 0-7923-2378-5
154. C.C. Gould and R.S. Cohen (eds.): *Artifacts, Representations, and Social Practice.* Essays for Marx W. Wartofsky. 1994 ISBN 0-7923-2481-1
155. R.E. Butts: *Historical Pragmatics.* Philosophical Essays. 1993
ISBN 0-7923-2498-6

# Boston Studies in the Philosophy of Science

156. R. Rashed: *The Development of Arabic Mathematics: Between Arithmetic and Algebra.* Translated from French by A.F.W. Armstrong. 1994 ISBN 0-7923-2565-6
157. I. Szumilewicz-Lachman (ed.): *Zygmunt Zawirski: His Life and Work.* With Selected Writings on Time, Logic and the Methodology of Science. Translations by Feliks Lachman. Ed. by R.S. Cohen, with the assistance of B. Bergo. 1994 ISBN 0-7923-2566-4
158. S.N. Haq: *Names, Natures and Things.* The Alchemist Jābir ibn Ḥayyān and His *Kitāb al-Aḥjār* (Book of Stones). 1994 ISBN 0-7923-2587-7
159. P. Plaass: *Kant's Theory of Natural Science.* Translation, Analytic Introduction and Commentary by Alfred E. and Maria G. Miller. 1994 ISBN 0-7923-2750-0

*Also of interest:*
R.S. Cohen and M.W. Wartofsky (eds.): *A Portrait of Twenty-Five Years Boston Colloquia for the Philosophy of Science, 1960-1985.* 1985 ISBN Pb 90-277-1971-3

*Previous volumes are still available.*

KLUWER ACADEMIC PUBLISHERS – DORDRECHT / BOSTON / LONDON

MIX
Papier aus verantwortungsvollen Quellen
Paper from responsible sources
FSC® C105338

If you have any concerns about our products,
you can contact us on
**ProductSafety@springernature.com**

In case Publisher is established outside the EU,
the EU authorized representative is:
**Springer Nature Customer Service Center GmbH**
**Europaplatz 3, 69115 Heidelberg, Germany**

Printed by Libri Plureos GmbH
in Hamburg, Germany